Identification and Control: Some New Challenges

CONTEMPORARY MATHEMATICS

757

Identification and Control: Some New Challenges

Summer School
Identification and Control: some challenges
June 18–20, 2019
Monastir, Tunisia

Kaïs Ammari
Stéphane Gerbi
Editors

AMERICAN MATHEMATICAL SOCIETY

2010 *Mathematics Subject Classification.* Primary 35R30, 35Q41, 35K65, 35P20, 35R02.

Library of Congress Cataloging-in-Publication Data

Names: Ammari, Kaïs, editor. | Gerbi, Stéphane, editor.

Title: Identification and control : some new challenges : Summer school on Identification and Control : some challenges, June 18-20, 2019, Monastir, Tunisia / Kaïs Ammari, Stéphane Gerbi, editors.

Description: Providence, Rhode Island : American Mathematical Society, [2020] | Series: Contemporary mathematics, 0271-4132 ; volume 757 | Includes bibliographical references.

Identifiers: LCCN 2020001921 | ISBN 9781470455477 (paperback) | ISBN 9781470456962 (ebook)

Subjects: LCSH: Control theory–Congresses. | Inverse problems (Differential equations)–Congresses. | Differential equations, Partial–Congresses. | Mathematical analysis–Congresses. | AMS: Partial differential equations – Inverse problems. | Partial differential equations – Equations of mathematical physics and other areas of application – Time-dependent Schrödinger equations, Dirac equations. | Partial differential equations – Parabolic equations and systems – Degenerate parabolic equations. | Partial differential equations – Spectral theory and eigenvalue problems – Asymptotic distribution of eigenvalues and eigenfunctions. | Partial differential equations.

Classification: LCC QA402.3 .I42 2020 | DDC 515/.642–dc23

LC record available at https://lccn.loc.gov/2020001921

Contemporary Mathematics ISSN: 0271-4132 (print); ISSN: 1098-3627 (online)

DOI: https://doi.org/10.1090/conm/757

∞ The paper used in this book is acid-free and falls within the guidelines
established to ensure permanence and durability.
Visit the AMS home page at `https://www.ams.org/`

10 9 8 7 6 5 4 3 2 1 25 24 23 22 21 20

Contents

Preface

This volume constitutes the proceedings of the Summer School on Identification and Control: some challenges. This summer school was held at the University of Monastir, Monastir, Tunisia during the period of June 18–20, 2019. It was organized by Kaïs Ammari, UR Analysis and Control of PDE, University of Monastir, Tunisia, and Stéphane Gerbi, Laboratoire de Mathématiques, Université Savoie Mont Blanc, France. (See `http://conf-identification-control.ur-acedp.org` for details.)

The summer school consisted of three mini-courses, six invited talks on the theme, and some short talks. This volume consists of the three mini-courses and the six contributions.

The first mini-course was held by Yavar Kian, Aix Marseille Université, France. The title was "Introduction to the problem of Calderón". The objective of this course was to present the problem of Calderón, which consists of determining a conductivity appearing in an elliptic equation from excitation and measurements on a part of the boundary of the domain. The study of this problem follows the approach initiated by Sylvester and Uhlmann based on complex geometric optics solutions.

The second mini-course was held by Éric Soccorsi, Aix Marseille Université, France. The title was "Introduction to Borg-Levinson inverse spectral theory". This course was an introduction to the mathematical analysis of inverse spectral problems of Borg-Levinson type. More precisely, they are concerned with the uniqueness and the stability issues in the inverse problem of determining the electric potential of the multidimensional Laplace operator by boundary spectral data.

The last mini-course was held by Louis Tebou, Florida International University, Miami, USA. The title was "Simultaneous and indirect control of waves: Some recent developments and open problems". The control of a single wave equation has seen a tremendous amount of activity since the pioneering works of Russell in the sixties and seventies, especially after the publication of Lions' monographs on controllability. However, the control of multi-component systems of wave equations can arguably be said to be at an early stage. Some recent work in this area, focusing on the notion of simultaneous control (using the same control scheme in all components of the system at hand) and indirect control (using a single control for a system consisting of two components), is presented. Some open problems are also presented.

The invited speakers agreed to write review papers related to their contributions to the summer school, while the others have written more traditional research papers; this constitutes the last part of this volume. They represent recent and new work on the topics of mathematical control theory as well as inverse problems such as data assimilation.

We believe that this volume therefore provides an accessible summary of a wide range of active research topics, along with some exciting new results, and we hope that it will serve as a useful resource for both graduate students new to the area and more established researchers.

The summer school brought together internationally leading researchers in the community of control theory and inverse problems and young researchers who came from all around the world. The organizers' intention was to provide a wide-angle snapshot of this exciting and fast-moving area and facilitate the exchange of ideas on recent advances in its various aspects. The numerous formal, informal, and sometimes lively discussions that resulted from this interaction were for us a sign that we achieved something in the direction of fulfilling this goal.

Our second aim was to ensure that the diffusion of these recent results was not limited to established researchers in the area and present at the summer school but were also available to newcomers and more junior members of the research community. This was reflected by the presence of many unfamiliar and/or young faces in the audience. The present proceedings should hopefully complete the fulfillment of our second aim.

This summer school would not have materialized without the help and support of the following institutions. We are very grateful to the GDRI LEM2I: "Laboratoire Euro-Maghrébin de Mathématiques et leurs Interactions" and to the Research Unity ACPDE, Analysis and Control of Partial Differential Equations of the University of Monastir for their financial support, without whom this summer school would not have be accessible without fees.

We would also like to thank all of the participants of the summer school who made this event a success, the contributors to these proceedings, and the reviewers for their hard work.

Finally, a special thanks is due to the American Mathematical Society for accepting our editorial proposal.

Kaïs Ammari and Stéphane Gerbi
Monastir and Chambéry, December 10, 2019

Contemporary Mathematics
Volume **757**, 2020
https://doi.org/10.1090/conm/757/15246

Lecture on Calderón problem

Yavar Kian

ABSTRACT. We consider the so called Calderón problem which corresponds to the determination of a conductivity appearing in an elliptic equation from boundary measurements. Using several known results we propose a simplified and self contained proof of this result.

CONTENTS

1. Introduction

A. P. Calderón published in 1980 the pioneer contribution [**C**] which motivated many applications in inverse problems. The problem considered by A. P. Calderón can be formulated as follows " can one determine the electrical conductivity of a medium by making voltage and current measurements at the boundary of the medium?". This problem is known as the Electrical Impedance Tomography (EIT in short). The work of Calderón was motivated by oil prospection. Since then, the EIT problem received a lot of attention among the mathematical community with several applications in medical imaging and geophysical prospection (see [**J**, **ZK**]). We refer to [**Uh**] for an overview of recent development for this problem. In this lecture we consider several results related to this problem.

We start with the first positive answer given to the question raised by Calderón in [**SU**]. Following several arguments borrowed in [**Ch**, **Ha**, **Ka**, **Ki**, **SU**] we propose a simple proof of this result which can be considered as an introduction to this field.

We consider also some partial data result associated with this problem corresponding to the same problem with inputs and measurements restricted to some portion of the boundary borrowed from [**AU**, **BU**].

2010 *Mathematics Subject Classification.* Primary 35R30.

2. The Calderón problem with full data

Let Ω be a bounded domain of $\mathbb{R}^n$, $n \geqslant 3$, with C^2 boundary $\partial\Omega$. We denote by $\nu(x)$ the outward unit normal to $\partial\Omega$ computed at $x \in \partial\Omega$. We consider $a \in W^{2,\infty}(\Omega)$ such that there exists $a_0 > 0$ for which the condition

$$\forall x \in \Omega, \quad a(x) \geqslant a_0 \tag{2.1}$$

is fulfilled. For all $\varphi \in H^{\frac{3}{2}}(\partial\Omega)$ we consider the boundary value problem

$$\begin{cases} -\mathrm{div}(a\nabla u) = 0 & \text{in } \Omega, \\ u = \varphi & \text{on } \partial\Omega. \end{cases} \tag{2.2}$$

It is well known that this problem admits a unique solution $u_{a,\varphi} \in H^2(\Omega)$. Therefore, we can define the Dirichlet-to-Neumann map (DN map in short) associated with (2.5) given by

$$\mathcal{N}_a : H^{\frac{3}{2}}(\partial\Omega) \ni \varphi \mapsto a\partial_\nu u_{a,\varphi} \in H^{\frac{1}{2}}(\partial\Omega)$$

which is a bounded operator. In this context, the Calderón problem corresponds to the determination of the conductivity a from the knowledge of $\mathcal{N}_a$. We consider here the unique recovery of a from $\mathcal{N}_a$ which can be stated as follows.

THEOREM 2.1. *For $j = 1, 2$, let $a_j \in W^{2,\infty}(\Omega)$ satisfy* (2.1) *and the condition*

$$a_1(x) = a_2(x), \quad \nabla a_1(x) = \nabla a_2(x), \quad x \in \partial\Omega. \tag{2.3}$$

Then we have

$$\mathcal{N}_{a_1} = \mathcal{N}_{a_2} \Longrightarrow a_1 = a_2. \tag{2.4}$$

REMARK 2.1. Note that applying the result of [**KV**], one can remove the condition (2.3) in Theorem 2.1. In these notes we will not consider the work of [**KV**] dealing with the unique recovery of the conductivity at the boundary from the knowledge of the DN map.

2.1. An associated inverse problem. Let $q \in L^\infty(\Omega)$ take values in $\mathbb{R}$ and consider for each $\varphi \in H^{\frac{3}{2}}(\partial\Omega)$ the boundary value problem

$$\begin{cases} -\Delta v + qv = 0 & \text{in } \Omega, \\ v = \varphi & \text{on } \partial\Omega. \end{cases} \tag{2.5}$$

Assuming that for all $w \in H^1_0(\Omega) \cap H^2(\Omega)$ the implication

$$-\Delta w + qw = 0 \Longrightarrow w = 0 \tag{2.6}$$

holds true, one can prove that problem (2.5) admits a unique solution $v_{q,\varphi} \in H^2(\Omega)$. We define $\mathcal{Q}$ as the subset of all $q \in L^\infty(\Omega)$ such that for all $w \in H^1_0(\Omega) \cap H^2(\Omega)$ the implication (2.6) is fulfilled. For all $q \in \mathcal{Q}$, we define the Dirichlet-to-Neumann map (DN map in short) associated with (2.5) given by

$$\Lambda_q : H^{\frac{3}{2}}(\partial\Omega) \ni \varphi \mapsto \partial_\nu v_{q,\varphi} \in H^{\frac{1}{2}}(\partial\Omega).$$

The Calderón problem is strongly connected with the determination of q from Λ_q. In terms of uniqueness this result can be stated as follows.

THEOREM 2.2. *For $j = 1, 2$, let $q_j \in \mathcal{Q}$. Then we have*

$$\Lambda_{q_1} = \Lambda_{q_2} \Longrightarrow q_1 = q_2. \tag{2.7}$$

Note that using the so called Liouville transform, one can deduce Theorem 2.1 from Theorem 2.2 in the following way.

Corollary 2.1. *Assume that the conditions of Theorem 2.1 are fulfilled and Theorem 2.2 holds true. Then we have (2.4).*

Proof. We assume that $\mathcal{N}_{a_1} = \mathcal{N}_{a_2}$ holds true and we will prove that $a_1 = a_2$. For $\varphi \in H^{\frac{3}{2}}(\partial\Omega)$ and $j = 1, 2$, fix $u_j \in H^2(\Omega)$ solving

$$\begin{cases} -\mathrm{div}(a_j \nabla u_j) = 0 & \text{in } \Omega, \\ u_j = a_1^{-\frac{1}{2}} \varphi & \text{on } \partial\Omega. \end{cases}$$

Consider $v_j = a_j^{\frac{1}{2}} u_j$. We have

$$\begin{aligned} -\Delta v_j &= -2\nabla\left(a_j^{\frac{1}{2}}\right) \cdot \nabla u_j - a_j^{\frac{1}{2}} \Delta u_j + (-\Delta a_j^{\frac{1}{2}}) u_j \\ &= a_j^{-\frac{1}{2}} [-\nabla a_j \cdot \nabla u_j - a_j \Delta u_j] - \left(a_j^{-\frac{1}{2}} \Delta a_j^{\frac{1}{2}}\right) v_j \\ &= a_j^{-\frac{1}{2}} [-\mathrm{div}(a_j \nabla u_j)] - \left(a_j^{-\frac{1}{2}} \Delta a_j^{\frac{1}{2}}\right) v_j \\ &= -\left(a_j^{-\frac{1}{2}} \Delta a_j^{\frac{1}{2}}\right) v_j. \end{aligned}$$

Thus, for $q_j = a_j^{-\frac{1}{2}} \Delta a_j^{\frac{1}{2}}$, v_j solves

$$-\Delta v_j + q_j v_j = 0, \quad \text{in } \Omega.$$

Moreover, (2.3) implies that v_j solves

$$\begin{cases} -\Delta v_j + q_j v_j = 0 & \text{in } \Omega, \\ v_j = \varphi & \text{on } \partial\Omega. \end{cases}$$

Note that due to this connection the uniqueness of the solution of (2.2) with $a = a_j$ implies that $q_j \in \mathcal{Q}$. In addition, from (2.3), for almost every $x \in \partial\Omega$ we get

$$\begin{aligned} \partial_\nu v_1(x) &= a_1^{\frac{1}{2}} \partial_\nu u_1(x) + \partial_\nu \left(a_1^{\frac{1}{2}}\right) u_1(x) \\ &= a_1^{-\frac{1}{2}} \mathcal{N}_{a_1} \varphi(x) + \frac{\nu \cdot \nabla a_1 \varphi(x)}{2 a_1^{\frac{1}{2}}(x)} \\ &= a_2^{-\frac{1}{2}} \mathcal{N}_{a_2} \varphi(x) + \frac{\nu \cdot \nabla a_2 \varphi(x)}{2 a_2^{\frac{1}{2}}(x)} = \partial_\nu v_2(x). \end{aligned}$$

From this identity we deduce that $\Lambda_{q_1} = \Lambda_{q_2}$ and from Theorem 2.2 we get

$$a_1^{-\frac{1}{2}} \Delta a_1^{\frac{1}{2}} = q_1 = q_2 = a_2^{-\frac{1}{2}} \Delta a_2^{\frac{1}{2}}. \tag{2.8}$$

Fix $y = a_1^{\frac{1}{2}} - a_2^{\frac{1}{2}}$. Using (2.8), we find

$$\begin{aligned} \Delta y &= \Delta\left(a_1^{\frac{1}{2}}\right) - \Delta\left(a_2^{\frac{1}{2}}\right) = a_1^{\frac{1}{2}} \left[a_1^{-\frac{1}{2}} \Delta\left(a_1^{\frac{1}{2}}\right) - a_1^{-\frac{1}{2}} \Delta\left(a_2^{\frac{1}{2}}\right)\right] \\ &= a_1^{\frac{1}{2}} \left[a_2^{-\frac{1}{2}} \Delta\left(a_2^{\frac{1}{2}}\right) - a_1^{-\frac{1}{2}} \Delta\left(a_2^{\frac{1}{2}}\right)\right] = a_2^{-\frac{1}{2}} \Delta\left(a_2^{\frac{1}{2}}\right) [a_1^{\frac{1}{2}} - a_2^{\frac{1}{2}}] = q_2 y. \end{aligned}$$

Then, (2.3) implies

$$\begin{cases} -\Delta y + q_2 y = 0 & \text{in } \Omega, \\ y = 0 & \text{on } \partial\Omega. \end{cases}$$

Using the fact that $q_2 \in \mathcal{Q}$, we deduce that $y = 0$ which implies that $a_1 = a_2$. □

In view of Corollary 2.1, it is clear that Theorem 2.1 follows from Theorem 2.2. From now on we will focus our attention on Theorem 2.2.

2.2. Construction of complex geometric optics solutions. In this subsection we will construct solutions of (2.5) suitably designed for our inverse problem. These solutions, called complex geometric optics solutions, will be the main ingredient in the proof of Theorem 2.2. This approach which has been initiated by [**SU**] is the main approach considered so far for proving Theorem 2.2. In this section we will use several arguments borrowed from [**Ch, Ha, Ka, Ki, SU**].

The complex geometric optics solutions under consideration here will depend on $\xi \in \mathbb{R}^n$. More precisely, for each $\xi \in \mathbb{R}^n$, we fix $\eta_1, \eta_2 \in \mathbb{S}^{n-1}$ such that

$$\xi \cdot \eta_1 = \xi \cdot \eta_2 = \eta_1 \cdot \eta_2 = 0.$$

Then, for $\rho > 1$ a large parameter, we consider solutions of $-\Delta v_j + q_j v_j = 0$ on Ω taking the form

$$v_1(x) = e^{\rho\eta_1\cdot x}\left(e^{i\rho\eta_2\cdot x}e^{-i\xi\cdot x} + w_1(x)\right), \quad x \in \Omega, \tag{2.9}$$

$$v_2(x) = e^{-\rho\eta_1\cdot x}\left(e^{-i\rho\eta_2\cdot x} + w_2(x)\right), \quad x \in \Omega, \tag{2.10}$$

with $w_j \in H^2(\Omega)$, $j = 1, 2$, satisfying

$$\|w_j\|_{H^k(\Omega)} \underset{\rho\to+\infty}{=} \mathcal{O}\left(\rho^{k-1}\right), \quad j = 1, 2, \ k = 0, 2. \tag{2.11}$$

Here we denote by $H^0(\Omega)$ the space $L^2(\Omega)$. The main point here will be the construction of the expression w_j satisfying (2.11). For this purpose we will use the approach of [**Ha**] based on construction of periodic solutions.

PROPOSITION 2.1. *For $j = 1, 2$ and for all $\xi \in \mathbb{R}^n$, there exists $\rho_1 > 0$ such that for all $\rho > \rho_1$ we can find a solution v_j of $-\Delta v_j + q_j v_j = 0$ on Ω taking the form* (2.9)-(2.10) *with $w_j \in H^2(\Omega)$ satisfying* (2.11).

PROOF. Since the proof of this result is similar for v_1 and v_2, we will only consider the construction of v_1. Note first that

$$\Delta\left(e^{\rho\eta_1\cdot x}e^{i\rho\eta_2\cdot x}\right) = \rho^2\left(|\eta_1|^2 - |\eta_2|^2\right)e^{\rho\eta_1\cdot x}e^{i\rho\eta_2\cdot x} = 0.$$

Therefore, we have

$$\begin{aligned} &-\Delta v_1 \\ &= e^{\rho\eta_1\cdot x}\left(e^{i\rho\eta_2\cdot x}|\xi|^2 e^{-i\xi\cdot x}\right) - \Delta\left(e^{\rho\eta_1\cdot x}w_1\right) \\ &= e^{\rho\eta_1\cdot x}\left(e^{i\rho\eta_2\cdot x}|\xi|^2 e^{-i\xi\cdot x}\right) + e^{\rho\eta_1\cdot x}(-\Delta w_1 - 2\rho\eta_1\cdot\nabla w_1 - \rho^2 w_1). \end{aligned}$$

It follows

$$-\Delta v_1 + q_1 v_1 = 0 \Longleftrightarrow -\Delta w_1 - 2\rho\eta_1\cdot\nabla w_1 - \rho^2 w_1 + q_1 w_1 = -e^{i\rho\eta_2\cdot x}(|\xi|^2 + q_1)e^{-i\xi\cdot x}.$$

Using this identity, it seems clear that we need to build $w_1 \in H^2(\Omega)$ satisfying

$$-\Delta w_1 - 2\rho\eta_1\cdot\nabla w_1 - \rho^2 w_1 + q_1 w_1 = -e^{i\rho\eta_2\cdot x}(|\xi|^2 + q_1)e^{-i\xi\cdot x} \tag{2.12}$$

and condition (2.11). For this purpose, we fix S an isometric operator of $\mathbb{R}^n$ such that $S\eta_1 = e_1 = (1, 0, \dots, 0)$ and $R > 0$ such that $\Omega \subset Q := S^*(-R, R)^n$. Here S^* denotes the adjoint of S with respect to the Euclidean scalar product of $\mathbb{R}^n$. Then, for $F \in L^2(Q)$ we consider a solution $z \in H^2(Q)$ of

$$-\Delta z - 2\rho\eta_1\cdot\nabla z - \rho^2 z = F, \quad \text{on } Q \tag{2.13}$$

taking the form

$$z(x) = \sum_{\alpha\in\mathbb{Z}^n} a_\alpha\varphi_\alpha, \quad \varphi_\alpha(x) := (2R)^{-\frac{n}{2}}e^{\frac{i\pi\eta_1\cdot x}{2R}}e^{\frac{i\pi\alpha\cdot S(x)}{R}}. \tag{2.14}$$

The construction of such solutions can be deduced from the following result.

LEMMA 2.1. *For all $F \in L^2(Q)$ the equation* (2.13) *admits a solution $z_F \in H^2(Q)$ taking the form* (2.14) *and satisfying*

$$\|z_F\|_{H^k(Q)} \leqslant C\rho^{k-1}\|F\|_{L^2(Q)}, \quad k=0,2, \tag{2.15}$$

with C independent of F and ρ.

The proof of this lemma will be postponed to the end of the present demonstration. Applying Lemma 2.1, we can define the operator

$$\mathcal{K}_\rho : L^2(Q) \ni F \mapsto z_F|_\Omega \in H^2(\Omega)$$

which is bounded. Here z_F denotes a solution of (2.13) satisfying (2.15). We have $\mathcal{K}_\rho \in \mathcal{B}(L^2(Q); H^2(\Omega))$ and

$$\|\mathcal{K}_\rho\|_{\mathcal{B}(L^2(Q);H^k(\Omega))} \leqslant C\rho^{k-1}, \quad k=0,2. \tag{2.16}$$

Fix $F_\rho \in L^2(Q)$ and $\widetilde{q_1} \in L^\infty(Q)$ defined by

$$F_\rho(x) := \begin{cases} -e^{i\rho\eta_2\cdot x}(|\xi|^2 + q_1(x))e^{-i\xi\cdot x} & \text{if } x \in \Omega \\ 0 & \text{if } x \notin \Omega, \end{cases}$$

$$\widetilde{q_1}(x) := \begin{cases} q_1(x) & \text{if } x \in \Omega \\ 0 & \text{if } x \notin \Omega. \end{cases}$$

Now consider the map

$$\begin{aligned} \mathcal{G}_\rho : L^2(\Omega) &\to L^2(\Omega), \\ w &\mapsto \mathcal{K}_\rho\left[F_\rho - \widetilde{q_1} w\right]. \end{aligned}$$

We will prove that there exists $\rho_1 > 0$ such that this map admits a fixed point w_1 in the closed set $B = \{h \in L^2(\Omega) : \|h\|_{L^2(\Omega)} \leqslant 1\}$ of $L^2(\Omega)$. Indeed, (2.16) implies

$$\begin{aligned} \forall y_1, y_2 \in B, \quad \|\mathcal{G}_\rho y_1 - \mathcal{G}_\rho y_2\|_{L^2(\Omega)} &= \|\mathcal{K}_\rho\left[\widetilde{q_1}(y_2 - y_1)\right]\|_{L^2(\Omega)} \\ &\leqslant C\rho^{-1}\|q_1\|_{L^\infty(\Omega)}\|y_1 - y_2\|_{L^2(\Omega)} \end{aligned}$$

and

$$\begin{aligned} \forall y \in B, \quad \|\mathcal{G}_\rho y\|_{L^2(\Omega)} &\leqslant C\rho^{-1}\left[2\|q_1\|_{L^\infty(\Omega)} + \left\|e^{i\rho\eta_2\cdot x}|\xi|^2 e^{-i\xi\cdot x}\right\|_{L^2(\Omega)}\right] \\ &\leqslant C\rho^{-1}(2\|q_1\|_{L^\infty(\Omega)} + |\Omega|^{\frac{1}{2}}|\xi|^2). \end{aligned}$$

Consequently, fixing $\rho_1 = 2C(2\|q_1\|_{L^\infty(\Omega)} + 2\|q_2\|_{L^\infty(\Omega)} + |\Omega|^{\frac{1}{2}}|\xi|^2)$, we deduce that, for all $\rho > \rho_1$, $\mathcal{G}_\rho$ is a contraction from B to B. Therefore, the Poincaré fixed point theorem implies that $\mathcal{G}_\rho$ admits a unique fixed point $w_1 \in B$. Moreover, we have $w_1 = \mathcal{K}_\rho\left[F_\rho - \widetilde{q_1} w_1\right]$ which proves that $w_1 \in H^2(\Omega)$ solves

$$\begin{aligned} &[-\Delta w_1 - 2\rho\eta_1 \cdot \nabla w_1 - \rho^2 w_1](x) \\ &= F_\rho(x) - \widetilde{q_1} w_1(x) \\ &= -e^{i\rho\eta_2\cdot x}(|\xi|^2 + q_1(x))e^{-i\xi\cdot x} - q_1(x)w_1(x), \quad x \in \Omega. \end{aligned}$$

Therefore, w_1 solves (2.12) and (2.16) implies

$$\|w_1\|_{H^k(\Omega)} \leqslant C\rho^{k-1}(2\|q_1\|_{L^\infty(\Omega)} + |\Omega|^{\frac{1}{2}}|\xi|^2), \ k=0,2,$$

which proves (2.11). □

Now that the proof of Proposition 2.1 is completed let us turn to Lemma 2.1.
Proof of Lemma 2.1. For all $\alpha\in\mathbb{Z}^n$, fix

$$\varphi_\alpha(x):=(2R)^{-\frac{n}{2}}e^{\frac{i\pi\eta_1\cdot x}{2R}}e^{\frac{i\pi\alpha\cdot Sx}{R}},\quad x\in Q.$$

Combining the fact that

$$\varphi_\alpha(S^*x)=\psi_\alpha(x)=(2R)^{-\frac{n}{2}}e^{\frac{i\pi\eta_1\cdot S^*x}{2R}}e^{\frac{i\pi\alpha\cdot x}{R}},\quad x\in(-R,R)^n,$$

with the fact that S is an isometry and $(\psi_\alpha)_{\alpha\in\mathbb{Z}^d}$ is an orthonormal basis of $L^2((-R,R)^n)$, we deduce that $(\varphi_\alpha)_{\alpha\in\mathbb{Z}^d}$ is an orthonormal basis of $L^2(Q)$. Thus, for all z of the form (2.14) and $k=0,2$, one can check that

$$\sum_{\alpha\in\mathbb{Z}^n}(1+|\alpha|^2)^k|a_\alpha|^2<+\infty\Longrightarrow z\in H^k(Q),\tag{2.17}$$

$$z\in H^k(Q)\Longrightarrow\|z\|^2_{H^k(Q)}\leqslant C\sum_{\alpha\in\mathbb{Z}^n}(1+|\alpha|^2)^k|a_\alpha|^2,\tag{2.18}$$

with $C>0$ independent of $z\in H^k(Q)$ in the last estimate. Note that since $S^*e_1=S^{-1}e_1=\eta_1$, one can check that

$$\forall\alpha\in\mathbb{Z}^n,\quad \eta_1\cdot S^*\alpha\in\mathbb{Z}$$

and we deduce that

$$\forall\alpha\in\mathbb{Z}^n,\quad \left|\eta_1\cdot S^*\alpha+\frac{1}{2}\right|\geqslant\frac{1}{2}.\tag{2.19}$$

Using (2.19), for $\alpha\in\mathbb{Z}^n$, we fix

$$a_\alpha:=\frac{\langle F,\varphi_\alpha\rangle_{L^2(Q)}}{\frac{\pi^2}{R^2}\left|S^*\alpha+\frac{\eta_1}{2}\right|^2-\rho^2-\frac{2i\pi\rho}{R}\left(S^*\alpha\cdot\eta_1+\frac{1}{2}\right)}.$$

Using the fact that $|S^*\alpha|=|\alpha|$ and (2.19), we get

$$\begin{aligned}&\left|\frac{\pi^2}{R^2}\left|S^*\alpha+\frac{\eta_1}{2}\right|^2-\rho^2-\frac{2i\pi\rho}{R}\left(S^*\alpha\cdot\eta_1+\frac{1}{2}\right)\right|\\ &\geqslant\max\left(\left|\frac{\pi^2}{R^2}\left|S^*\alpha+\frac{\eta_1}{2}\right|^2-\rho^2\right|,\frac{2\pi\rho}{R}\left|S^*\alpha\cdot\eta_1+\frac{1}{2}\right|\right)\\ &\geqslant\max\left(\left|\frac{\pi^2}{2R^2}|\alpha|^2-\frac{\pi^2}{4R^2}-\rho^2\right|,\frac{\pi\rho}{R}\right).\end{aligned}\tag{2.20}$$

It follows that

$$\begin{aligned}
&\sum_{\alpha\in\mathbb{Z}^n}(1+|\alpha|^2)^2|a_\alpha|^2\\
&\leqslant \sum_{|\alpha|^2\geqslant 2+\frac{4R^2\rho^2}{\pi^2}}(1+|\alpha|^2)^2|a_\alpha|^2+\sum_{|\alpha|^2\leqslant 2+\frac{4R^2\rho^2}{\pi^2}}(1+|\alpha|^2)^2|a_\alpha|^2\\
&\leqslant \sum_{|\alpha|^2\geqslant 2+\frac{4R^2\rho^2}{\pi^2}}\frac{(1+|\alpha|^2)^2\left|\langle F,\varphi_\alpha\rangle_{L^2(Q)}\right|^2}{\left|\frac{\pi^2}{2R^2}|\alpha|^2\right|^2}+\sum_{|\alpha|^2\leqslant 2+\frac{4R^2\rho^2}{\pi^2}}\frac{(1+|\alpha|^2)^2\left|\langle F,\varphi_\alpha\rangle_{L^2(Q)}\right|^2}{\frac{\pi^2\rho^2}{R^2}}\\
&\leqslant \sum_{|\alpha|^2\geqslant 2+\frac{4R^2\rho^2}{\pi^2}}\frac{16R^4}{\pi^4}\left|\langle F,\varphi_\alpha\rangle_{L^2(Q)}\right|^2+\frac{\left(3+\frac{4R^2\rho^2}{\pi^2}\right)^2}{\frac{\pi^2\rho^2}{R^2}}\sum_{|\alpha|^2\leqslant 2+\frac{4R^2\rho^2}{\pi^2}}\left|\langle F,\varphi_\alpha\rangle_{L^2(Q)}\right|^2\\
&\leqslant C\rho^2\sum_{\alpha\in\mathbb{Z}^n}\left|\langle F,\varphi_\alpha\rangle_{L^2(Q)}\right|^2=C\rho^2\|F\|^2_{L^2(Q)}.
\end{aligned}$$

Here $C>0$ depends only on R. Combining this estimate with (2.17)-(2.18) we deduce that $z\in H^2(Q)$ fulfills (2.15) for $k=2$. In particular, for all $\beta=(\beta_1,\dots,\beta_n)\in\mathbb{N}^n$, $|\beta|\leqslant 2$, one can check that

$$\partial_x^\beta z=\sum_{\alpha\in\mathbb{Z}^n}\left(\frac{i\pi}{R}\right)^{|\beta|}\left[\prod_{j=1}^n\left(\alpha\cdot Se_j+\frac{\eta_1\cdot e_j}{2}\right)^{\beta_j}\right]a_\alpha\varphi_\alpha,$$

where

$$e_j=(0,\dots,0,\underbrace{1}_{\text{position } j},0,\dots 0).$$

It follows

$$\begin{aligned}
-\Delta z&=\sum_{\alpha\in\mathbb{Z}^n}\frac{\pi^2}{R^2}\left[\sum_{j=1}^n\left(\alpha\cdot Se_j+\frac{\eta_1\cdot e_j}{2}\right)^2\right]a_\alpha\varphi_\alpha\\
&=\sum_{\alpha\in\mathbb{Z}^n}\frac{\pi^2}{R^2}\left[\sum_{j=1}^n\left(S^*\alpha\cdot e_j+\frac{\eta_1\cdot e_j}{2}\right)^2\right]a_\alpha\varphi_\alpha\\
&=\sum_{\alpha\in\mathbb{Z}^n}\frac{\pi^2}{R^2}\left|S^*\alpha+\frac{\eta_1}{2}\right|^2a_\alpha\varphi_\alpha.
\end{aligned}$$

In the same way, we find

$$-\eta_1\cdot\nabla z=\sum_{\alpha\in\mathbb{Z}^n}\frac{-i\pi}{R}\left[S^*\alpha\cdot\eta_1+\frac{1}{2}\right]a_\alpha\varphi_\alpha.$$

From these two identities we deduce that

$$\begin{aligned}
-\Delta z-2\rho\eta_1\cdot\nabla z-\rho^2z&=\sum_{\alpha\in\mathbb{Z}^n}\left[\frac{\pi^2}{R^2}\left|S^*\alpha+\frac{\eta_1}{2}\right|^2-\frac{2i\pi\rho}{R}\left(S^*\alpha\cdot\eta_1+\frac{1}{2}\right)-\rho^2\right]a_\alpha\varphi_\alpha\\
&=\sum_{\alpha\in\mathbb{Z}^n}\langle F,\varphi_\alpha\rangle_{L^2(Q)}\varphi_\alpha=F.
\end{aligned}$$

Therefore z is a solution of (2.13). Using (2.20), we obtain

$$|a_\alpha| \leqslant \frac{R\left|\langle F, \varphi_\alpha\rangle_{L^2(Q)}\right|}{\pi\rho}$$

and, applying (2.18), we get

$$\|z\|_{L^2(Q)} \leqslant C\,\|F\|_{L^2(Q)}\,\rho^{-1},$$

with C independent of F and ρ. This proves (2.15) for $k=0$ and it completes the proof of the lemma. □

2.3. Proof of Theorem 2.2. In this section we will complete the proof of Theorem 2.2. For this purpose, let us assume that $\Lambda_{q_1} = \Lambda_{q_2}$. We fix $\xi \in \mathbb{R}^n$ and applying Proposition 2.1 we deduce the existence of $v_j \in H^2(\Omega)$, $j=1,2$, solving $-\Delta v_j + q_j v_j = 0$ of the form (2.9)-(2.10) with $w_j \in H^2(\Omega)$ satisfying (2.11). Consider $y_2 \in H^2(\Omega)$ solving

$$\begin{cases} -\Delta y_2 + q_2 y_2 = 0 & \text{in } \Omega, \\ y_2 = v_1 & \text{on } \partial\Omega. \end{cases}$$

Then, $v = y_2 - v_1$ solves

$$\begin{cases} -\Delta v + q_2 v = (q_1 - q_2)v_1 & \text{in } \Omega, \\ v = 0 & \text{on } \partial\Omega. \end{cases}$$

Moreover, fixing $\varphi \in H^{\frac{3}{2}}(\partial\Omega)$ defined by

$$\varphi(x) = v_1(x), \quad x \in \partial\Omega,$$

we find

$$\partial_\nu v = \partial_\nu y_2 - \partial_\nu v_1 = \Lambda_{q_2}\varphi - \Lambda_{q_1}\varphi = 0.$$

Thus, v satisfies the condition

$$\begin{cases} -\Delta v + q_2 v = (q_1 - q_2)v_1 & \text{in } \Omega, \\ v = \partial_\nu v = 0 & \text{on } \partial\Omega. \end{cases}$$

Multiplying, this equation by v_2 and integrating by parts, we get

$$\int_\Omega (q_1 - q_2)v_1 v_2 dx = \int_\Omega (-\Delta v + q_2 v)v_2 dx = \int_\Omega v(-\Delta v_2 + q_2 v_2)dx = 0.$$

We obtain the orthogonality identity

$$\int_\Omega (q_1 - q_2)v_1 v_2 dx = 0. \tag{2.21}$$

In addition, (2.9)-(2.10) imply

$$\int_\Omega (q_1 - q_2)v_1 v_2 dx = \int_\Omega (q_1 - q_2)e^{-ix\cdot\xi}dx + \int_\Omega Z_\rho dx,$$

with

$$Z_\rho(x) = (q_1(x) - q_2(x))\left[e^{-i\rho\eta_2\cdot x}w_1(x) + e^{i\rho\eta_2\cdot x}e^{-i\xi\cdot x}w_2(x) + w_1(x)w_2(x)\right], \ x \in \Omega.$$

Fixing $q \in L^\infty(\mathbb{R}^n) \cap L^1(\mathbb{R}^n)$ defined by

$$q(x) := \begin{cases} q_1(x) - q_2(x), & \text{if } x \in \Omega, \\ 0, & \text{if } x \in \mathbb{R}^n \setminus \Omega \end{cases}$$

and applying (2.11) we get

$$\lim_{\rho\to+\infty}\int_\Omega (q_1-q_2)v_1v_2dx=\int_\Omega qe^{-ix\cdot\xi}dx=\int_{\mathbb{R}^n} qe^{-ix\cdot\xi}dx.$$

Combining this with (2.21) we obtain

$$\int_{\mathbb{R}^n} qe^{-ix\cdot\xi}dx=0.$$

Since $\xi\in\mathbb{R}^n$ is arbitrary chosen from the injectivity of the Fourier transform we deduce that this condition implies $q=0$ and $q_1=q_2$. This completes the proof of Theorem 2.2.

3. Partial data results

Since the pioneer work of [**SU**], several authors considered the Calderón problem with measurements restricted to some portion of the boundary instead of the full boundary $\partial\Omega$ as stated in Theorem 2.1, 2.2. These class of inverse problems are called inverse problem with partial data. The goal of these inverse problems is to reduce as much as possible the portion of the boundary where the input are imposed to the system (corresponding to the support of the Dirichlet input) and the portion where the measurements are made (corresponding to restriction on the knowledge of the Neumann boundary values of solutions). In this section, we consider two class of partial data results associated with the Calderón problem. The first class of partial data results corresponds to partial data results under the assumption of Theorem 2.1, 2.2. The second class of inverse problems requires additional assumption. Namely, they need the knowledge of the coefficients on the neighborhood of the boundary. In this section we give an example of each of these two types of partial data results.

3.1. Partial data result without restriction. In this subsection we prove that the result of Theorem 2.1 and 2.2 is still true if one restrict the Neumann boundary measurement to, roughly speaking, half of the boundary. This approach which goes back to [**BU**], combines the complex geometric optics solutions of Section 3 with Carleman estimates with linear weight.

Let us first state this result. We start by fixing $\eta\in\mathbb{S}^{n-1}$ and we consider the decomposition of $\partial\Omega$ into the η-illuminated face $\partial\Omega_{-,\eta}:=\{x\in\partial\Omega:\ \nu(x)\cdot\eta\leqslant 0\}$ and the η-shadowed face $\partial\Omega_{+,\eta}:=\{x\in\partial\Omega:\ \nu(x)\cdot\eta\geqslant 0\}$. Then, for $\epsilon>0$, we introduce $V:=\{x\in\partial\Omega:\ \nu(x)\cdot\eta<2\epsilon\}$ and we consider the partial DN map

$$\Lambda_q^*: H^{\frac{3}{2}}(\partial\Omega)\ni\varphi\mapsto\partial_\nu v_{q,\varphi}|_V\in H^{\frac12}(V),$$

with $v_{q,\varphi}$ solving (2.5). The knowledge of this map corresponds to the knowledge of DN map Λ_q with restriction of the measurements to V which is a neighborhood of the η-illuminated face of $\partial\Omega$. The main result of this section can be stated as follows.

THEOREM 3.1. *For $j=1,2$, let $q_j\in\mathcal{Q}$. Then we have*

$$\Lambda_{q_1}^*=\Lambda_{q_2}^*\Longrightarrow q_1=q_2. \tag{3.1}$$

Let us also consider the partial DN map associated with (2.2) given by

$$\mathcal{N}_a^*: H^{\frac{3}{2}}(\partial\Omega)\ni\varphi\mapsto a\partial_\nu u_{a,\varphi}|_V\in H^{\frac12}(V).$$

Then in a similar way to Corollary 2.1, we can deduce from Theorem 3.1 the following result.

COROLLARY 3.1. *Assume that the conditions of Theorem 2.1 are fulfilled and Theorem 3.1 holds true. Then we have*

$$\mathcal{N}_{a_1}^* = \mathcal{N}_{a_2}^* \Longrightarrow a_1 = a_2.$$

From now on we will focus our attention on Theorem 3.1. We start with a Carleman estimate which will be the key ingredient for our problem.

PROPOSITION 3.1. *Let $q \in L^\infty(\Omega)$, $\eta_1 \in \mathbb{S}^{n-1}$ and $v \in H_0^1(\Omega) \cap H^2(\Omega)$. Then there exists $C > 0$ and $\rho_2 > \rho_1$[1] depending only on Ω, η_1 and $\|q\|_{L^\infty(\Omega)}$ such that for all $\rho > \rho_2$ we have*

$$\begin{aligned} &\rho^2 \int_\Omega e^{-2\rho x\cdot\eta_1} |v(x)|^2 dx + \rho \int_{\partial\Omega_{+,\eta_1}} e^{-2\rho x\cdot\eta_1} |\partial_\nu v(x)|^2 |\eta_1 \cdot \nu| d\sigma(x) \\ &\leqslant C \left[\int_\Omega e^{-2\rho x\cdot\eta_1} |(-\Delta + q) v(x)|^2 dx + \rho \int_{\partial\Omega_{-,\eta_1}} e^{-2\rho x\cdot\eta_1} |\partial_\nu v(x)|^2 |\eta_1 \cdot \nu| d\sigma(x) \right]. \end{aligned} \tag{3.2}$$

PROOF. We start by proving (3.2) for $q = 0$. Without lost of generality we may assume that v is real valued. We fix $w(x) = e^{-\rho x\cdot\eta_1} v(x)$ and $P = e^{-\rho x\cdot\eta_1} \Delta e^{\rho x\cdot\eta_1} = \Delta + 2\rho\eta_1\cdot\nabla + \rho^2$. Using the fact that $v \in H_0^1(\Omega)$ and the fact that $e^{-2\rho x\cdot\eta_1} |\Delta v(x)|^2 = |Pw(x)|^2$, we deduce that (3.2), with $q = 0$, follows from

$$\begin{aligned} &\rho^2 \int_\Omega |w(x)|^2 dx + \rho \int_{\partial\Omega_{+,\eta_1}} |\partial_\nu w(x)|^2 |\eta_1 \cdot \nu| d\sigma(x) \\ &\leqslant C \left[\int_\Omega |Pw(x)|^2 dx + \rho \int_{\partial\Omega_{-,\eta_1}} |\partial_\nu w(x)|^2 |\eta_1 \cdot \nu| d\sigma(x) \right], \end{aligned} \tag{3.3}$$

with $C > 0$ depending only on Ω. We start with this last estimate. For this purpose, we decompose the differential operator P into $P_+ + P_-$ with

$$P_+ = \Delta + \rho^2, \quad P_- = 2\rho\eta_1 \cdot \nabla.$$

We have

$$\begin{aligned} \|Pw\|_{L^2(\Omega)}^2 &\geqslant \|P_- w\|_{L^2(\Omega)}^2 + 2 \int_\Omega P_+ w P_- w dx \\ &\geqslant \|P_- w\|_{L^2(\Omega)}^2 + 2I_1 + 2I_2, \end{aligned} \tag{3.4}$$

with

$$I_1 = 2\rho \int_\Omega \Delta w(\eta_1 \cdot \nabla w) dx, \quad I_2 = 2\rho^3 \int_\Omega w(\eta_1 \cdot \nabla w) dx.$$

[1]Here ρ_1 denotes the constant of Proposition 2.1.

Using the fact that $w_{|\partial\Omega} = 0$ we deduce that $\nabla w = (\partial_\nu w)\nu$ on $\partial\Omega$ and integrating by parts, we obtain

$$
\begin{aligned}
I_1 &= 2\rho\int_{\partial\Omega}\partial_\nu w(\eta_1\cdot\nabla w)d\sigma(x) - 2\rho\int_\Omega \nabla w\cdot\nabla(\eta_1\cdot\nabla w)dx\\
&= 2\rho\int_{\partial\Omega}|\partial_\nu w|^2(\eta_1\cdot\nu)d\sigma(x) - \rho\int_\Omega \eta_1\cdot\nabla|\nabla w|^2 dx\\
&= 2\rho\int_{\partial\Omega}|\partial_\nu w|^2(\eta_1\cdot\nu)d\sigma(x) - \rho\int_\Omega \mathrm{div}\left(|\nabla w|^2\eta_1\right)dx\\
&= \rho\int_{\partial\Omega}|\partial_\nu w|^2(\eta_1\cdot\nu)d\sigma(x).
\end{aligned}
\tag{3.5}
$$

In the same way, we obtain

$$I_2 = \rho^3\int_\Omega \mathrm{div}\left(w^2\eta_1\right)dx = \rho^3\int_{\partial\Omega} w^2(\eta_1\cdot\nu)d\sigma(x) = 0.$$

Combining this with (3.4)-(3.5), we obtain

$$
\begin{aligned}
&\rho^2\int_\Omega|\eta_1\cdot\nabla w(x)|^2dx + 2\rho\int_{\partial\Omega_{+,\eta_1}}|\partial_\nu w(x)|^2|\eta_1\cdot\nu|d\sigma(x)\\
&\leqslant \int_\Omega|Pw(x)|^2dx + 2\rho\int_{\partial\Omega_{-,\eta_1}}|\partial_\nu w(x)|^2|\eta_1\cdot\nu|d\sigma(x).
\end{aligned}
\tag{3.6}
$$

In view of this estimate, the proof of (3.3) will be completed if we can estimate the left hand side of (3.3) with the one of (3.6). This will be possible thanks to the following Poincaré type of inequality.

LEMMA 3.1. *There exists a constant $C > 0$ depending only on Ω such that for all $w \in H^1_0(\Omega)$ we have*

$$\int_\Omega |w(x)|^2dx \leqslant C\int_\Omega|\eta_1\cdot\nabla w(x)|^2dx. \tag{3.7}$$

We postpone the proof of this lemma to the end of the present demonstration. Combining (3.7) with (3.6), we deduce easily (3.3) and (3.2) for $q = 0$. Now let us consider the case $q \neq 0$. For this purpose, note first that

$$
\begin{aligned}
\int_\Omega e^{-2\rho x\cdot\eta_1}|(-\Delta+q)v(x)|^2dx &\geqslant \int_\Omega e^{-2\rho x\cdot\eta_1}\left(\frac{|\Delta v|^2}{2} - |q|^2|v|^2\right)dx\\
&\geqslant \frac{1}{2}\int_\Omega e^{-2\rho x\cdot\eta_1}|\Delta v(x)|^2dx\\
&\qquad - \|q\|^2_{L^\infty(\Omega)}\int_\Omega e^{-2\rho x\cdot\eta_1}|v(x)|^2dx.
\end{aligned}
$$

Combining this with (3.2) for $q = 0$, we obtain

$$
\begin{aligned}
&\left(\rho^2 - 2C\|q\|^2_{L^\infty(\Omega)}\right)\int_\Omega e^{-2\rho x\cdot\eta_1}|v(x)|^2dx + \rho\int_{\partial\Omega_{+,\eta_1}} e^{-2\rho x\cdot\eta_1}|\partial_\nu v(x)|^2|\eta_1\cdot\nu|d\sigma(x)\\
&\leqslant C\left[2\int_\Omega e^{-2\rho x\cdot\eta_1}|(-\Delta+q)v(x)|^2dx + \rho\int_{\partial\Omega_{-,\eta_1}} e^{-2\rho x\cdot\eta_1}|\partial_\nu v(x)|^2|\eta_1\cdot\nu|d\sigma(x)\right].
\end{aligned}
$$

Therefore, fixing $\rho_2 = 2\sqrt{C}\,\|q\|_{L^\infty(\Omega)} + \rho_1$ we deduce easily (3.2) from this estimate.

□

Now that the proof of Proposition 3.1 is completed, let us consider Lemma 3.1.

Proof of Lemma 3.1. By density we only need to show this result for $w \in \mathcal{C}_0^\infty(\Omega)$. We fix $R > 0$ such that $\Omega \subset B_R := \{x \in \mathbb{R}^n : |x| < R\}$. Fixing $x \in \Omega$ and $h : s \mapsto w[(x - (x \cdot \eta_1)\eta_1) + s\eta_1]$, we deduce that

$$\begin{aligned} w(x) = w[(x - (x \cdot \eta_1)\eta_1) + (x \cdot \eta_1)\eta_1] &= h(x \cdot \eta_1) \\ &= \int_{-\infty}^{x\cdot\eta_1} h'(s)ds \\ &= \int_{-\infty}^{x\cdot\eta_1} \eta_1 \cdot \nabla w[(x - (x \cdot \eta_1)\eta_1) + s\eta_1]ds. \end{aligned}$$

Using the fact that

$$\forall s \in (-\infty, -R), \quad |(x - (x \cdot \eta_1)\eta_1) + s\eta_1|^2 = |x - (x \cdot \eta_1)\eta_1|^2 + s^2 \geqslant R^2,$$

and the fact that $\mathrm{supp}(w) \subset B_R$, we get

$$w(x) = \int_{-R}^{x\cdot\eta_1} \eta_1 \cdot \nabla w[(x - (x \cdot \eta_1)\eta_1) + s\eta_1]ds.$$

Therefore, integrating this last expression with respect to $x \in \mathbb{R}^n$ and making a change of variable with any coordinates having an orthonormal basis containing η_1, we obtain

$$\begin{aligned} \int_{\mathbb{R}^n} |w(x)|^2 dx &= \int_{\eta_1^\perp} \int_{-R}^{R} |w(x' + t\eta_1)|^2 dt dx' \\ &= \int_{\eta_1^\perp} \int_{-R}^{R} \left| \int_{-R}^{t} \eta_1 \cdot \nabla w(x' + s\eta_1) ds \right|^2 dx' dt. \end{aligned}$$

Here and from now on, for any $y \in \mathbb{R}^n$, we denote by $y^\perp$ the set $\{x \in \mathbb{R}^n : x \cdot y = 0\}$. Applying Hölder and Cauchy-Schwarz inequality, we obtain

$$\begin{aligned} \int_{\mathbb{R}^n} |w(x)|^2 dx &\leqslant \int_{\eta_1^\perp} \int_{-R}^{R} (t+R) \int_{-R}^{t} |\eta_1 \cdot \nabla w(x' + s\eta_1)|^2 ds dt dx' \\ &\leqslant \int_{\eta_1^\perp} \int_{-R}^{R} (t+R) \int_{-R}^{R} |\eta_1 \cdot \nabla w(x' + s\eta_1)|^2 ds dt dx' \\ &\leqslant 4R^2 \int_{\eta_1^\perp} \int_{-R}^{R} |\eta_1 \cdot \nabla w(x' + s\eta_1)|^2 ds dx' \\ &\leqslant 4R^2 \int_{\mathbb{R}^n} |\eta_1 \cdot \nabla w(x)|^2 dx. \end{aligned}$$

Combining this with the fact that w is supported in Ω we deduce (3.7) from this estimate. □

Armed with Proposition 3.1, we are now in position to complete the proof of Theorem 3.1.

Proof of Theorem 3.1. Let us assume that $\Lambda^*_{q_1} = \Lambda^*_{q_2}$. We fix $\eta_1 \in \{\theta \in \mathbb{S}^{n-1} : |\eta_1 - \eta| < \epsilon\}$ and we consider $\xi \in \eta_1^\perp$, $\eta_2 \in \eta_1^\perp \cap \mathbb{S}^{n-1}$. Applying Proposition 2.1, for $\rho > \rho_2$, we deduce the existence of $v_j \in H^2(\Omega)$, $j = 1, 2$, solving $-\Delta v_j + q_j v_j = 0$ of the form (2.9)-(2.10) with $w_j \in H^2(\Omega)$ satisfying (2.11). Consider $y_2 \in H^2(\Omega)$ solving

$$\begin{cases} -\Delta y_2 + q_2 y_2 = 0 & \text{in } \Omega, \\ y_2 = v_1 & \text{on } \partial\Omega. \end{cases}$$

Then, $v = y_2 - v_1$ solves

$$\begin{cases} -\Delta v + q_2 v = (q_1 - q_2)v_1 & \text{in } \Omega, \\ v = 0 & \text{on } \partial\Omega. \end{cases} \tag{3.8}$$

Moreover, fixing $\varphi \in H^{\frac{3}{2}}(\partial\Omega)$ defined by

$$\varphi(x) = v_1(x), \quad x \in \partial\Omega,$$

we find

$$\partial_\nu v = \partial_\nu y_2 - \partial_\nu v_1 = \Lambda_{q_2}^* \varphi - \Lambda_{q_1}^* \varphi = 0, \quad \text{on } V. \tag{3.9}$$

Therefore, multiplying (3.8) by v_2 and integrating by parts we get

$$\begin{aligned} \int_\Omega (q_1 - q_2) v_1 v_2 dx &= \int_\Omega (-\Delta v + q_2 v) v_2 dx \\ &= \int_\Omega v(-\Delta v_2 + q_2 v_2) dx + \int_U \partial_\nu v v_2 d\sigma(x) \\ &= \int_U \partial_\nu v v_2 d\sigma(x), \end{aligned} \tag{3.10}$$

where

$$U = \{x \in \partial\Omega : \ \nu(x) \cdot \eta \geqslant 2\epsilon\} = \partial\Omega \setminus V.$$

Now let us show that

$$\lim_{\rho \to +\infty} \int_U \partial_\nu v v_2 d\sigma(x) = 0. \tag{3.11}$$

For this purpose, we fix $U_1 = \{x \in \partial\Omega : \ \nu(x) \cdot \eta_1 \geqslant \epsilon\}$ and we remark that

$$\forall x \in U, \quad |\nu(x) - \eta_1| \geqslant |\nu(x) - \eta| - |\eta_1 - \eta| \geqslant \epsilon,$$

which proves that $U \subset U_1$. Therefore, we have

$$\left| \int_U \partial_\nu v v_2 d\sigma(x) \right| \leqslant \int_{U_1} |\partial_\nu v v_2| \, d\sigma(x).$$

Using the fact that v_2 takes the form (2.10) and applying the Cauchy-Schwarz inequality we get

$$\begin{aligned} \left| \int_U \partial_\nu v v_2 d\sigma(x) \right| &\leqslant \int_{U_1} e^{-\eta_1 \cdot x} |\partial_\nu v| (1 + |w_2|) d\sigma(x) \\ &\leqslant C \left(\int_{U_1} e^{-2\rho\eta_1 \cdot x} |\partial_\nu v|^2 d\sigma(x) \right)^{\frac{1}{2}} \left(1 + \|w_2\|_{L^2(\partial\Omega)} \right). \end{aligned} \tag{3.12}$$

Recall that using local coordinates, one can check that there exists $C > 0$ depending only on Ω such that

$$\|w_2\|_{L^2(\partial\Omega)} \leqslant C \|w_2\|_{L^2(\Omega)}^{\frac{1}{2}} \|\nabla w_2\|_{L^2(\Omega)}^{\frac{1}{2}} \leqslant C \|w_2\|_{L^2(\Omega)}^{\frac{1}{2}} \|w_2\|_{H^2(\Omega)}^{\frac{1}{2}}.$$

Therefore, applying (2.11), we deduce that there exists a constant $C > 0$ independent of $\rho > 0$ such that $\|w_2\|_{L^2(\partial\Omega)} \leqslant C$. Combining this with (3.12), we obtain

$$\left| \int_U \partial_\nu v v_2 d\sigma(x) \right| \leqslant C \left(\int_{U_1} e^{-2\rho\eta_1 \cdot x} |\partial_\nu v|^2 d\sigma(x) \right)^{\frac{1}{2}}. \tag{3.13}$$

On the other hand, using the fact that

$$\forall x \in U_1, \quad \nu(x) \cdot \eta_1 \geqslant \epsilon,$$

we obtain

$$\begin{aligned}\int_{U_1} e^{-2\rho\eta_1\cdot x}\left|\partial_\nu v\right|^2 d\sigma(x) &\leqslant \int_{U_1} e^{-2\rho\eta_1\cdot x}\left|\partial_\nu v\right|^2 \frac{\eta_1\cdot\nu(x)}{\epsilon} d\sigma(x)\\ &\leqslant \epsilon^{-1}\int_{U_1} e^{-2\rho\eta_1\cdot x}\left|\partial_\nu v\right|^2 |\eta_1\cdot\nu(x)| d\sigma(x)\\ &\leqslant \epsilon^{-1}\int_{\partial\Omega_{+,\eta_1}} e^{-2\rho\eta_1\cdot x}\left|\partial_\nu v\right|^2 |\eta_1\cdot\nu(x)| d\sigma(x).\end{aligned}$$

Therefore, using the fact that $v \in H_0^1(\Omega)\cap H^2(\Omega)$ we can apply the Carleman estimate of Proposition 3.1, with $q = q_2$, to obtain

$$\begin{aligned}&\int_{U_1} e^{-2\rho\eta_1\cdot x}\left|\partial_\nu v\right|^2 d\sigma(x)\\ &\leqslant \rho^{-1}\epsilon^{-1}\left(\rho\int_{\partial\Omega_{+,\eta_1}} e^{-2\rho\eta_1\cdot x}\left|\partial_\nu v\right|^2 |\eta_1\cdot\nu(x)| d\sigma(x)\right)\\ &\leqslant C\rho^{-1}\epsilon^{-1}\left(\int_\Omega e^{-2\rho\eta_1\cdot x}\left|-\Delta v+q_2 v\right|^2 dx+\rho\int_{\partial\Omega_{-,\eta_1}} e^{-2\rho\eta_1\cdot x}\left|\partial_\nu v\right|^2 |\eta_1\cdot\nu(x)| d\sigma(x)\right).\end{aligned}$$

Applying the fact that $\partial\Omega_{-,\eta_1}\subset V$, $-\Delta v + q_2 v = (q_1-q_2)v_1$ and (3.9), we get

$$\int_{U_1} e^{-2\rho\eta_1\cdot x}\left|\partial_\nu v\right|^2 d\sigma(x) \leqslant \left(\|q_1\|_{L^\infty(\Omega)}+\|q_2\|_{L^\infty(\Omega)}\right) C\rho^{-1}\epsilon^{-1}\int_\Omega e^{-2\rho\eta_1\cdot x}\left|v_1\right|^2 dx$$

and using (2.9) and (2.11), we find

$$\begin{aligned}\int_{U_1} e^{-2\rho\eta_1\cdot x}\left|\partial_\nu v\right|^2 d\sigma(x) &\leqslant C\rho^{-1}\int_\Omega e^{-2\rho\eta_1\cdot x}\left|e^{\rho\eta_1\cdot x}(1+|w_1|)\right|^2 dx\\ &\leqslant C\rho^{-1},\end{aligned}$$

with C independent of ρ. This proves (3.11). Applying (3.11) and the arguments used in the proof of Theorem 2.2, for $q = q_1 - q_2$ extended by zero to a function of $\mathbb{R}^n$, we get

$$\int_{\mathbb{R}^n} q(x)e^{-ix\cdot\xi}dx = 0. \tag{3.14}$$

Note that, following the above arguments, this property is true for any $\xi\in\eta_1^\perp$ while $\eta_1\in\{y\in\mathbb{S}^{n-1}:\ |y-\eta_1|<\epsilon\}$. Let us consider the following property.

Lemma 3.2. *Let $h\in L^1(\mathbb{R}^n)$ be compactly supported and fix the sequence $(\theta_k)_{k\in\mathbb{N}}$ of $\mathbb{S}^{n-1}$ such that*

$$\forall k,\ell\in\mathbb{N},\quad (k\neq\ell \Longrightarrow \theta_k\neq\theta_\ell). \tag{3.15}$$

Then the condition

$$\forall k\in\mathbb{N},\ \forall\xi\in\theta_k^\perp,\quad \int_{\mathbb{R}^n} h(x)e^{-ix\cdot\xi}dx = 0 \tag{3.16}$$

implies that $h = 0$.

We postpone the proof of Lemma 3.2 to the end of the present demonstration. According to Lemma 3.2, the fact that (3.14) is fulfilled for all $\xi\in\eta_1^\perp$ for $\eta_1\in\{y\in\mathbb{S}^{n-1}:\ |y-\eta_1|<\epsilon\}$ implies that $q=0$. This proves that $q_1=q_2$ which completes the proof of Theorem 3.1. □

Now that the proof of Theorem 3.1 is completed, let us consider Lemma 3.2.

Proof of Lemma 3.2. Fix $R > 0$ such that $\mathrm{supp}(h) \subset \{y \in \mathbb{R}^n : \ |y| < R\}$. Note first that for all $\xi \in \mathbb{R}^n$, we have

$$\mathcal{F}h(\xi) = \int_{\mathbb{R}^n} h(x)e^{-ix\cdot\xi}dx = \int_{|x|<R} h(x)\sum_{k=0}^{\infty}\frac{(-ix\cdot\xi)^k}{k!}dx.$$

Using the fact that

$$\forall x \in \{y \in \mathbb{R}^n : \ |y| < R\},\ \forall \xi \in \mathbb{R}^n, \qquad \left|\frac{(-ix\cdot\xi)^k}{k!}\right| \leqslant \frac{(R|\xi|)^k}{k!}$$

and

$$\sum_{k=0}^{\infty}\frac{(R|\xi|)^k}{k!} = e^{R|\xi|} < \infty,$$

we deduce that

$$\forall \xi \in \mathbb{R}^n, \quad \mathcal{F}h(\xi) = \sum_{k=0}^{\infty}\left(\int_{|x|<R} h(x)\frac{(-ix\cdot\xi)^k}{k!}dx\right). \tag{3.17}$$

Combining this representation with (3.15)-(3.16), we can show by iteration that for all $\ell \geqslant 1$ we can find a complex valued real-analytic function $g_\ell : \mathbb{R}^n \longrightarrow \mathbb{C}$ such that

$$\forall \xi \in \mathbb{R}^n, \quad \mathcal{F}h(\xi) = \left[\prod_{j=1}^{\ell}(\xi\cdot\theta_j)\right] g_\ell(\xi). \tag{3.18}$$

Now let us fix $\xi \in \mathbb{R}^n$ and consider the function $P : \mathbb{R} \ni t \mapsto \mathcal{F}h(t\xi)$. It is clear that

$$P(t) = \sum_{k=0}^{\infty}\left(\int_{|x|<R} h(x)\frac{(-ix\cdot\xi)^k}{k!}dx\right)t^k. \tag{3.19}$$

Moreover, applying (3.18), we deduce that

$$\begin{aligned}\forall \ell \geqslant 1,\ P(t) &= \left[\prod_{j=1}^{\ell}(t\xi\cdot\theta_j)\right] g_\ell(t\xi)\\ &= \left[\prod_{j=1}^{\ell}(\xi\cdot\theta_j)\right] g_\ell(t\xi)t^\ell = \underset{t\to 0}{\mathcal{O}}(t^\ell).\end{aligned}$$

Combining this with (3.19) we obtain

$$\forall k \geqslant 0, \quad \int_{|x|<R} h(x)\frac{(-ix\cdot\xi)^k}{k!}dx = 0$$

and combining this with (3.17) we deduce that $\mathcal{F}h(\xi) = 0$. Since ξ is arbitrary chosen we get

$$\forall \xi \in \mathbb{R}^n, \quad \mathcal{F}h(\xi) = 0$$

and using the injectivity of the Fourier transform we deduce that $h = 0$. This completes the proof of the lemma. □

3.2. Partial data result for coefficients known close to the boundary. In this subsection we consider an improvement of Theorem 3.1 for coefficients known on the neighborhood of the boundary of Ω. The approach that we consider is borrowed from [**AU**]. We fix γ_1, γ_2 two open sets of $\partial\Omega$ and we consider the subspace $H^{\frac{3}{2}}_{\gamma_1}(\partial\Omega)$ of $H^{\frac{3}{2}}(\partial\Omega)$ given by

$$H^{\frac{3}{2}}_{\gamma_1}(\partial\Omega) := \{h \in H^{\frac{3}{2}}(\partial\Omega) : \ \mathrm{supp}(h) \subset \gamma_1\}.$$

For $q \in \mathcal{Q}$, we consider the partial Dirichlet-to-Neumann map

$$\Lambda_{\gamma_1,\gamma_2,q} : H^{\frac{3}{2}}_{\gamma_1}(\partial\Omega) \ni \varphi \mapsto \partial_\nu v_{q,\varphi}|_{\gamma_2},$$

with $v_{q,\varphi}$ the solution of (2.5). Then, we consider the following result.

THEOREM 3.2. *Let Ω be connected. For $j = 1, 2$, let $q_j \in \mathcal{Q}$ be such that there exists an open set $\Omega_1 \subset \Omega$, corresponding to a connected neighborhood of the boundary (i.e $\partial\Omega \subset \partial\Omega_1$), such that*

$$q_1(x) = q_2(x), \quad x \in \Omega_1. \tag{3.20}$$

Then we have

$$\Lambda_{\gamma_1,\gamma_2,q_1} = \Lambda_{\gamma_1,\gamma_2,q_2} \Longrightarrow q_1 = q_2. \tag{3.21}$$

REMARK 3.1. In contrast to Theorem 3.1, Theorem 3.2 is stated with arbitrary restriction on the support of the Dirichlet input and on the location of the Neumann measurements. For instance, we can consider here problems with inputs and measurements on disjoint sets which is an interesting setting for several applications. However, in contrast to Theorem 3.1, Theorem 3.2 requires (3.20) to be fulfilled which corresponds to the knowledge of the coefficients under consideration on a neighborhood of the boundary.

For γ an open set of $\partial\Omega$ and $q \in \mathcal{Q}$ we introduce the following sets

$$S_q := \{v \in H^2(\Omega) : \ -\Delta v + qv = 0\}, \quad S_{q,\gamma} := \{v \in S_q : \ \mathrm{supp}(v_{|\partial\Omega}) \subset \gamma\}.$$

For Theorem 3.2, we need to consider first the following density result.

LEMMA 3.3. *The space $\{v_{|\Omega\setminus\Omega_1} : \ v \in S_{q,\gamma}\}$ is dense in $\{v_{|\Omega\setminus\Omega_1} : \ v \in S_q\}$ in the sense of $L^2(\Omega \setminus \Omega_1)$.*

PROOF. Assume the contrary. Then, the Hahn Banach theorem implies that there exists a continuous linear form on $L^2(\Omega\setminus\Omega_1)$ vanishing on $\{v_{|\Omega\setminus\Omega_1} : v \in S_{q,\gamma}\}$ but non uniformly vanishing on $\{v_{|\Omega\setminus\Omega_1} : \ v \in S_q\}$. This is equivalent to the existence of $y \in L^2(\Omega \setminus \Omega_1)$ and $v_0 \in S_q$ such that

$$\forall v \in S_{q,\gamma}, \quad \langle y, v\rangle_{L^2(\Omega\setminus\Omega_1)} = 0, \tag{3.22}$$

$$\langle y, v_0\rangle_{L^2(\Omega\setminus\Omega_1)} = 1. \tag{3.23}$$

We extend y by zero to Ω and we consider $w \in H^2(\Omega)$ solving

$$\begin{cases} -\Delta w + qw = y & \text{in } \Omega, \\ w = 0 & \text{on } \partial\Omega. \end{cases} \tag{3.24}$$

Applying (3.22), we get

$$\forall v \in S_{q,\gamma}, \quad \int_\Omega (-\Delta + q)w\overline{v}dx = \langle y, v\rangle_{L^2(\Omega\setminus\Omega_1)} = 0$$

and, integrating by parts, for all $v \in S_{q,\gamma}$, we obtain

$$\int_\Omega w\overline{(-\Delta + q)v}dx - \int_{\partial\Omega} \partial_\nu w\overline{v}d\sigma(x) = \int_\Omega (-\Delta + q)w\overline{v}dx = 0.$$

It follows that

$$\forall v \in S_{q,\gamma}, \quad \int_\gamma \partial_\nu w\overline{v}d\sigma(x) = 0. \tag{3.25}$$

Since $q \in \mathcal{Q}$, for any $\varphi \in \mathcal{C}_0^\infty(\gamma)$ there exists a unique $v_{q,\varphi} \in H^2(\Omega)$ solving (2.5). Using the fact that $v_{q,\varphi} \in S_{q,\gamma}$, we deduce from (3.25) that

$$\forall \varphi \in \mathcal{C}_0^\infty(\gamma), \quad \int_\gamma \partial_\nu w\overline{\varphi}d\sigma(x) = \int_\gamma \partial_\nu w\overline{v_{q,\varphi}}d\sigma(x) = 0.$$

This proves that $\partial_\nu w_{|\gamma} = 0$ and combining this with (3.24) and the fact that $y_{|\Omega_1} = 0$, we deduce that

$$\begin{cases} -\Delta w + qw = 0 & \text{in } \Omega_1, \\ w = \partial_\nu w = 0 & \text{on } \gamma. \end{cases}$$

Therefore, applying results of unique continuation for elliptic equations (e.g. [**GL**, Theorem 1.1] and [**SS**, Theorem 1]), we obtain $w = 0$ on Ω_1. In particular, we have $w = \partial_\nu w = 0$ on $\partial\Omega$. Therefore, integrating by parts, we find

$$\langle y, v_0\rangle_{L^2(\Omega\setminus\Omega_1)} = \int_\Omega (-\Delta + q)w\overline{v_0}dx = \int_\Omega w\overline{(-\Delta + q)v_0}dx = 0,$$

which contradicts (3.23). This completes the proof of the lemma. □

Armed with this lemma, we are now in position to complete the proof of Theorem 3.2.

Proof of Theorem 3.2. For $j = 1, 2$, consider $v_j \in S_{q_j,\gamma_j}$. We consider $y_2 \in H^2(\Omega)$ solving

$$\begin{cases} -\Delta y_2 + q_2 y_2 = 0 & \text{in } \Omega, \\ y_2 = v_1 & \text{on } \partial\Omega. \end{cases}$$

Then, $v = y_2 - v_1$ solves

$$\begin{cases} -\Delta v + q_2 v = (q_1 - q_2)v_1 & \text{in } \Omega, \\ v = 0 & \text{on } \partial\Omega. \end{cases} \tag{3.26}$$

Moreover, fixing $\varphi \in H^{\frac{3}{2}}_{\gamma_1}(\partial\Omega)$ defined by $\varphi(x) = v_1(x)$, $x \in \partial\Omega$, we find

$$\partial_\nu v_{|\gamma_2} = \Lambda_{q_2,\gamma_1,\gamma_2}\varphi - \Lambda_{q_1,\gamma_1,\gamma_2}\varphi = 0.$$

Therefore, multiplying (3.26) by v_2 and integrating by parts we get

$$\begin{aligned} \int_\Omega (q_1 - q_2)v_1 v_2 dx &= \int_\Omega (-\Delta v + q_2 v)v_2 dx \\ &= \int_\Omega v(-\Delta v_2 + q_2 v_2)dx - \int_{\gamma_2} \partial_\nu v v_2 d\sigma(x) \\ &= 0. \end{aligned}$$

Applying (3.20), we deduce that, for any $v_j \in S_{q_j,\gamma_j}$, $j = 1, 2$, we have

$$\int_{\Omega\setminus\Omega_1} (q_1 - q_2)v_1 v_2 dx = \int_\Omega (q_1 - q_2)v_1 v_2 dx = 0.$$

Using the density result of Lemma 3.3, we deduce that, for any $v_j \in S_{q_j}$, $j = 1,2$, we have

$$\int_\Omega (q_1 - q_2)v_1v_2dx = \int_{\Omega\setminus\Omega_1} (q_1 - q_2)v_1v_2dx = 0.$$

Thus, choosing v_j, $j = 1,2$, of the form (2.9)-(2.10), with $w_j \in H^2(\Omega)$ satisfying (2.11), in this last identity and repeating the argument used at the end of the proof of Theorem 2.2, we deduce that $q_1 = q_2$. □

References

[AU] H. Ammari and G. Uhlmann, *Reconstruction of the potential from partial Cauchy data for the Schrödinger equation*, Indiana Univ. Math. J. **53** (2004), no. 1, 169–183, DOI 10.1512/iumj.2004.53.2299. MR2048188

[BU] A. L. Bukhgeim and G. Uhlmann, *Recovering a potential from partial Cauchy data*, Comm. Partial Differential Equations **27** (2002), no. 3-4, 653–668, DOI 10.1081/PDE-120002868. MR1900557

[C] A.-P. Calderón, *On an inverse boundary value problem*, Seminar on Numerical Analysis and its Applications to Continuum Physics (Rio de Janeiro, 1980), Soc. Brasil. Mat., Rio de Janeiro, 1980, pp. 65–73. MR590275

[Ch] M. Choulli, *Une introduction aux problèmes inverses elliptiques et paraboliques* (French), Mathématiques & Applications (Berlin) [Mathematics & Applications], vol. 65, Springer-Verlag, Berlin, 2009. MR2554831

[GL] N. Garofalo and F.-H. Lin, *Unique continuation for elliptic operators: a geometric-variational approach*, Comm. Pure Appl. Math. **40** (1987), no. 3, 347–366, DOI 10.1002/cpa.3160400305. MR882069

[Ha] P. Hähner, *A periodic Faddeev-type solution operator*, J. Differential Equations **128** (1996), no. 1, 300–308, DOI 10.1006/jdeq.1996.0096. MR1392403

[J] J. Jossinet, *The impedivity of freshly excised human breast tissue*, Physiol. Meas., **19** (1998), 61-75.

[Ka] O. Kavian, *Lectures on parameter identification*, Three courses on partial differential equations, IRMA Lect. Math. Theor. Phys., vol. 4, de Gruyter, Berlin, 2003, pp. 125–162. MR2046906

[Ki] Y. Kian, *Recovery of non compactly supported coefficients of elliptic equations on an infinite waveguide*, J. Inst. Math. Jussieu **19** (2020), no. 5, 1573–1600, DOI 10.1017/s1474748018000488. MR4138952

[KV] R. Kohn and M. Vogelius, *Determining conductivity by boundary measurements*, Comm. Pure Appl. Math. **37** (1984), no. 3, 289–298, DOI 10.1002/cpa.3160370302. MR739921

[SS] J.-C. Saut and B. Scheurer, *Sur l'unicité du problème de Cauchy et le prolongement unique pour des équations elliptiques à coefficients non localement bornés* (French), J. Differential Equations **43** (1982), no. 1, 28–43, DOI 10.1016/0022-0396(82)90072-9. MR645635

[SU] J. Sylvester and G. Uhlmann, *A global uniqueness theorem for an inverse boundary value problem*, Ann. of Math. (2) **125** (1987), no. 1, 153–169, DOI 10.2307/1971291. MR873380

[Uh] G. Uhlmann, *Electrical impedance tomography and Calderón's problem*, Inverse Problems **25** (2009), no. 12, 123011, 39, DOI 10.1088/0266-5611/25/12/123011. MR3460047

[ZK] M. S. Zhdanov and G. V. Keller, *The geoelectrical methods in geophysical exploration*, Methods in Geochemistry and Geophysics, **31** (1994), Elsevier.

Aix Marseille Univ, Université de Toulon, CNRS, CPT, Marseille, France
Email address: yavar.kian@univ-amu.fr

Contemporary Mathematics
Volume **757**, 2020
https://doi.org/10.1090/conm/757/15248

Multidimensional Borg-Levinson inverse spectral problems

Éric Soccorsi

ABSTRACT. This text deals with multidimensional Borg-Levinson inverse theory. Its main purpose is to establish that the Dirichlet eigenvalues and Neumann boundary data of the operator $-\Delta+q$, acting in a bounded domain of $\mathbb{R}^d$ with $d \geqslant 2$, uniquely determine the real-valued bounded potential q. We first address the case of incomplete spectral data, where finitely many boundary spectral eigen-pairs remain unknown. Under suitable summability condition on the Neumann data, we also consider the case where only the asymptotic behavior of the eigenvalues is known. Finally, we use the multidimensional Borg-Levinson theory for solving parabolic inverse coefficient problems.

CONTENTS

1. A short introduction to inverse spectral problems

Let $\Omega \subset \mathbb{R}^d$, where $d \in \mathbb{N} := \{1, 2, \ldots\}$, be a bounded domain with $C^{1,1}$ boundary $\partial\Omega$. In the particular case where $d = 1$, we set $\Omega := (0, 1)$. Given $q \in L^\infty(\Omega)$, real-valued, we perturb the Dirichlet Laplacian in $L^2(\Omega)$ by q, i.e. we consider the operator acting in $L^2(\Omega)$ as $-\Delta+q$, that is endowed with homogeneous Dirichlet boundary conditions.

We investigate the inverse problem of determining the operator $-\Delta + q$, that is of determining the perturbation potential q, from knowledge of partial spectral data of $-\Delta + q$. More precisely, we are interested in two types of results:

- A *uniqueness result*, expressing that every two admissible potentials q_j, $j = 1, 2$, are equal whenever the spectral data of $-\Delta + q_1$ coincide with the ones of $-\Delta + q_2$, i.e. we seek the following implication:

$$(\text{Spectral data of } -\Delta + q_1 = \text{Spectral data of } -\Delta + q_2) \Longrightarrow (q_1 = q_2).$$

2010 *Mathematics Subject Classification.* Primary 35R30, 35J10, 35K10.
This work was partially supported by the Agence Nationale de la Recherche under grant ANR-17- CE40-0029.

- A *stability result*, claiming that any unknown admissible potential q is not only uniquely determined (in the sense of the above implication) by the spectral data of $-\Delta + q$, but also that it depends continuously on these data.

1.1. Self-adjointness, spectral data and all that. For $M \in (0,+\infty)$ fixed, let $q \in L^{\infty}(\Omega,\mathbb{R})$ fulfill

$$\|q\|_{L^{\infty}(\Omega)} \leqslant M. \tag{1.1}$$

We define A_q as the operator in $L^2(\Omega)$, associated with the closed sesquilinear form

$$a_q(u,v) := \int_{\Omega} \left(\nabla u(x) \cdot \overline{\nabla v(x)} + q(x) u(x) \overline{v(x)} \right) dx, \ u, v \in D(a_q) := H_0^1(\Omega), \tag{1.2}$$

where $H_0^1(\Omega)$ denotes the closure of $C_0^{\infty}(\Omega)$, the set of infinitely differentiable and compactly supported functions in Ω, for the topology of the first-order Sobolev space $H^1(\Omega)$. The operator A_q is self-adjoint in $L^2(\Omega)$ and acts on its domain[1] as

$$A_q u = (-\Delta + q) u, \ u \in D(A_q) = H_0^1(\Omega) \cap H^2(\Omega). \tag{1.3}$$

Here, the notation $H^2(\Omega)$ stands for the usual second-order Sobolev space in Ω, and we recall that the graph norm of A_q is equivalent to the one of $H^2(\Omega)$, i.e.

$$c^{-1} \|u\|_{H^2(\Omega)} \leqslant \|u\|_{D(A_q)} := \|u\|_{L^2(\Omega)} + \|A_q u\|_{L^2(\Omega)} \leqslant c \|u\|_{H^2(\Omega)}, \ u \in D(A_q), \tag{1.4}$$

for some constant $c \in (1,+\infty)$ that depends only on Ω and M.

Next, since the injection $H_0^1(\Omega) \hookrightarrow L^2(\Omega)$ is compact, then the same is true for the resolvent[2] of A_q, and the spectrum of A_q is discrete. We denote by $\{\lambda_n, \ n \in \mathbb{N}\}$ the non-decreasing sequence of eigenvalues of A_q, repeated with their multiplicity,

$$\lambda_1 \leqslant \ldots \leqslant \lambda_n \leqslant \lambda_{n+1} \leqslant \ldots, \ n \in \mathbb{N}.$$

In view of (1.1)-(1.2), we infer from the Min-Max principle that

$$\lambda_1 \geqslant -M,$$

and we recall for further use that

$$\lim_{n \to \infty} \lambda_n = +\infty. \tag{1.5}$$

Let $\{\varphi_n, \ n \in \mathbb{N}\}$ be an orthonormal basis in $L^2(\Omega)$ of eigenfunctions of A_q, such that

$$A_q \varphi_n = \lambda_n \varphi_n, \ n \in \mathbb{N}.$$

With reference to (1.4)-(1.5), there exist two constants $n_M \in \mathbb{N}$ and $c \in (0,+\infty)$, both of them depending only on Ω and M, such that

$$c^{-1} \lambda_n \leqslant \|\varphi_n\|_{H^2(\Omega)} \leqslant c \lambda_n, \ n \geqslant n_M. \tag{1.6}$$

Put

$$\psi_n := \partial_{\nu} \varphi_n, \ n \in \mathbb{N},$$

[1] The assumption $\partial\Omega \in C^{1,1}$ is needed for applying the classical elliptic regularity theory that establishes that $D(A_q) \subset H^2(\Omega)$.

[2] This is provided 0 is in the resolvent set of A_q, but since q is bounded as in (1.1), one can assume that this is the case without restricting the generality of the reasoning, upon possibly substituting $A_q + M$ for A_q.

where ν denotes the outward normal vector to $\partial\Omega$ and $\partial_\nu u := \nabla u \cdot \nu$ is the normal derivative of φ. Then, it follows from (1.6) and the continuity of the trace operator $\tau_1 : u \mapsto (\partial_\nu u)_{|\partial\Omega}$ from $H^2(\Omega)$ into $H^{1/2}(\partial\Omega)$, that we have

$$\|\psi_n\|_{H^{1/2}(\partial\Omega)} \leqslant c\lambda_n,\ n \geqslant n_M, \tag{1.7}$$

for some positive constant c that depends only on Ω and M.

1.2. Review of the one-dimensional case. Fix $d = 1$ and recall that we have $\Omega = (0,1)$ with

$$A_q = -\frac{d^2}{dx^2} + q(x),\ D(A_q) = \{u \in H^2(0,1),\ u(0) = u(1) = 0\},$$

in this case.

1.2.1. *An obstruction to identifiability.* A very natural question that arises in this context is to know whether q can be determined by knowledge of $\mathrm{Sp}(A_q) = \{\lambda_n,\ n \in \mathbb{N}\}$. But the answer is negative as the spectrum does not discriminate between symmetric potentials. This can be seen by noticing that we have

$$UA_qU^{-1} = A_{Uq}, \tag{1.8}$$

where we have set $(Uf)(x) := f(1-x)$ for all $f \in L^2(\Omega)$ and a.e. $x \in \Omega$. Since U is unitary in $L^2(\Omega)$, then the operators A_q and A_{Uq} are unitarily equivalent, by (1.8). Hence they are iso-spectral: $\mathrm{Sp}(A_{Uq}) = \mathrm{Sp}(A_q)$. Thus, one cannot distinguish between the potentials q and Uq, from knowledge of the two spectra $\mathrm{Sp}(A_q)$ and $\mathrm{Sp}(A_{Uq})$, despite of the fact that $q \neq Uq$ when q is not symmetric about the midpoint $x = 1/2$ of the interval Ω.

Therefore, the spectrum of A_q does not uniquely determine q, and some additional spectral data is needed for identifying the potential.

1.2.2. *One-dimensional Borg-Levinson theorem.* Assuming that $\varphi_n'(0) = \frac{d\varphi_n}{dx}(0) = 1$ for all $n \in \mathbb{N}$, G. Borg [**4**] and N. Levinson [**16**] established when $\mathrm{Sp}(A_q)$ is known, that additional knowledge of $\{\|\varphi_n\|_{L^2(\Omega)},\ n \in \mathbb{N}\}$ uniquely determines q.

THEOREM 1.1 (Borg (1946) and Levinson (1949)). *For $\lambda \in \mathbb{R}$ and for $q_j \in L^\infty(0,1;\mathbb{R})$, $j = 1,2$, let $u_j(\cdot,\lambda)$ be the $H^2(0,1)$-solution to the initial values problem*

$$\begin{cases} (-\frac{d^2}{dx^2} + q_j(x))u_j(x,\lambda) = \lambda u_j(x,\lambda), & x \in (0,1) \\ u_j(0,\lambda) = 0,\ u_j'(0,\lambda) = 1. \end{cases} \tag{1.9}$$

Denote by $\{\lambda_{j,n},\ n \in \mathbb{N}\}$ the non-decreasing sequence of the Dirichlet eigenvalues associated with A_{q_j}, obtained by imposing:

$$u_j(1,\lambda_{j,n}) = 0,\ n \in \mathbb{N}.$$

Then, we have the implication:

$$\left(\lambda_{1,n} = \lambda_{2,n} \text{ and } \|u_1(\cdot,\lambda_{1,n})\|_{L^2(0,1)} = \|u_2(\cdot,\lambda_{2,n})\|_{L^2(0,1)},\ n \in \mathbb{N}\right) \Longrightarrow (q_1 = q_2 \text{ in } (0,1)).$$

Later on, I. M. Gel'fand and B. M. Levitan proved that uniqueness is still valid upon substituting $u_j'(1,\lambda_{j,n})$ for $\|u_j(\cdot,\lambda_{j,n})\|_{L^2(0,1)}$, $j=1,2$, in Theorem 1.1:

THEOREM 1.2 (Gel'fand-Levitan (1951)). *Under the conditions of Theorem 1.1 we have:*

$$(\lambda_{1,n}=\lambda_{2,n} \text{ and } u_1'(1,\lambda_{1,n})=u_2'(1,\lambda_{2,n}),\ n\in\mathbb{N}) \Longrightarrow (q_1=q_2 \text{ in } (0,1)).$$

REMARK 1.3. Let p, q, ρ be real-valued and bounded functions in $(0,1)$, with p and ρ positive. Introduce the operator

$$A_{p,q,\rho}=-\rho^{-1}\frac{d}{dx}\left(p\frac{d}{dx}\right)+q,\ D(A_{p,q,\rho})=\{u\in H_0^1(0,1),\ pu'\in H^1(0,1)\},$$

which is self-adjoint in $L^2_\rho(0,1)$, the usual $L^2(0,1)$-space endowed with the weighted scalar product $\langle u,v\rangle_{L^2_\rho(0,1)}=\int_0^1\rho(x)u(x)\overline{v(x)}dx$. Denote by $\{\lambda_n,\ n\in\mathbb{N}\}$ the sequence of the eigenvalues of $A_{p,q,\rho}$ and by $\{u_n,\ n\in\mathbb{N}\}$ a $L^2_\rho(0,1)$-orthonormal basis of eigenfunctions of $A_{p,q,\rho}$, obeying $A_{p,q,\rho}u_n=\lambda_n u_n$. If p and ρ are $C^{1,1}(0,1)$, then the boundary spectral data

$$\mathrm{BSD}(p,q,\rho)=\{(\lambda_n,u_n'(1)),\ n\in\mathbb{N}\}$$

uniquely determine either of the three coefficients p, q and ρ, when the two others are known. Indeed, one can check by using the Liouville transformation $y(x)=L^{-1}\int_0^x p^{-1/2}(t)\rho^{1/2}(t)dt$, $x\in(0,1)$, where $L=\int_0^1 p^{-1/2}(t)\rho^{1/2}(t)dt$, as a coordinate transformation, that the equation $-(pu')'+qu-\lambda\rho u=0$ in $(0,1)$ reduces to its normal form $-u''+Vu=\lambda u$ in $(0,1)$, where $V=V_{p,q,\rho}$ is expressed in terms of p, q and ρ, while the boundary spectral data is preserved, i.e.

$$\mathrm{BSD}(1,V,1)=\mathrm{BSD}(p,q,\rho).$$

Thus, Theorem 1.2 yields recovery of V from $\mathrm{BSD}(p,q,\rho)$, hence the result. Notice that this change of coordinates is no longer valid for discontinuous p and ρ. We refer the reader to [**7**] for a specific treatment of this problem.

All the approaches from G. Borg, N. Levinson or I. M. Gel'fand and B. M. Levitan, were based on highly one-dimensional techniques, but two great ideas emerged in the 80', that paved the way toward solving the multidimensional Borg-Levinson inverse spectral problem. The first one is called the C-property, see [**21**]. It is due to A. G. Ramm who showed that the set $\{u_1(\cdot,\lambda)u_2(\cdot,\lambda),\ \lambda\in(0,+\infty)\}$ is dense in $L^1(0,1)$. The second one is called the boundary control method, see [**3**]. It was established by M. I. Belishev upon applying the boundary controllability theory to the hyperbolic equation associated with the Sturm-Liouville system (1.9). One common nice feature of these two great ideas is that they apply to higher dimensions $d\geqslant 2$ as well.

1.3. Multidimensional identification results.

1.3.1. *Boundary spectral data.* Let us recall that $\{\lambda_n,\ n\in\mathbb{N}\}$ is the non-decreasing sequence of the eigenvalues of A_q (repeated with the multiplicity), that $\{\varphi_n,\ n\in\mathbb{N}\}$ is a $L^2(\Omega)$-orthonormal basis of eigenfunctions of A_q such that $A_q\varphi_n=\lambda_n\varphi_n$, and that $\psi_n=\partial_\nu\varphi_n$. We define the boundary spectral data (BSD) of A_q, or the BSD associated with q, as:

$$\mathrm{BSD}(q):=\{(\lambda_n,\psi_n),\ n\in\mathbb{N}\}.$$

REMARK 1.4. For all $n \in \mathbb{N}$, one may replace φ_n by $e^{i\theta_n}\varphi_n$ with $\theta_n \in \mathbb{R}$, in the above definition. Thus it is clear that the BSD are not defined in a unique way: they depend on the choice of the $L^2(\Omega)$-orthonormal basis $\{\varphi_n,\ n \in \mathbb{N}\}$ of eigenfunctions of A_q.

1.3.2. *Multidimensional identifiability.* In 1988, it was proved for $d \geqslant 2$ by A. Nachman, J. Sylvester and G. Uhlmann in [**17**], and independently by R. Novikov in [**18**], that the potential q is uniquely determined by BSD(q), i.e. that the following implication

$$(\mathrm{BSD}(q_1) = \mathrm{BSD}(q_2)) \Longrightarrow (q_1 = q_2),$$

holds for any two suitable potentials q_j, $j = 1, 2$. This result has been improved in several ways by various authors.

Firstly, H. Isozaki [**11**] (see also M. Choulli [**8**]) extended the result of [**17**, **18**] when finitely many eigenpairs remain unknown.

THEOREM 1.5. *For $j=1,2$, let $q_j \in L^\infty(\Omega, \mathbb{R})$ and write*[3] $\mathrm{BSD}(q_j)=\{(\lambda_{j,n}, \psi_{j,n}),\ n \in \mathbb{N}\}$. *Then, for all $N \in \mathbb{N}$, we have the following implication:*

$$((\lambda_{1,n}, \psi_{1,n}) = (\lambda_{2,n}, \psi_{2,n}),\ n \geqslant N) \Longrightarrow (q_1 = q_2).$$

Recently, uniqueness in the determination of q was proved in [**9**, **13**] from the knowledge of the asymptotic behavior of BSD(q) when $n \to +\infty$.

THEOREM 1.6. *Let q_j for $j = 1, 2$, and the notations, be the same as in Theorem 1.5. Assume that the asymptotics of* BSD(q_1) *and* BSD(q_2) *coincide, in the sense that*

$$\lim_{n \to \infty} (\lambda_{1,n} - \lambda_{2,n}) = 0 \text{ and } \sum_{n=1}^{+\infty} \|\psi_{1,n} - \psi_{2,n}\|^2_{L^2(\partial\Omega)} < \infty.$$

Then, we have $q_1 = q_2$ in Ω.

The multidimensional Borg-Levinson theorem has been studied in many different kinds of settings[4] and it is not quite possible to give an extensive survey of this here, but we shall mention a few results which are relevant for the problem under investigation in this text. In [**19**], L. Päivärinta and V. Serov proved identifiability of unbounded potentials $q \in L^p(\Omega, \mathbb{R})$ for $p > d/2$, by BSD(q). The case of $p = d/2$, $d \geqslant 3$, has been studied by V. Pohjola in [**20**]. As for Borg-Levinson inverse spectral theory with partial Neumann data, we refer the reader to M. Bellassoued, M. Choulli and M. Yamamoto's article [**2**], where a log-stability estimate for electric potentials which are known in a neighborhood of the boundary, is established with respect to the BSD measured on an arbitrary non-empty open subset of the boundary[5].

---

[3]That is to say that $\{\lambda_{j,n},\ n \in \mathbb{N}\}$ is the non-decreasing sequence of the eigenvalues of A_{q_j} and that $\psi_{j,n} = \partial_\nu \varphi_{j,n}$ for all $n \in \mathbb{N}$, where $\{\varphi_{j,n},\ n \in \mathbb{N}\}$ is a $L^2(\Omega)$-orthonormal basis of eigenvectors of A_{q_j} such that $A_{q_j}\varphi_{j,n} = \lambda_{j,n}\varphi_{j,n}$.

[4]Such as operators in the divergence form, see e.g. B. Canuto and O. Kavian's paper [**6**], where two unknown coefficients out of three are simultaneously identified by the BSD, or magnetic Schrödinger operators, see e.g. Y. Kian's article [**14**].

[5]The strategy that is used in this paper is quite the opposite of the one we apply in the last section of this text for solving parabolic inverse coefficient problems by means of the Borg-Levinson theorem, in the sense that the authors rather derive their spectral stability result from a hyperbolic stability inequality.

1.3.3. *Outline.* The paper is organized as follows. In Section 2 we prove Theorems 1.5 and 1.6. In Subsection 2.1 we express the strong solution to the Dirichlet problem for $-\Delta+q-\lambda$, $\lambda\in\mathbb{C}\setminus\mathrm{Sp}(A_q)$, in terms of BSD$(q)$. Subsection 2.2 contains the proof Isozaki's formula, which is useful for the derivation of Theorems 1.5 and 1.6, presented in Subsections 2.3 and 2.4, respectively. In Subsection 2.5 we examine the stability issue of the Borg-Levinson inverse problem under study. Finally, we derive a parabolic identification result in Section 3, by means of Theorem 1.5.

2. Multidimensional Borg-Levinson theory

This section contains the proof of the incomplete Borg-Levinson theorem stated in Theorem 1.5 and the asymptotic Borg-Levinson theorem stated in Theorem 1.6. We start by establishing several technical results that are needed by the derivation of Theorems 1.5 and 1.6.

2.1. Preliminaries. For $q\in L^\infty(\Omega,\mathbb{R})$, $f\in H^{3/2}(\partial\Omega)$ and $\lambda\in\mathbb{C}$, we consider the boundary value problem (BVP)

$$\begin{cases} (-\Delta+q-\lambda)u=0 & \text{in } \Omega \\ u=f & \text{on } \partial\Omega. \end{cases} \tag{2.1}$$

First, we establish that there exists a unique strong solution[6] to the Cauchy problem (2.1) that can be expressed in terms of BSD(q).

LEMMA 2.1. *Let $q\in L^\infty(\Omega,\mathbb{R})$ and $f\in H^{3/2}(\partial\Omega)$. Then, for each $\lambda\in\mathbb{C}\setminus\mathrm{Sp}(A_q)$ there exists a unique solution $u_\lambda\in H^2(\Omega)$ to (2.1). Moreover u_λ reads*

$$u_\lambda=\sum_{n=1}^{+\infty}\frac{\langle f,\psi_n\rangle_{L^2(\partial\Omega)}}{\lambda-\lambda_n}\varphi_n \text{ in } L^2(\Omega), \tag{2.2}$$

and we have

$$\lim_{\lambda\to-\infty}\|u_\lambda\|^2_{L^2(\Omega)}=\lim_{\lambda\to-\infty}\left(\sum_{n=1}^{+\infty}\left|\frac{\langle f,\psi_n\rangle_{L^2(\partial\Omega)}}{\lambda-\lambda_n}\right|^2\right)=0. \tag{2.3}$$

PROOF. We split the proof into three steps.

STEP 1 (Existence and uniqueness of the solution to (2.1)). Since $f\in H^{3/2}(\partial\Omega)$ and since the trace operator $\tau_0: v\mapsto v_{|\partial\Omega}$ is surjective from $H^2(\Omega)$ onto $H^{3/2}(\partial\Omega)$, then there exists $F\in H^2(\Omega)$ such that $\tau_0F=f$. Thus, u_λ is a solution to (2.1) iff $v_\lambda:=u_\lambda-F$ solves

$$\begin{cases} (-\Delta+q-\lambda)v=G & \text{in } \Omega \\ v=0 & \text{on } \partial\Omega, \end{cases} \tag{2.4}$$

with $G:=-(-\Delta+q-\lambda)F\in L^2(\Omega)$. Next, λ being in the resolvent set of A_q, we see that (2.4) admits a unique solution $v_\lambda=(A_q-\lambda)^{-1}G$. Therefore, $u_\lambda=v_\lambda+F$ is the unique solution to (2.1), and $u_\lambda\in H^2(\Omega)$.

STEP 2 (Proof of (2.2)). For all $n\in\mathbb{N}$, we have

$$0=\langle(-\Delta+q-\lambda)u_\lambda,\varphi_n\rangle_{L^2(\Omega)}=\int_\Omega(-\Delta+q(x)-\lambda)u_\lambda(x)\overline{\varphi_n(x)}dx,$$

[6] A strong solution to (2.1) is a solution in $H^2(\Omega)$, which satisfies the equation a. e. in Ω.

whence

$$\begin{aligned} 0 &= -\int_{\partial\Omega} \partial_\nu u_\lambda(x)\overline{\varphi_n(x)}dx + \int_{\partial\Omega} u_\lambda(x)\overline{\psi_n(x)}dx + \int_\Omega u_\lambda(x)\overline{(A_q-\overline{\lambda})\varphi_n(x)}dx \\ &= \int_{\partial\Omega} f(x)\overline{\psi_n(x)}dx + (\lambda_n-\lambda)\int_\Omega u_\lambda(x)\overline{\varphi_n(x)}dx, \end{aligned}$$

by integrating by parts. As a consequence we have $\langle u_\lambda, \varphi_n\rangle_{L^2(\Omega)} = \frac{\langle f,\psi_n\rangle_{L^2(\partial\Omega)}}{\lambda-\lambda_n}$, so (2.2) follows readily from this and the $L^2(\Omega)$-decomposition $u_\lambda = \sum_{n=1}^{+\infty}\langle u_\lambda, \varphi_n\rangle_{L^2(\Omega)}\varphi_n$.

STEP 3 (Proof of (2.3)). With reference to (1.1)-(1.2), we have for all $n\in\mathbb{N}$,

$$\lambda_n = \langle A_q\varphi_n, \varphi_n\rangle_{L^2(\Omega)} = a_q(\varphi_n,\varphi_n) = \int_\Omega |\nabla\varphi_n(x)|^2\,dx + \int_\Omega q(x)\,|\varphi_n(x)|^2\,dx \geqslant -M,$$

hence $\mathrm{Sp}(A_q)\subset[-M,+\infty)$. Thus,we see that every $\lambda\in(-\infty,-(1+M)]$ lies in the resolvent set of A_q, and that

$$\left|\frac{\langle f,\psi_n\rangle_{L^2(\partial\Omega)}}{\lambda-\lambda_n}\right|^2 \leqslant \left|\frac{\langle f,\psi_n\rangle_{L^2(\partial\Omega)}}{1+M+\lambda_n}\right|^2,\ n\in\mathbb{N}. \tag{2.5}$$

Further, since $\sum_{n=1}^{+\infty}\left|\frac{\langle f,\psi_n\rangle_{L^2(\partial\Omega)}}{1+M+\lambda_n}\right|^2 = \left\|u_{-(1+M)}\right\|^2_{L^2(\Omega)} < \infty$, by (2.2) and the Parseval theorem, and since $\lim_{\lambda\to-\infty}\left|\frac{\langle f,\psi_n\rangle_{L^2(\partial\Omega)}}{\lambda-\lambda_n}\right|^2 = 0$ for all $n\in\mathbb{N}$, we infer from (2.5) and the Lebesgue dominated convergence theorem that

$$\lim_{\lambda\to-\infty}\left(\sum_{n=1}^{+\infty}\left|\frac{\langle f,\psi_n\rangle_{L^2(\partial\Omega)}}{\lambda-\lambda_n}\right|^2\right) = \sum_{n=1}^{+\infty}\left(\lim_{\lambda\to-\infty}\left|\frac{\langle f,\psi_n\rangle_{L^2(\partial\Omega)}}{\lambda-\lambda_n}\right|^2\right) = 0.$$

Putting this, together with (2.2) and the Parseval formula, we obtain (2.3). □

Notice that the series in (2.2) converges in $L^2(\Omega)$ and not in $H^2(\Omega)$. Therefore, the normal derivative $\partial_\nu u_\lambda$ of the solution u_λ to (2.1) cannot be obtained directly from (2.2), by substituting ψ_n for φ_n on the right hand side. To achieve this, we need to introduce an additional specific spectral parameter μ, and consider the difference $u_\lambda - u_\mu$, as follows.

LEMMA 2.2. *Let q and f be the same as in Lemma 2.1. Then, for all λ and μ in $\mathbb{C}\setminus\mathrm{Sp}(A_q)$, we have*

$$\partial_\nu(u_\lambda-u_\mu) = (\mu-\lambda)\sum_{n=1}^{+\infty}\frac{\langle f,\psi_n\rangle_{L^2(\partial\Omega)}}{(\lambda-\lambda_n)(\mu-\lambda_n)}\psi_n \ \textit{in}\ H^{1/2}(\partial\Omega). \tag{2.6}$$

Here u_λ (resp., u_μ) denotes the $H^2(\Omega)$-solution to (2.1) (resp., (2.1) where λ is replaced by μ), given by Lemma 2.1.

PROOF. In view of (2.1), we see that $v := u_\lambda - u_\mu$ solves

$$\begin{cases} (-\Delta+q-\lambda)v = (\lambda-\mu)u_\mu & \text{in } \Omega \\ v = 0 & \text{on } \partial\Omega. \end{cases} \tag{2.7}$$

Since λ is in the resolvent set of A_q, (2.7) yields

$$v = (\lambda-\mu)(A_q-\lambda)^{-1}u_\mu = (\lambda-\mu)\sum_{n=1}^{+\infty}\frac{\langle u_\mu,\varphi_n\rangle_{L^2(\Omega)}}{\lambda_n-\lambda}\varphi_n, \tag{2.8}$$

the series being convergent in $L^2(\Omega)$. Recall that we have

$$u_\mu = \sum_{n=1}^{+\infty} \frac{\langle f, \psi_n \rangle_{L^2(\partial\Omega)}}{\mu - \lambda_n} \varphi_n \text{ in } L^2(\Omega), \tag{2.9}$$

upon substituting μ for λ in (2.2). Putting this together with (2.8), we get that

$$v = (\lambda - \mu) \sum_{n=1}^{+\infty} \frac{\langle f, \psi_n \rangle_{L^2(\partial\Omega)}}{(\lambda_n - \lambda)(\mu - \lambda_n)} \varphi_n \text{ in } L^2(\Omega). \tag{2.10}$$

Next, since $v \in D(A_q)$ and $A_q v = (\lambda - \mu) u_\mu + \lambda v$, we deduce from (2.9)-(2.10) that

$$A_q v = (\lambda - \mu) \sum_{n=1}^{+\infty} \frac{\lambda_n \langle f, \psi_n \rangle_{L^2(\partial\Omega)}}{(\lambda_n - \lambda)(\mu - \lambda_n)} \varphi_n \text{ in } L^2(\Omega).$$

Therefore, the series in (2.10) converges for the topology of the norm of A_q, hence it converges in $H^2(\Omega)$, according to (1.4). Finally, we obtain (2.6) from this by invoking the continuity of the trace operator $\tau_1 : u \mapsto (\partial_\nu u)_{|\partial\Omega}$ from $H^2(\Omega)$ into $H^{1/2}(\partial\Omega)$. □

The next lemma claims for any two real-valued bounded potentials q_1 and q_2, that the solutions to (2.1) associated with either $q = q_1$ or $q = q_2$, are closed as $\lambda \to -\infty$: in some sense the influence of the potential is dimmed when the spectral parameter λ goes to $-\infty$.

LEMMA 2.3. *Let $f \in H^{3/2}(\partial\Omega)$ and let $q_j \in L^\infty(\Omega, \mathbb{R})$, $j = 1, 2$. For $\lambda \in \mathbb{C} \setminus (\mathrm{Sp}(A_{q_1}) \cup \mathrm{Sp}(A_{q_2}))$, let $u_{j,\lambda}$ be the solution to (2.1) where q_j is substituted for q, which is given by Lemma 2.1. Then, we have*

$$\lim_{\lambda \to -\infty} \|\partial_\nu u_{1,\lambda} - \partial_\nu u_{2,\lambda}\|_{L^2(\partial\Omega)} = 0. \tag{2.11}$$

PROOF. Set $w_\lambda := u_{1,\lambda} - u_{2,\lambda}$, so we have

$$\begin{cases} (-\Delta + q_1 - \lambda) w_\lambda = (q_2 - q_1) u_{2,\lambda} & \text{in } \Omega \\ w_\lambda = 0 & \text{on } \partial\Omega, \end{cases}$$

from (2.1), and hence $w_\lambda = (A_{q_1} - \lambda)^{-1}(q_2 - q_1) u_{2,\lambda}$. Bearing in mind for all real number $\lambda < -\|q_1\|_{L^\infty(\Omega)}$, that $\|(A_{q_1} - \lambda)^{-1}\|_{\mathcal{B}(L^2(\Omega))} = \mathrm{dist}^{-1}(\lambda, \mathrm{Sp}(A_{q_1})) \leqslant 1/(\|q_1\|_{L^\infty(\Omega)} + \lambda)$, we find that

$$\|w_\lambda\|_{L^2(\Omega)} \leqslant \frac{\|q_2 - q_1\|_{L^\infty(\Omega)} \|u_{2,\lambda}\|_{L^2(\Omega)}}{-\lambda - \|q_1\|_{L^\infty(\Omega)}}, \quad \lambda \in \left(-\infty, -\|q_1\|_{L^\infty(\Omega)}\right).$$

Here and in the remaining part of this text, $\mathcal{B}(L^2(\Omega))$ denotes the space of linear bounded operators[7] in $L^2(\Omega)$. From this and (2.3) it then follows that

$$\lim_{\lambda \to -\infty} \lambda \|w_\lambda\|_{L^2(\Omega)} = 0. \tag{2.12}$$

Next, since $A_{q_1} w_\lambda = (q_2 - q_1) u_{2,\lambda} + \lambda w_\lambda$, it holds true for every $\lambda < -\|q_1\|_{L^\infty(\Omega)}$ that

$$\|A_{q_1} w_\lambda\|_{L^2(\Omega)} \leqslant \|q_2 - q_1\|_{L^\infty(\Omega)} \|u_{2,\lambda}\|_{L^2(\Omega)} - \lambda \|w_\lambda\|_{L^2(\Omega)},$$

[7] The usual norm of $T \in \mathcal{B}(L^2(\Omega))$ is defined by $\|T\|_{\mathcal{B}(L^2(\Omega))} = \sup_{f \in L^2(\Omega) \setminus \{0\}} \frac{\|Tf\|_{L^2(\Omega)}}{\|f\|_{L^2(\Omega)}}$.

so we get $\lim_{\lambda\to-\infty}\|A_{q_1}w_\lambda\|_{L^2(\Omega)}=0$, from (2.3) and (2.12). As a consequence we have

$$\lim_{\lambda\to-\infty}\left(\|w_\lambda\|_{L^2(\Omega)}+\|A_{q_1}w_\lambda\|_{L^2(\Omega)}\right)=0.$$

This and (1.4) entail

$$\lim_{\lambda\to-\infty}\|u_{1,\lambda}-u_{2,\lambda}\|_{H^2(\Omega)}=0$$

which together with the continuity of the trace operator $\tau_1 : u \mapsto (\partial_\nu u)_{|\partial\Omega}$ from $H^2(\Omega)$ into $H^{1/2}(\partial\Omega)$, yield (2.11). □

2.2. Isozaki's asymptotic representation formula. Let $q_j \in L^\infty(\Omega,\mathbb{R})$ satisfy

$$\|q_j\|_{L^\infty(\Omega)}\leqslant M,\ j=1,2, \tag{2.13}$$

for some *a priori* fixed constant $M\in(0,+\infty)$. In [**11**], H. Isozaki gives a simple representation formula, expressing the difference q_1-q_2 in terms of the Dirichlet-to-Neumann (DN) operator associated with the BVP obtained by substituting q_j for q in (2.1). More precisely, adapting the argument of [**11**] to fit our aim in this text, we fix $\tau\in(1,+\infty)$ and we consider the BVP (2.1) with $\lambda=\lambda_\tau^+:=(\tau+i)^2$ and $q=q_j$, i.e.

$$\begin{cases}(-\Delta+q_j-\lambda_\tau^+)u=0 & \text{in } \Omega\\ u=f & \text{on } \partial\Omega.\end{cases} \tag{2.14}$$

We denote by u_{j,λ_τ^+} the $H^2(\Omega)$-solution to (2.14) (for the sake of notational simplicity we drop the dependence of u_{j,λ_τ^+} on f). Let us introduce the DN map associated with (2.14), as

$$\begin{array}{rccc}\Lambda_{j,\lambda_\tau^+}: & H^{3/2}(\partial\Omega) & \to & H^{1/2}(\partial\Omega)\\ & f & \mapsto & \left(\partial_\nu u_{j,\lambda_\tau^+}\right)_{|\partial\Omega}.\end{array} \tag{2.15}$$

Given two test functions $f_\tau^\pm$, we shall make precise below, we aim to link the difference q_1-q_2 to the asymptotic behavior of

$$S_\tau:=S_{1,\tau}-S_{2,\tau},\ \text{where } S_{j,\tau}:=\langle\Lambda_{j,\lambda_\tau^+}f_\tau^+,f_\tau^-\rangle_{L^2(\partial\Omega)}, \tag{2.16}$$

as $\tau\to+\infty$.

2.2.1. *Test functions.* For $\xi\in\mathbb{R}^d$ fixed, and for every $\tau\in(|\xi|+1,+\infty)$, we set $\lambda_\tau^\pm:=(\tau\pm i)^2$, and we seek two functions $f_\tau^\pm$ such that

$$\begin{cases}(-\Delta-\lambda_\tau^\pm)f_\tau^\pm=0 \text{ in } \Omega\\ \lim_{\tau\to+\infty}f_\tau^+(x)\overline{f_\tau^-(x)}=e^{-i\xi\cdot x},\ x\in\Omega\\ \sup_{\tau\in(|\xi|+1,+\infty)}\|f_\tau^\pm\|_{C(\overline{\Omega})}<\infty.\end{cases} \tag{2.17}$$

Here and in the remaining part of this text, the notation $\cdot$ (resp., $|\cdot|$) stands for the Euclidian product (resp., norm) in $\mathbb{R}^d$.

Pick $\eta\in\mathbb{S}^{d-1}$ such that $\xi\cdot\eta=0$, and put

$$\beta_\tau:=\sqrt{1-\frac{|\xi|^2}{4\tau^2}}\ \text{and}\ \eta_\tau^\pm:=\beta_\tau\eta\mp\frac{\xi}{2\tau}, \tag{2.18}$$

in such a way that $|\eta_\tau^\pm|=1$. Then, the two following functions

$$f_\tau^\pm(x):=e^{i(\tau\pm i)\eta_\tau^\pm\cdot x},\ x\in\Omega, \tag{2.19}$$

fulfill the conditions of (2.17). As a matter of fact, it can be checked through direct computation from (2.18)-(2.19), that $\Delta f_\tau^\pm = -\lambda_\tau^\pm |\eta_\tau^\pm|^2 f_\tau^\pm = -\lambda_\tau^\pm f_\tau^\pm$ in Ω, that $f_\tau^+(x)\overline{f_\tau^-(x)} = e^{-i\frac{\tau+i}{\tau}\xi\cdot x}$ for all $x \in \Omega$, and that

$$|f_\tau^\pm(x)| \leqslant e^{|\eta_\tau^\pm||x|} \leqslant e^{|x|},\ x \in \overline{\Omega}. \tag{2.20}$$

We notice for further use from (2.20) that the estimate

$$\|f_\tau^\pm\|_{L^p(X)} \leqslant c_* := \left(1 + |\Omega|^{1/2} + |\partial\Omega|^{1/2}\right) \sup_{x\in\overline{\Omega}} e^{|x|}, \tag{2.21}$$

holds with $X = \Omega$ or $X = \partial\Omega$, and with $p = 2$ or $p = \infty$. Here $|\Omega|$ (resp., $|\partial\Omega|$) denotes the diameter of Ω (resp., the length of $\partial\Omega$).

2.2.2. *Probing* (2.1) *with* $f_\tau^\pm$. For $j = 1, 2$ and $z \in \mathbb{C} \setminus \mathrm{Sp}(A_{q_j})$, we denote by $u_{j,z}^\pm$ the $H^2(\Omega)$-solution to the BVP (2.1), where $(q_j, z, f_\tau^\pm)$ is substituted for (q, λ, f). The function $u_{j,z}^\pm$ is characterized by

$$\begin{cases} (-\Delta + q_j - z)u_{j,z}^\pm = 0 & \text{in } \Omega \\ u_{j,z}^\pm = f_\tau^\pm & \text{on } \partial\Omega, \end{cases} \tag{2.22}$$

hence $v_{j,z}^\pm := u_{j,z}^\pm - f_\tau^\pm$ solves

$$\begin{cases} (-\Delta + q_j - z)v_{j,z}^\pm = -(-\Delta + q_j - z)f_\tau^\pm & \text{in } \Omega \\ v_{j,z}^\pm = 0 & \text{on } \partial\Omega. \end{cases} \tag{2.23}$$

Moreover, since $(-\Delta + q_j - z)f_\tau^\pm = (q_j + \lambda_\tau^\pm - z)f_\tau^\pm$, by the first line in (2.17), it follows from (2.23) that

$$v_{j,z}^\pm = -(A_{q_j} - z)^{-1}(q_j + \lambda_\tau^\pm - z)f_\tau^\pm. \tag{2.24}$$

Let us now examine the case where $z = \lambda_\tau^\pm$, which is permitted since $\lambda_\tau^\pm$ belongs to the resolvent set of the self-adjoint operator A_{q_j}, as we have:

$$\mathrm{Im}\left(\lambda_\tau^\pm\right) = \pm 2\tau \neq 0. \tag{2.25}$$

We shall establish that the $L^2(\Omega)$-norm of $v_{j,\lambda_\tau^\pm}^\pm$ scales like τ^{-1} as τ becomes large, whereas the one of $u_{j,\lambda_\tau^\pm}^\pm$ is bounded uniformly in $\tau \in (1 + |\xi|, +\infty)$. To do that, we substitute $\lambda_\tau^\pm$ for z in (2.24) and get that $v_{j,\lambda_\tau^\pm}^\pm = -(A_{q_j} - \lambda_\tau^\pm)^{-1} q_j f_\tau^\pm$. Next, using that $\|(A_{q_j} - \lambda_\tau^\pm)^{-1}\|_{\mathcal{B}(L^2(\Omega))} = \mathrm{dist}^{-1}(\lambda_\tau^\pm, \mathrm{Sp}(A_{q_j})) \leqslant (2\tau)^{-1}$, according to (2.25), we obtain

$$\left\|v_{j,\lambda_\tau^\pm}^\pm\right\|_{L^2(\Omega)} \leqslant \frac{\|q_j\|_{L^\infty(\Omega)} \|f_\tau^\pm\|_{L^2(\Omega)}}{2\tau} \leqslant \frac{Mc_*}{2\tau},\ j = 1, 2, \tag{2.26}$$

upon applying (2.13) and (2.21) with $(X, p) = (\Omega, 2)$. Now, bearing in mind that $\tau \geqslant 1$ and recalling that $u_{j,z}^\pm = v_{j,z}^\pm + f_\tau^\pm$, we derive from (2.26) that

$$\left\|u_{j,\lambda_\tau^\pm}^\pm\right\|_{L^2(\Omega)} \leqslant \frac{M+2}{2} c_*,\ j = 1, 2. \tag{2.27}$$

2.2.3. *Isozaki's formula.*

PROPOSITION 2.4. *For $j = 1, 2$, let $q_j \in L^\infty(\Omega, \mathbb{R})$ satisfy* (2.13). *Then, for all $\xi \in \mathbb{R}^d$, we have*

$$\int_\Omega (q_1(x) - q_2(x))e^{-i\xi\cdot x}dx = \lim_{\tau\to+\infty} S_\tau, \tag{2.28}$$

where S_τ is defined by (2.15)-(2.16).

PROOF. For $j = 1, 2$, we consider the $H^2(\Omega)$-solution u^+_{j,λ^+_τ} to the BVP (2.14) with $f = f^+_\tau$:

$$\begin{cases} (-\Delta + q_j - \lambda^+_\tau)u^+_{j,\lambda^+_\tau} = 0 & \text{in } \Omega \\ u^+_{j,\lambda^+_\tau} = f^+_\tau & \text{on } \partial\Omega. \end{cases} \tag{2.29}$$

Upon left-multiplying the first line of (2.29) by $\overline{f^-_\tau}$, integrating over Ω, and applying the Green formula, we obtain with the aid of (2.17) that

$$\begin{aligned} 0 &= \int_\Omega (-\Delta + q_j - \lambda^+_\tau)u^+_{j,\lambda^+_\tau}(x)\overline{f^-_\tau(x)}dx \\ &= \langle f^+_\tau, \partial_\nu f^-_\tau\rangle_{L^2(\partial\Omega)} - \langle \partial_\nu u^+_{j,\lambda^+_\tau}, f^-_\tau\rangle_{L^2(\partial\Omega)} + \int_\Omega u^+_{j,\lambda^+_\tau}(x)\overline{(-\Delta + q_j - \lambda^-_\tau)f^-_\tau(x)}dx \\ &= \langle f^+_\tau, \partial_\nu f^-_\tau\rangle_{L^2(\partial\Omega)} - S_{j,\tau} + \int_\Omega q_j(x)u^+_{j,\lambda^+_\tau}(x)\overline{f^-_\tau(x)}dx,\ j = 1, 2. \end{aligned}$$

Thus, we have $S_{j,\tau} = \langle f^+_\tau, \partial_\nu f^-_\tau\rangle_{L^2(\partial\Omega)} + \int_\Omega q_j(x)u^+_{j,\lambda^+_\tau}(x)\overline{f^-_\tau(x)}dx$ for $j = 1, 2$, and consequently

$$S_\tau = S_{1,\tau} - S_{2,\tau} = \int_\Omega \left(q_1(x)u^+_{1,\lambda^+_\tau}(x) - q_2(x)u^+_{2,\lambda^+_\tau}(x)\right)\overline{f^-_\tau(x)}dx. \tag{2.30}$$

Next, taking into account that $u^+_{j,\lambda^+_\tau} = v^+_{j,\lambda^+_\tau} + f^+_\tau$ for $j = 1, 2$, we deduce from (2.30) that

$$S_\tau - \int_\Omega (q_1(x) - q_2(x))f^+_\tau(x)\overline{f^-_\tau(x)}dx = \int_\Omega (q_1(x)v^+_{1,\lambda^+_\tau}(x) - q_2(x)v^+_{2,\lambda^+_\tau}(x))\overline{f^-_\tau(x)}dx.$$

Therefore, by applying (2.21) with $(X, p) = (\Omega, 2)$ and (2.26), we get

$$\begin{aligned} \left|S_\tau - \int_\Omega (q_1(x) - q_2(x))f^+_\tau(x)\overline{f^-_\tau(x)}dx\right| &\leqslant \left(\sum_{j=1}^2 \|q_j\|_{L^\infty(\Omega)}\left\|v^+_{j,\lambda^+_\tau}\right\|_{L^2(\Omega)}\right)\|f^-_\tau\|_{L^2(\Omega)} \\ &\leqslant \frac{M^2c_*^2}{\tau}, \end{aligned}$$

which leads to:

$$\lim_{\tau\to+\infty}\left(S_\tau - \int_\Omega (q_1(x) - q_2(x))f^+_\tau(x)\overline{f^-_\tau(x)}dx\right) = 0. \tag{2.31}$$

Finally, as we have

$$\lim_{\tau\to+\infty}\int_\Omega (q_1(x) - q_2(x))f^+_\tau(x)\overline{f^-_\tau(x)}dx = \int_\Omega (q_1(x) - q_2(x))e^{-i\xi\cdot x}dx,$$

by the second line of (2.17), (2.21) with $(X, p) = (\Omega, +\infty)$, (2.13), and the dominated convergence theorem, the desired result follows directly from this and from (2.31). □

Armed with Proposition 2.4 we turn now to proving Theorems 1.5 and 1.6.

2.3. Proof of the incomplete Borg-Levinson theorem. In this section we prove Theorem 1.5. In view of Proposition 2.4, we have to show that

$$\lim_{\tau\to+\infty} S_\tau = 0,\ \xi \in \mathbb{R}^d. \tag{2.32}$$

Indeed, by combining the Isozaki formula (2.28) with (2.32), we get for every $\xi \in \mathbb{R}^d$ that the Fourier transform

$$(\mathcal{F}q)(\xi) := \frac{1}{(2\pi)^{d/2}} \int_{\mathbb{R}^d} q(x) e^{-i\xi\cdot x} dx \tag{2.33}$$

of the following function

$$q(x) := \begin{cases} q_1(x) - q_2(x) & \text{if } x \in \Omega \\ 0 & \text{if } x \in \mathbb{R}^d \setminus \Omega, \end{cases} \tag{2.34}$$

reads $(\mathcal{F}q)(\xi) = \frac{1}{(2\pi)^{d/2}} \int_\Omega (q_1(x) - q_2(x)) e^{-i\xi\cdot x} dx = \frac{1}{(2\pi)^{d/2}} \lim_{\tau\to+\infty} S_\tau = 0$. By the injectivity of Fourier transform $\mathcal{F}$, this entails that $q = 0$ in $\mathbb{R}^d$, i.e. that $q_1 = q_2$ in Ω.

We turn now to establishing (2.32). To this purpose, we fix $\xi \in \mathbb{R}^d$, pick $\tau \in (|\xi| + 1, +\infty)$, and for $j = 1, 2$ and all $z \in \mathbb{C} \setminus \mathrm{Sp}(A_{q_j})$, we consider the $H^2(\Omega)$-solution $u_{j,z}^+$ to the BVP (2.1), where (q_j, z, f_τ^+) is substituted for (q, λ, f), i.e.

$$\begin{cases} (-\Delta + q_j - z) u_{j,z}^+ = 0 & \text{in } \Omega \\ u_{j,z}^+ = f_\tau^+ & \text{on } \partial\Omega. \end{cases}$$

For $z_j \in \mathbb{C} \setminus \mathrm{Sp}(A_{q_j})$, $j = 1, 2$, we put $u_{j,z_1,z_2}^+ := u_{j,z_1}^+ - u_{j,z_2}^+$ and recall from (2.16) that

$$S_\tau = \langle \Lambda_{1,\lambda_\tau^+} f_\tau^+, f_\tau^- \rangle_{L^2(\partial\Omega)} - \langle \Lambda_{2,\lambda_\tau^+} f_\tau^+, f_\tau^- \rangle_{L^2(\partial\Omega)} = \langle \partial_\nu u_{1,\lambda_\tau^+}^+ - \partial_\nu u_{2,\lambda_\tau^+}^+, f_\tau^- \rangle_{L^2(\partial\Omega)}.$$

Thus, for every $\mu \in (-\infty, -M)$ we have

$$S_\tau = \langle \partial_\nu u_{1,\lambda_\tau^+,\mu}^+, f_\tau^- \rangle_{L^2(\partial\Omega)} - \langle \partial_\nu u_{2,\lambda_\tau^+,\mu}^+, f_\tau^- \rangle_{L^2(\partial\Omega)} + \langle \partial_\nu u_{1,\mu}^+ - \partial_\nu u_{2,\mu}^+, f_\tau^- \rangle_{L^2(\partial\Omega)}. \tag{2.35}$$

In view of (2.11), we have $\lim_{\mu\to-\infty} \left\| \partial_\nu u_{1,\mu}^+ - \partial_\nu u_{2,\mu}^+ \right\|_{L^2(\partial\Omega)} = 0$ so we get

$$S_\tau = \lim_{\mu\to-\infty} \langle \partial_\nu u_{1,\lambda_\tau^+,\mu}^+ - \partial_\nu u_{2,\lambda_\tau^+,\mu}^+, f_\tau^- \rangle_{L^2(\partial\Omega)}, \tag{2.36}$$

upon sending μ to $-\infty$ in (2.35).

Next, we introduce

$$\kappa_{\tau,\mu}(t) := \frac{\mu - \lambda_\tau^+}{(\lambda_\tau^+ - t)(\mu - t)},\ t \in \mathbb{R} \setminus \{\mu\}, \tag{2.37}$$

and set

$$\zeta_\tau(\psi, \varphi) := \langle f_\tau^+, \psi \rangle_{L^2(\partial\Omega)} \overline{\langle f_\tau^-, \varphi \rangle}_{L^2(\partial\Omega)},\ \psi,\ \varphi \in L^2(\partial\Omega). \tag{2.38}$$

In light of Lemma 2.2, the scalar product on the right hand side of (2.36) decomposes as

$$\langle \partial_\nu u_{1,\lambda_\tau^+,\mu}^+ - \partial_\nu u_{2,\lambda_\tau^+,\mu}^+, f_\tau^- \rangle_{L^2(\partial\Omega)} = \sum_{n=1}^{+\infty} \left(\kappa_{\tau,\mu}(\lambda_{1,n}) \zeta_\tau(\psi_{1,n}) - \kappa_{\tau,\mu}(\lambda_{2,n}) \zeta_\tau(\psi_{2,n}) \right), \tag{2.39}$$

where the notation $\zeta_\tau(\psi)$ is a shorthand for $\zeta_\tau(\psi,\psi)$. Further, as $(\lambda_{1,n},\psi_{1,n}) = (\lambda_{2,n},\psi_{2,n})$ for every $n \geqslant N$, by assumption, (2.39) becomes
(2.40)

$$\langle \partial_\nu u^+_{1,\lambda_\tau^+,\mu} - \partial_\nu u^+_{2,\lambda_\tau^+,\mu}, f_\tau^- \rangle_{L^2(\partial\Omega)} = \sum_{n=1}^{N-1} \left(\kappa_{\tau,\mu}(\lambda_{1,n})\zeta_\tau(\psi_{1,n}) - \kappa_{\tau,\mu}(\lambda_{2,n})\zeta_\tau(\psi_{2,n})\right),$$

the sum on the right hand side of the above equality being taken equal to zero when $N=1$. Further, taking into account that

$$\lim_{\mu\to-\infty} \kappa_{\tau,\mu}(\lambda_{j,n}) = 1/(\lambda_\tau^+ - \lambda_{j,n}),\ j=1,2,\ n=1,\dots,N-1,$$

we deduce from (2.36) and (2.40), that

$$S_\tau = \sum_{n=1}^{N-1} \left(\frac{\zeta_\tau(\psi_{1,n})}{\lambda_\tau^+ - \lambda_{1,n}} - \frac{\zeta_\tau(\psi_{2,n})}{\lambda_\tau^+ - \lambda_{2,n}} \right).$$

Next, bearing in mind that $\mathrm{Im}\,(\lambda_\tau^+) = 2\tau$, we see that $|\lambda_\tau^+ - \lambda_{j,n}| \geqslant 2\tau$ for $j=1,2$ and for all $n=1,\dots,N-1$, and hence that

$$\begin{aligned} |S_\tau| &\leqslant \sum_{n=1}^{N-1} \left(|\zeta_\tau(\psi_{1,n})| + |\zeta_\tau(\psi_{2,n})|\right)(2\tau)^{-1} \\ &\leqslant \|f_\tau^+\|_{L^2(\partial\Omega)} \|f_\tau^-\|_{L^2(\partial\Omega)} \sum_{n=1}^{N-1} \left(\|\psi_{1,n}\|^2_{L^2(\partial\Omega)} + \|\psi_{2,n}\|^2_{L^2(\partial\Omega)} \right)(2\tau)^{-1}, \end{aligned}$$

from (2.37)-(2.38). Now, applying (2.21) with $X=\partial\Omega$ and $p=2$, we obtain that

$$|S_\tau| \leqslant c_*^2 \sum_{n=1}^{N-1} \left(\|\psi_{1,n}\|^2_{L^2(\partial\Omega)} + \|\psi_{2,n}\|^2_{L^2(\partial\Omega)} \right)\tau^{-1},$$

which immediately entails (2.32).

2.4. Proof of the asymptotic Borg-Levinson theorem. In this section we prove Theorem 1.6. We stick with the notations of Section 2.3 and recall from (2.36) and (2.39) that
(2.41)

$$S_\tau = \lim_{\mu\to-\infty} \langle \partial_\nu u^+_{1,\lambda_\tau^+,\mu} - \partial_\nu u^+_{2,\lambda_\tau^+,\mu}, f_\tau^- \rangle_{L^2(\partial\Omega)} = \lim_{\mu\to-\infty} \sum_{n=1}^{+\infty} \left(A_{n,\tau,\mu} + B_{n,\tau,\mu}\right),$$

where

$$A_{n,\tau,\mu} := \left(\kappa_{\tau,\mu}(\lambda_{1,n}) - \kappa_{\tau,\mu}(\lambda_{2,n})\right)\zeta_\tau(\psi_{1,n}), \tag{2.42}$$

and

$$B_{n,\tau,\mu} := \kappa_{\tau,\mu}(\lambda_{2,n})\left(\zeta_\tau(\psi_{1,n}-\psi_{2,n},\psi_{1,n}) + \zeta_\tau(\psi_{2,n},\psi_{1,n}-\psi_{2,n})\right). \tag{2.43}$$

We split the proof into three steps. The first one, presented in Section 2.4.1, is to show that

$$\lim_{\mu\to-\infty} \sum_{n=1}^{+\infty} A_{n,\tau,\mu} = \sum_{n=1}^{+\infty} A_{n,\tau,*},\ \text{where } A_{n,\tau,*} := \frac{(\lambda_{1,n}-\lambda_{2,n})\zeta_\tau(\psi_{1,n})}{(\lambda_\tau^+ - \lambda_{1,n})(\lambda_\tau^+ - \lambda_{2,n})}, \tag{2.44}$$

while the second one, given in Section 2.4.2, establishes that

$$\text{(2.45)}\qquad \lim_{\mu\to-\infty}\sum_{n=1}^{+\infty}B_{n,\tau,\mu}=\sum_{n=1}^{+\infty}B_{n,\tau,*},$$

$$\text{where } B_{n,\tau,*}:=\frac{\zeta_\tau(\psi_{1,n}-\psi_{2,n},\psi_{1,n})+\zeta_\tau(\psi_{2,n},\psi_{1,n}-\psi_{2,n})}{\lambda_\tau^+-\lambda_{2,n}}.$$

Finally, the end of the proof is displayed in Section 2.4.3.

2.4.1. *Step 1: Proof of* (2.44). Let us start by noticing that

$$|\kappa_{\tau,\mu}(\lambda_{1,n})-\kappa_{\tau,\mu}(\lambda_{2,n})|\leqslant 2\,|\lambda_{1,n}-\lambda_{2,n}|\max_{t\in[\lambda_{1,n},\lambda_{2,n}]}\left(\frac{1}{|\lambda_\tau^+-t|^2}+\frac{1}{|\mu-t|^2}\right),\ n\in\mathbb{N}.$$

This can be seen from the identity $\kappa_{\tau,\mu}(\lambda_{1,n})-\kappa_{\tau,\mu}(\lambda_{2,n})=-\int_{\lambda_{1,n}}^{\lambda_{2,n}}\kappa_{\tau,\mu}'(t)dt$, which yields

$$\begin{aligned}|\kappa_{\tau,\mu}(\lambda_{1,n})&-\kappa_{\tau,\mu}(\lambda_{2,n})|\\ &\leqslant|\lambda_{1,n}-\lambda_{2,n}|\max_{t\in[\lambda_{1,n},\lambda_{2,n}]}\left(\frac{|\lambda_\tau^+-\mu|}{|\lambda_\tau^+-t|\,|\mu-t|^2}+\frac{|\lambda_\tau^+-\mu|}{|\lambda_\tau^+-t|^2\,|\mu-t|}\right),\end{aligned}$$

and from the basic estimate $|\lambda_\tau^+-\mu|\leqslant|\lambda_\tau^+-t|+|\mu-t|$, entailing:

$$\begin{aligned}\frac{|\lambda_\tau^+-\mu|}{|\lambda_\tau^+-t|\,|\mu-t|^2}+\frac{|\lambda_\tau^+-\mu|}{|\lambda_\tau^+-t|^2\,|\mu-t|}&\leqslant\frac{1}{|\mu-t|^2}+\frac{2}{|\lambda_\tau^+-t|\,|\mu-t|}+\frac{1}{|\lambda_\tau^+-t|^2}\\ &\leqslant\frac{2}{|\mu-t|^2}+\frac{2}{|\lambda_\tau^+-t|^2}.\end{aligned}$$

Denote by $\lambda_{*,n}$ a real number between $\lambda_{1,n}$ and $\lambda_{2,n}$, where the maximum of the function $t\mapsto|\lambda_\tau^+-t|^{-2}+|\mu-t|^{-2}$ is achieved, in such a way that we have
(2.46)

$$|\kappa_{\tau,\mu}(\lambda_{1,n})-\kappa_{\tau,\mu}(\lambda_{2,n})|\leqslant 2\,|\lambda_{1,n}-\lambda_{2,n}|\left(\frac{1}{|\lambda_\tau^+-\lambda_{*,n}|^2}+\frac{1}{|\mu-\lambda_{*,n}|^2}\right),\ n\in\mathbb{N}.$$

Next, bearing in mind that $\lim_{n\to+\infty}\lambda_{1,n}=+\infty$, we pick $N_0\in\mathbb{N}$ so large, that

$$\text{(2.47)}\qquad \lambda_{1,N_0}\geqslant|\lambda_\tau^+|+4M.$$

Since $\lambda_{1,n}\geqslant\lambda_{1,N_0}$ for all $n\geqslant N_0$, we have $|\lambda_\tau^+-\lambda_{1,n}|\geqslant\lambda_{1,n}-|\lambda_\tau^+|\geqslant 4M$ in this case, whence $|\lambda_\tau^+-\lambda_{*,n}|\geqslant|\lambda_\tau^+-\lambda_{1,n}|-|\lambda_{1,n}-\lambda_{*,n}|\geqslant|\lambda_\tau^+-\lambda_{1,n}|-2M$. Here, we used the basic inequality $|\lambda_{1,n}-\lambda_{*,n}|\leqslant|\lambda_{1,n}-\lambda_{2,n}|$ and the estimate

$$\text{(2.48)}\qquad |\lambda_{1,n}-\lambda_{2,n}|\leqslant\|q_1-q_2\|_{L^\infty(\Omega)}\leqslant 2M,\ n\in\mathbb{N},$$

arising from the Min-Max principle and the operator identity $A_{q_2}=A_{q_1}+q_2-q_1$. Therefore, we have

$$\text{(2.49)}\qquad |\lambda_\tau^+-\lambda_{*,n}|\geqslant\frac{|\lambda_\tau^+-\lambda_{1,n}|}{2},\ n\geqslant N_0.$$

Similarly, taking $\mu\in(-\infty,-(1+5M))$, we have $|\mu-\lambda_{1,n}|\geqslant-\mu-M\geqslant 4M$. Since $|\lambda_{1,n}-\lambda_{*,n}|\leqslant 2M$, by (2.48), we get that $|\lambda_{1,n}-\lambda_{*,n}|\leqslant|\mu-\lambda_{1,n}|/2$, and hence

$$|\mu-\lambda_{*,n}|\geqslant|\mu-\lambda_{1,n}|-|\lambda_{1,n}-\lambda_{*,n}|\geqslant\frac{|\mu-\lambda_{1,n}|}{2},\ n\in\mathbb{N}.$$

Putting this together with (2.42), (2.46) and (2.49), we obtain that

$$|A_{n,\tau,\mu}| \leqslant 8\delta_1 \left(\frac{|\zeta_\tau(\psi_{1,n})|}{\left|\lambda_\tau^+ - \lambda_{1,n}\right|^2} + \frac{|\zeta_\tau(\psi_{1,n})|}{\left|\mu - \lambda_{1,n}\right|^2} \right), \ n \geqslant N_0, \tag{2.50}$$

where $\delta_1 := \sup_{n\in\mathbb{N}} |\lambda_{1,n} - \lambda_{2,n}| < \infty$.

Further, in light of (2.38), we deduce from (2.3) that

$$\begin{aligned} \sum_{n=1}^{+\infty} \frac{|\zeta_\tau(\psi_{1,n})|}{\left|\ell - \lambda_{1,n}\right|^2} &\leqslant \left(\sum_{n=1}^{+\infty} \left| \frac{\langle f_\tau^+, \psi_{1,n}\rangle_{L^2(\partial\Omega)}}{\ell - \lambda_{1,n}} \right|^2 \right)^{1/2} \left(\sum_{n=1}^{+\infty} \left| \frac{\langle f_\tau^-, \psi_{1,n}\rangle_{L^2(\partial\Omega)}}{\overline{\ell} - \lambda_{1,n}} \right|^2 \right)^{1/2} \\ &\leqslant \left\| u_{1,\ell}^+ \right\|_{L^2(\Omega)} \left\| u_{1,\overline{\ell}}^- \right\|_{L^2(\Omega)}, \ \ell = \lambda_\tau^+, \ \mu. \end{aligned}$$

Here, $u_{1,\ell}^\pm$ denotes the $H^2(\Omega)$-solution to (2.1) where $(\ell, q_1, f_\tau^\pm)$ is substituted for (λ, q, f). Thus, bearing in mind that $\overline{\lambda_\tau^+} = \lambda_\tau^-$, we derive from (2.50) that

$$\sum_{n=N_0}^{+\infty} |A_{n,\tau,\mu}| \leqslant 8\delta_1 \left(\left\| u_{1,\lambda_\tau^+}^+ \right\|_{L^2(\Omega)} \left\| u_{1,\lambda_\tau^-}^- \right\|_{L^2(\Omega)} + \left\| u_{1,\mu}^+ \right\|_{L^2(\Omega)} \left\| u_{1,\mu}^- \right\|_{L^2(\Omega)} \right).$$

With reference to (2.3), we assume upon possibly enlarging $(-\mu)$, that $\left\| u_{1,\mu}^\pm \right\|_{L^2(\Omega)} \leqslant 1$, and obtain

$$\sum_{n=N_0}^{+\infty} |A_{n,\tau,\mu}| \leqslant 2\delta_1 \left((M+2)^2 c_*^2 + 4 \right), \tag{2.51}$$

with the aid of (2.27). Now, since $\lim_{\mu\to-\infty} A_{n,\tau,\mu} = A_{n,\tau,*}$ for all $n \in \mathbb{N}$, we deduce (2.44) from this and (2.51) by invoking the Lebesgue dominated convergence theorem.

2.4.2. *Step 2: Proof of* (2.45). For all $n \geqslant N_0$, we infer from (2.47)-(2.48) that

$$\lambda_{2,n} \geqslant \lambda_{1,n} - |\lambda_{1,n} - \lambda_{2,n}| \geqslant \lambda_{1,N_0} - 2M \geqslant \left|\lambda_\tau^+\right|,$$

whence $|\mu - \lambda_{2,n}| = \lambda_{2,n} - \mu \geqslant |\lambda_\tau^+| - \mu \geqslant |\lambda_\tau^+ - \mu|$. Therefore, we get

$$|\kappa_{\tau,\mu}(\lambda_{2,n})| \leqslant \frac{1}{\left|\lambda_\tau^+ - \lambda_{2,n}\right|}, \ n \geqslant N_0,$$

from (2.37). This and (2.38) entail

$$\begin{aligned} |B_{n,\tau,\mu}| &\leqslant \|\psi_{1,n} - \psi_{2,n}\|_{L^2(\partial\Omega)} \\ &\quad \left(\left\| f_\tau^+ \right\|_{L^2(\partial\Omega)} \left| \frac{\langle f_\tau^-, \psi_{1,n}\rangle_{L^2(\partial\Omega)}}{\lambda_\tau^+ - \lambda_{2,n}} \right| + \left\| f_\tau^- \right\|_{L^2(\partial\Omega)} \left| \frac{\langle f_\tau^+, \psi_{2,n}\rangle_{L^2(\partial\Omega)}}{\lambda_\tau^+ - \lambda_{2,n}} \right| \right) \\ &\leqslant c_* \|\psi_{1,n} - \psi_{2,n}\|_{L^2(\partial\Omega)} \left(\left| \frac{\langle f_\tau^-, \psi_{1,n}\rangle_{L^2(\partial\Omega)}}{\lambda_\tau^+ - \lambda_{2,n}} \right| + \left| \frac{\langle f_\tau^+, \psi_{2,n}\rangle_{L^2(\partial\Omega)}}{\lambda_\tau^+ - \lambda_{2,n}} \right| \right), \end{aligned} \tag{2.52}$$

by applying (2.21) with $(X, p) = (\partial\Omega, 2)$.

Further, recalling from (2.47) that $|\lambda_\tau^+ - \lambda_{1,n}| \geqslant 2\|q_1 - q_2\|_{L^\infty(\Omega)}$ for all $n \geqslant N_0$, and using the estimate $|\lambda_\tau^+ - \lambda_{2,n}| \geqslant |\lambda_\tau^+ - \lambda_{1,n}| - \|q_1 - q_2\|_{L^\infty(\Omega)}$ arising from (2.48), we find that

$$\left|\lambda_\tau^+ - \lambda_{2,n}\right| \geqslant \frac{|\lambda_\tau^+ - \lambda_{1,n}|}{2} = \frac{|\lambda_\tau^- - \lambda_{1,n}|}{2}, \ n \geqslant N_0. \tag{2.53}$$

Putting this together with (2.52), we get for every $\mu \in (-\infty, -(1+5M)]$, that

$$\sum_{n=N_0}^{+\infty} |B_{n,\tau,\mu}|$$
$$\leqslant 2c_* \varepsilon_1 \left(\left(\sum_{n=N_0}^{+\infty} \left| \frac{\langle f_\tau^-, \psi_{1,n} \rangle_{L^2(\partial\Omega)}}{\lambda_\tau^- - \lambda_{1,n}} \right|^2 \right)^{1/2} + \left(\sum_{n=N_0}^{+\infty} \left| \frac{\langle f_\tau^+, \psi_{2,n} \rangle_{L^2(\partial\Omega)}}{\lambda_\tau^+ - \lambda_{2,n}} \right|^2 \right)^{1/2} \right),$$

with $\varepsilon_1 := \left(\sum_{n=1}^{+\infty} \| \psi_{1,n} - \psi_{2,n} \|_{L^2(\partial\Omega)}^2 \right)^{1/2}$. This leads to

$$\sum_{n=N_0}^{+\infty} |B_{n,\tau,\mu}| \leqslant 2c_* \varepsilon_1 \left(\left\| u_{1,\lambda_\tau^-}^- \right\|_{L^2(\Omega)} + \left\| u_{2,\lambda_\tau^+}^+ \right\|_{L^2(\Omega)} \right) \leqslant 8(M+2)c_*^2,$$

with the aid of Lemma 2.1 and (2.27). Now, (2.45) follows from this and the identities

$$\lim_{\mu \to -\infty} B_{n,\tau,\mu} = B_{n,\tau,*},\ n \in \mathbb{N},$$

by applying Lebesgue's dominated convergence theorem.

2.4.3. *Step 3: End of the proof.* Putting (2.41) and (2.44)-(2.45) together, we obtain that

$$S_\tau = \sum_{n=1}^{+\infty} (A_{n,\tau,*} + B_{n,\tau,*}). \tag{2.54}$$

Further, since $\mathrm{Im}\,(\lambda_\tau^+ - \lambda_{j,n}) = 2\tau$ for $j = 1, 2$ and all $n \in \mathbb{N}$, by (2.25), we infer from (2.21) with $(X, p) = (\partial\Omega, 2)$, (2.38), and (2.44)-(2.45) that

$$|A_{n,\tau,*}| \leqslant c_*^2 \frac{|\lambda_{1,n} - \lambda_{2,n}| \, \|\psi_{1,n}\|_{L^2(\partial\Omega)}^2}{\tau^2}$$

and

$$|B_{n,\tau,*}| \leqslant c_*^2 \frac{\left(\|\psi_{1,n}\|_{L^2(\partial\Omega)} + \|\psi_{2,n}\|_{L^2(\partial\Omega)} \right) \|\psi_{1,n} - \psi_{2,n}\|_{L^2(\partial\Omega)}}{\tau}.$$

Therefore, it holds true for all $n \in \mathbb{N}$ that $\lim_{\tau \to +\infty} A_{n,\tau,*} = \lim_{\tau \to +\infty} B_{n,\tau,*} = 0$, so it follows from (2.54) that

$$\limsup_{\tau \to +\infty} |S_\tau| \leqslant \limsup_{\tau \to +\infty} \sum_{n=N}^{+\infty} |A_{n,\tau,*}| + \limsup_{\tau \to +\infty} \sum_{n=N}^{+\infty} |B_{n,\tau,*}|,\ N \in \mathbb{N}. \tag{2.55}$$

Moreover, setting $\delta_N := \sup_{n \geqslant N} |\lambda_{1,n} - \lambda_{2,n}|$, we infer from (2.38) and (2.44) that

$$\begin{aligned} \sum_{n=N}^{+\infty} |A_{n,\tau,*}| &\leqslant \delta_N \sum_{n=N}^{+\infty} \left| \frac{\langle f_\tau^-, \psi_{1,n} \rangle_{L^2(\partial\Omega)}}{\lambda_\tau^+ - \lambda_{1,n}} \right| \left| \frac{\langle f_\tau^+, \psi_{1,n} \rangle_{L^2(\partial\Omega)}}{\lambda_\tau^+ - \lambda_{2,n}} \right| \\ &\leqslant 2\delta_N \left(\sum_{n=N}^{+\infty} \left| \frac{\langle f_\tau^-, \psi_{1,n} \rangle_{L^2(\partial\Omega)}}{\lambda_\tau^- - \lambda_{1,n}} \right|^2 \right)^{1/2} \left(\sum_{n=N}^{+\infty} \left| \frac{\langle f_\tau^+, \psi_{1,n} \rangle_{L^2(\partial\Omega)}}{\lambda_\tau^+ - \lambda_{1,n}} \right|^2 \right)^{1/2}, \end{aligned}$$

$N \geqslant N_0$. In the last line we used the Cauchy-Schwarz inequality, the estimate (2.53) and the identity $|\lambda_\tau^- - \lambda_{1,n}| = |\lambda_\tau^+ - \lambda_{1,n}|$. Therefore, applying Lemma 2.1

and (2.27), we obtain for all $N \geqslant N_0$, that

$$\sum_{n=N}^{+\infty} |A_{n,\tau,*}| \leqslant 2\delta_N \left\| u^-_{1,\lambda^-_\tau} \right\|_{L^2(\partial\Omega)} \left\| u^+_{1,\lambda^+_\tau} \right\|_{L^2(\partial\Omega)} \leqslant \frac{(M+2)^2}{2} c_*^2 \delta_N,$$

which entails

$$\limsup_{\tau\to+\infty} \sum_{n=N}^{+\infty} |A_{n,\tau,*}| \leqslant \frac{(M+2)^2}{2} c_*^2 \left(\sup_{n\geqslant N} |\lambda_{1,n} - \lambda_{2,n}| \right), \ N \geqslant N_0. \tag{2.56}$$

Similarly, using (2.38) and (2.45), we can upper bound $\sum_{n=N}^{+\infty} |B_{n,\tau,*}|$ by

$$\begin{aligned}
&\sum_{n=N}^{+\infty} \|\psi_{1,n} - \psi_{2,n}\|_{L^2(\partial\Omega)} \\
&\quad \left(\|f^+_\tau\|_{L^2(\partial\Omega)} \left| \frac{\langle f^-_\tau, \psi_{1,n}\rangle_{L^2(\partial\Omega)}}{\lambda^+_\tau - \lambda_{2,n}} \right| + \|f^-_\tau\|_{L^2(\partial\Omega)} \left| \frac{\langle f^+_\tau, \psi_{2,n}\rangle_{L^2(\partial\Omega)}}{\lambda^+_\tau - \lambda_{2,n}} \right| \right) \\
&\qquad \leqslant \varepsilon_N \left(\|f^+_\tau\|_{L^2(\partial\Omega)} \left(\sum_{n=N}^{+\infty} \left| \frac{\langle f^-_\tau, \psi_{1,n}\rangle_{L^2(\partial\Omega)}}{\lambda^+_\tau - \lambda_{2,n}} \right|^2 \right)^{1/2} \right. \\
&\qquad\qquad \left. + \|f^-_\tau\|_{L^2(\partial\Omega)} \left(\sum_{n=N}^{+\infty} \left| \frac{\langle f^+_\tau, \psi_{2,n}\rangle_{L^2(\partial\Omega)}}{\lambda^+_\tau - \lambda_{2,n}} \right|^2 \right)^{1/2} \right),
\end{aligned}$$

where $\varepsilon_N := \left(\sum_{n=N}^{+\infty} \|\psi_{1,n} - \psi_{2,n}\|^2_{L^2(\partial\Omega)} \right)^{1/2}$. Next, applying Lemma 2.1 we get for all $N \in \mathbb{N}$ that $\sum_{n=N}^{+\infty} \left| \frac{\langle f^+_\tau, \psi_{2,n}\rangle_{L^2(\partial\Omega)}}{\lambda^+_\tau - \lambda_{2,n}} \right|^2 \leqslant \left\| u^+_{2,\lambda^+_\tau} \right\|^2_{L^2(\Omega)}$, and for every $N \geqslant N_0$ that

$$\sum_{n=N}^{+\infty} \left| \frac{\langle f^-_\tau, \psi_{1,n}\rangle_{L^2(\partial\Omega)}}{\lambda^+_\tau - \lambda_{2,n}} \right|^2 \leqslant 4 \sum_{n=N}^{+\infty} \left| \frac{\langle f^-_\tau, \psi_{1,n}\rangle_{L^2(\partial\Omega)}}{\lambda^+_\tau - \lambda_{1,n}} \right|^2 \leqslant 4 \left\| u^-_{1,\lambda^-_\tau} \right\|^2_{L^2(\Omega)},$$

by virtue of (2.53). Therefore, in light of (2.21) with $(X, p) = (\partial\Omega, 2)$ and (2.27), we have

$$\begin{aligned}
\sum_{n=N}^{+\infty} |B_{n,\tau,*}| &\leqslant \varepsilon_N \left(2 \|f^+_\tau\|_{L^2(\partial\Omega)} \left\| u^-_{1,\lambda^-_\tau} \right\|_{L^2(\Omega)} + \|f^-_\tau\|_{L^2(\partial\Omega)} \left\| u^+_{2,\lambda^+_\tau} \right\|_{L^2(\Omega)} \right) \\
&\leqslant \frac{3(M+2)}{2} c_*^2 \varepsilon_N,
\end{aligned}$$

provided $N \geqslant N_0$, and hence

$$\limsup_{\tau\to+\infty} \sum_{n=N}^{+\infty} \left| B_{n,\lambda^+_\tau,*} \right| \leqslant \frac{3(M+2)}{2} c_*^2 \left(\sum_{n=N}^{+\infty} \|\psi_{1,n} - \psi_{2,n}\|^2_{L^2(\partial\Omega)} \right)^{1/2}, \ N \geqslant N_0.$$

Putting this together with (2.55)-(2.56), we obtain

$$\limsup_{\tau\to+\infty} |S_\tau| \leqslant c \left(\left(\sum_{n=N}^{+\infty} \|\psi_{1,n} - \psi_{2,n}\|^2_{L^2(\partial\Omega)} \right)^{1/2} + \sup_{n\geqslant N} |\lambda_{1,n} - \lambda_{2,n}| \right), \ N \geqslant N_0, \tag{2.57}$$

where the constant $c := (M+2)(M+5)c_*^2/2$ is independent of N. Now, by sending N to $+\infty$ on the right hand side of the above estimate, we get that

$\limsup_{\tau\to+\infty}|S_\tau| = 0$. Thus, we have $\lim_{\tau\to+\infty} S_\tau = 0$, by virtue of Proposition 2.4. This entails in the same way as in Section 2.3 that $q_1 = q_2$ in Ω, which terminates the proof of Theorem 1.6

2.5. The stability issue. The stability issue for the Borg-Levinson inverse problem was first examined by G. Alessandrini and J. Sylvester in [**1**], who proved Hölder stable determination of q by BSD(q) (see also [**8**, Theorem 2.31] for a reformulation of their result). We shall establish in this section, at the expense of stronger regularity on q, that it can be Hölder-stably determined by the asymptotic behavior of its BSD, provided q is known on the boundary $\partial\Omega$.

2.5.1. *Notations and stability inequality.* We stick with the notations of Section 2. In particular, given two real-valued potentials q_j, $j = 1, 2$, we denote by $\{\lambda_{j,n},\ n \in \mathbb{N}\}$ the sequence of the eigenvalues of A_{q_j}, arranged in non-decreasing order (and repeated with the multiplicity), and we write $\psi_{j,n}$ instead of $\partial_\nu \varphi_{j,n}$ for all $n \in \mathbb{N}$, where $\{\varphi_{j,n},\ n \in \mathbb{N}\}$ is a $L^2(\Omega)$-orthonormal basis of eigenvectors of A_{q_j}, such that $A_{q_j}\varphi_{j,n} = \lambda_{j,n}\varphi_{j,n}$.

THEOREM 2.5. *For $M \in (0, +\infty)$ fixed, pick q_1 and q_2 in $L^\infty(\Omega, \mathbb{R}) \cap H^1(\Omega)$, such that*

$$\|q_j\|_{L^\infty(\Omega)} + \|q_j\|_{H^1(\Omega)} \leqslant M,\ j = 1, 2, \tag{2.58}$$

and

$$q_1 = q_2 \text{ on } \partial\Omega. \tag{2.59}$$

Assume moreover that

$$\sum_{n=1}^{+\infty} \|\psi_{1,n} - \psi_{2,n}\|^2_{L^2(\partial\Omega)} < \infty. \tag{2.60}$$

Then, the following stability estimate

$$\|q_1 - q_2\|_{L^2(\Omega)} \leqslant C \limsup_{n\to+\infty} |\lambda_{1,n} - \lambda_{2,n}|^{2/(d+2)},$$

holds for some positive constant C that depends only on Ω and M.

2.5.2. *Proof of Theorem* 2.5. Let us recall from (2.57) that for all $N \geqslant N_0$, we have

$$\limsup_{\tau\to+\infty} |S_\tau| \leqslant c \left(\left(\sum_{n=N}^{+\infty} \|\psi_{1,n} - \psi_{2,n}\|^2_{L^2(\partial\Omega)} \right)^{1/2} + \sup_{n\geqslant N} |\lambda_{1,n} - \lambda_{2,n}| \right),$$

for some positive constant c that is independent of N and ξ. Thus, in light of (2.60) we get upon sending N to infinity, that

$$\limsup_{\tau\to+\infty} |S_\tau| \leqslant c \limsup_{n\to\infty} |\lambda_{1,n} - \lambda_{2,n}|. \tag{2.61}$$

Further, we recall from Proposition 2.4 that

$$\lim_{\tau\to+\infty} S_\tau = \int_\Omega q(x) e^{-x\cdot\xi} dx = (2\pi)^{d/2} \widehat{q}(\xi),$$

where q is the same as in (2.34) and $\widehat{q}$ stands for the Fourier transform $\mathcal{F}q$ of q, defined by (2.33). This, (2.61) and the basic estimate $|\lim_{\tau\to+\infty} S_\tau| \leqslant \limsup_{\tau\to+\infty} |S_\tau|$, yield $|\widehat{q}(\xi)| \leqslant (2\pi)^{-d/2} c \limsup_{n \to +\infty} |\lambda_{1,n} - \lambda_{2,n}|$, uniformly in $\xi \in \mathbb{R}^d$. Thus, we obtain

$$\|\widehat{q}\|_{L^\infty(\mathbb{R}^d)} \leqslant c \limsup_{n \to +\infty} |\lambda_{1,n} - \lambda_{2,n}| . \tag{2.62}$$

upon substituting $(2\pi)^{-d/2}c$ for c.

On the other hand, we infer from (2.34) and the Plancherel theorem that

$$\|q_1 - q_2\|^2_{L^2(\Omega)} = \|q\|^2_{L^2(\mathbb{R}^d)} = \|\widehat{q}\|^2_{L^2(\mathbb{R}^d)} = \int_{\mathbb{R}^d} |\widehat{q}(\xi)|^2 \, d\xi. \tag{2.63}$$

For $R \in (1, +\infty)$ fixed, set $B_R := \{\xi \in \mathbb{R}^d,\ |\xi| \leqslant R\}$ and notice from (2.63) that

$$\|q_1 - q_2\|^2_{L^2(\Omega)} = \int_{B_R} |\widehat{q}(\xi)|^2 \, d\xi + \int_{\mathbb{R}^d \setminus B_R} |\widehat{q}(\xi)|^2 \, d\xi. \tag{2.64}$$

The first term on the right hand side of (2.64) is easily treated, as we have

$$\int_{B_R} |\widehat{q}(\xi)|^2 \, d\xi \leqslant \tilde{c} R^d \, \|\widehat{q}\|^2_{L^\infty(B_R)} , \tag{2.65}$$

for some positive constant $\tilde{c}$ that is independent of R. Further, since $q_1 - q_2 \in H^1_0(\Omega)$ from (2.59), we see that $q \in H^1(\mathbb{R}^d)$. Thus we may write

$$\begin{aligned}\int_{\mathbb{R}^d \setminus B_R} (1+|\xi|^2) \, |\widehat{q}(\xi)|^2 \, d\xi &= \|q\|^2_{H^1(\mathbb{R}^d)} \\ &= \|q_1 - q_2\|^2_{H^1(\Omega)} \leqslant \left(\|q_1\|_{H^1(\Omega)} + \|q_2\|_{H^1(\Omega)} \right)^2 \leqslant 4M^2,\end{aligned}$$

from (2.58), and consequently

$$\int_{\mathbb{R}^d \setminus B_R} |\widehat{q}(\xi)|^2 \, d\xi \leqslant R^{-2} \int_{\mathbb{R}^d \setminus B_R} (1 + |\xi|^2) \, |\widehat{q}(\xi)|^2 \, d\xi \leqslant 4M^2 R^{-2}.$$

Putting this and (2.64)-(2.65) together, we find that

$$\|q_1 - q_2\|^2_{L^2(\Omega)} \leqslant \tilde{c} \left(R^d \, \|\widehat{q}\|^2_{L^\infty(B_R)} + R^{-2} \right), \tag{2.66}$$

upon possibly substituting $\max(\tilde{c}, 4M^2)$ for $\tilde{c}$.

Set $\delta := \limsup_{n \to +\infty} |\lambda_{1,n} - \lambda_{2,n}|$. We shall examine the two cases $\delta \in (0, 1)$ and $\delta \in [1, +\infty)$ separately. In the first case we plug the estimate $\|\widehat{q}\|_{L^\infty(B_R)} \leqslant c\delta$, arising from (2.62), in (2.66), choose $R = \delta^{-2/(d+2)}$, and get

$$\|q_1 - q_2\|_{L^2(\Omega)} \leqslant C \delta^{2/(d+2)}, \tag{2.67}$$

with $C := \left(\tilde{c}(1 + c^2) \right)^{1/2}$. In the second case we have obviously

$$\|q_1 - q_2\|_{L^2(\Omega)} \leqslant \|q_1\|_{L^2(\Omega)} + \|q_2\|_{L^2(\Omega)} \leqslant 2M \leqslant 2M\delta^{2/(d+2)},\ \delta \in [1, +\infty),$$

so the desired result follows from this and (2.67).

3. Application to parabolic inverse coefficient problems

Let $T \in (0, +\infty)$, let Ω be as in the preceding sections, that is $\Omega \subset \mathbb{R}^d$, $d \geqslant 2$, is a bounded domain with boundary $\partial\Omega \in C^{1,1}$. We consider the diffusion equation

$$\left\{ \begin{array}{rcll} (\partial_t - \Delta + q)u & = & 0 & \text{in } Q := \Omega \times (0,T) \\ u & = & f & \text{on } \Sigma := \partial\Omega \times (0,T) \\ u(\cdot,0) & = & 0 & \text{in } \Omega, \end{array} \right. \tag{3.1}$$

where q is a real-valued bounded potential and f fulfills the compatibility condition:

$$f(\cdot,0) = 0 \text{ on } \partial\Omega.$$

The inverse problem we examine in this section can be stated as follows. Given $M \in (0, +\infty)$ and two open subsets Γ_{in} and Γ_{out} of $\partial\Omega$, determine

$$q \in Q_M := \{q \in L^\infty(\Omega, \mathbb{R}),\ \|q\|_{L^\infty(\Omega)} \leqslant M\}$$

by knowledge of the parabolic partial DN map at one fixed time $T_0 \in (0,T)$:

$$\Lambda_q : f \in \mathscr{H}_{\text{in}} \mapsto \partial_\nu(\cdot, T_0)_{|\Gamma_{\text{out}}}.$$

Here, we have set $\mathscr{H}_{\text{in}} := \{f \in \mathscr{H},\ \text{supp} f \subset \Gamma_{\text{in}} \times (0,T_0)\}$, with $\mathscr{H} := C^{1,\alpha}([0,T], H^{3/2}(\partial\Omega))$ for some $\alpha \in (0,1]$.

REMARK 3.1. Since $\text{supp} f \subset (0,T_0) \times \Gamma_{\text{in}}$ for any $f \in \mathscr{H}_{\text{in}}$, then the compatibility condition $f(\cdot,0) = 0$ holds on $\partial\Omega$.

Evidently, the inverse problem under investigation can be reformulated as whether the mapping $q \in Q_M \mapsto \Lambda_q$ is injective.

3.1. Parabolic Dirichlet-to-Neumann map and identifiability.

We start by recalling the following uniqueness and existence result (see e.g. [**8**, Section 3.5]).

PROPOSITION 3.2. *For all $q \in Q_M$ and all $f \in \mathscr{H}$, there exists a unique solution*

$$u \in \mathscr{Z} := C^1((0,T], H^2(\Omega)) \cap C^0([0,T], L^2(\Omega))$$

to (3.1).

Thus, by continuity of the trace operator $g \to (\partial_\nu g)_{|\Gamma_{\text{out}}}$ from $H^2(\Omega)$ to $H^{1/2}(\Gamma_{\text{out}})$, the map

$$\begin{array}{rccl} \Lambda_q : & \mathscr{H}_{\text{in}} & \to & H^{1/2}(\Gamma_{\text{out}}) \\ & f & \mapsto & \partial_\nu u(\cdot, T_0)_{|\Gamma_{\text{out}}} \end{array}$$

is well-defined.

The main result if this section is as follows.

THEOREM 3.3. *Assume that $\Gamma_{\text{in}} \cup \Gamma_{\text{out}} = \partial\Omega$ and that $\Gamma_{\text{in}} \cap \Gamma_{\text{out}} \neq \emptyset$. For $j = 1,2$, let $q_j \in Q_M$ and put $\Lambda_j := \Lambda_{q_j}$. Then, we have the implication:*

$$(\forall f \in \mathscr{H}_{\text{in}},\ \Lambda_1(f) = \Lambda_2(f)) \Longrightarrow (q_1 = q_2).$$

REMARK 3.4. This result was proved by B. Canuto and O. Kavian in [**5**]. Recently, in [**15**], it was extended to the case of time-fractional diffusion equations $(\partial_t^\alpha - \Delta + q)u = 0$ in Q, with $\alpha \in (0,1) \cup (1,2)$, and where ∂_t^α denotes the Caputo fractional derivative of order α.

3.2. Technical tools. The proof of Theorem 3.3 consists of two steps.

- The first one is to show that knowledge of Λ_q uniquely determines[8] $\mathrm{BSD}(q)$:

$$(\forall f \in \mathscr{H}_{\rm in},\ \Lambda_1(f) = \Lambda_2(f)) \Longrightarrow (\mathrm{BSD}(q_1) = \mathrm{BSD}(q_2))\,. \tag{3.2}$$

- The second step is to identify q through $\mathrm{BSD}(q)$, with the aid of Theorem 1.5.

In [**12**], A. Katchalov, Y. Kurylev, M. Lassas and C. Mandache have established the equivalence between the full parabolic DN map and the BSD. Their statement is quite similar to the claim of the first step, except that this is the partial data Λ_q (and not the full parabolic DN map) that is considered here and that we only seek determination of $\mathrm{BSD}(q)$ by Λ_q (and not equivalence of these two data).

3.2.1. *Some notations and useful properties.* We stick with the notations of Section 2. That is to say that for $j = 1, 2$, we write $\mathrm{BSD}(q_j) = \{(\lambda_{j,n}, \psi_{j,n}),\ n \in \mathbb{N}\}$ with $\psi_{j,n} = \partial_\nu \varphi_{j,n}$.

Weyl's law. It is well known (see e.g. [**22**, Section XIII.15]) that there exist two constants $n_M \in \mathbb{N}$ and $c \in (1, +\infty)$, both of them depending only on Ω and M, such that we have

$$c^{-1} n^{2/d} \leqslant \lambda_n \leqslant c n^{2/d},\ n \geqslant n_M.$$

This entails for all $k \in \mathbb{N}$ and all $\varepsilon > 0$, that the series

$$\sum_{n=n_M}^{+\infty} \lambda_n^k e^{-\varepsilon\lambda_n} \leqslant c^k \sum_{n=n_M}^{+\infty} n^{2k/d} e^{-c^{-1}\varepsilon n^{2/d}} < \infty. \tag{3.3}$$

Linear independence of the Neumann data. Given a non-empty open subset Γ of $\partial\Omega$, the family $\{\psi_{n|\Gamma},\ n \in \mathbb{N}\}$ is, in general, not linearly independent in $L^2(\Gamma)$, but the normal derivatives of the eigenfunctions associated with one eigenvalue are linearly independent. More precisely, if m_n denotes the geometric multiplicity of λ_n, let $\{\varphi_{n,i},\ i = 1, \ldots, m_n\}$ be an orthonormal basis of the $L^2(\Omega)$-subspace $\ker(A_q - \lambda_n)$. Then, we have

$$\dim\{\psi_{n,i|\Gamma},\ i = 1, \ldots, m_n\} = m_n. \tag{3.4}$$

The proof of (3.4) essentially relies on the following unique continuation principle for local Cauchy data.

LEMMA 3.5. *For $j, k = 1, \ldots, d$, let $a_{j,k} = a_{k,j} \in W^{1,\infty}(\Omega)$, $b_j \in L^\infty(\Omega)$ and $c \in L^\infty(\Omega)$, and suppose that the differential operator*

$$P := -\sum_{j,k=1}^{d} a_{j,k}(x)\partial_{j,k}^2 + \sum_{j=1}^{d} b_j(x)\partial_j + c(x),$$

fulfills the ellipticity condition:

$$\exists \lambda \in (0, +\infty),\ \sum_{j,k=1}^{d} a_{j,k}(x)\xi_j\xi_k \geqslant \lambda\, |\xi|^2\,,\ x \in \Omega,\ \xi \in \mathbb{R}^d.$$

Then, for all $u \in H^2(\Omega)$, we have:

$$(Pu = 0,\ u_{|\Gamma} = \partial_\nu u_{|\Gamma} = 0) \Longrightarrow (u = 0 \ in\ \Omega).$$

[8]Or, equivalently, the BSD associated with the Dirichlet Laplacian A_q, defined by (1.2).

To show (3.4), we pick m_n constants c_i, $i=1,\dots,m_n$, such that

$$\sum_{i=1}^{m_n} c_i \psi_{n,i} = 0 \text{ on } \Gamma.$$

Putting $\varphi := \sum_{i=1}^{m_n} c_i \varphi_{n,i}$, we see that φ solves

$$\begin{cases} A_q \varphi &= \lambda_n \varphi & \text{in } \Omega \\ \varphi &= 0 & \text{on } \Gamma \\ \partial_\nu \varphi &= \sum_{i=1}^{m_n} c_i \psi_{n,i} = 0 & \text{on } \Gamma. \end{cases}$$

Thus, upon applying Lemma 3.5 with $a_{jk} = \delta_{jk}$ and $b_j = 0$ for $j,k = 1,\dots,d$, and $c = q$, we get that $\varphi = 0$ in Ω. Here and below, δ denotes the Kronecker symbol, i.e.

$$\delta_{jk} := \begin{cases} 1 & \text{if } j = k \\ 0 & \text{otherwise.} \end{cases}$$

Next, since the family $\{\varphi_{n,i},\ i=1,\dots,m_n\}$ is linearly independent in $L^2(\Omega)$, we obtain that $c_i = 0$ for all $i = 1,\dots,m_n$. Therefore, the $\psi_{n,i|\Gamma}$, $i=1,\dots,m_n$, are linearly independent in $L^2(\Gamma)$, which establishes (3.4).

We may now prove the:

LEMMA 3.6. *Let Γ and Γ' be two non-empty open subsets of $\partial\Omega$. Then, the function*

$$\theta_n(\sigma,\sigma') = \sum_{i=1}^{m_n} \psi_{n,i}(\sigma)\psi_{n,i}(\sigma'),\ (\sigma,\sigma') \in \Gamma \times \Gamma',$$

is not identically zero in $\Gamma \times \Gamma'$.

PROOF. We prove Lemma 3.6 by contradiction. If we assume that $\theta_n(\sigma,\sigma') = 0$ for a.e. $(\sigma,\sigma') \in \Gamma \times \Gamma'$, then we would have

$$\sum_{i=1}^{m_n} \psi_{n,i}(\sigma)\psi_{n,i} = 0 \text{ on } \Gamma', \sigma \in \Gamma,$$

and hence

$$\psi_{n,i} = 0 \text{ on } \Gamma,\ i = 1,\dots,m_n,$$

from (3.4), which is in contradiction with (3.4). Therefore, θ_n is not identically zero in $\Gamma \times \Gamma'$. □

3.3. Proof of the parabolic uniqueness result. In this section we prove Theorem 3.3. In light of Theorem 1.5, it is enough to show (3.2). More precisely, we shall establish the:

THEOREM 3.7. *Under the assumptions of Theorem 3.3, we have*

$$(\forall f \in \mathscr{H}_{\text{in}},\ \Lambda_1(f) = \Lambda_2(f)) \Longrightarrow (\lambda_{1,n} = \lambda_{2,n} \textit{ and } \psi_{1,n} = \psi_{2,n} \textit{ on } \Gamma_{\text{in}} \cup \Gamma_{\text{out}},\ n \in \mathbb{N}),$$

up to an appropriate choice of the eigenfunctions $\varphi_{2,n}$ of A_2.

Prior to proving Theorem 3.7, we establish a representation formula of the normal derivative of the solution to (3.1).

3.3.1. *A representation formula of the DN map.* We start by expressing the solution to (3.1) in terms of the BSD.

LEMMA 3.8. *For all $f \in \mathscr{H}$, the solution u to (3.1), given by Proposition 3.2, reads*

$$u(\cdot,t) = \sum_{n\geqslant 1} u_n(t)\varphi_n \ \textit{in}\ L^2(\Omega),\ t\in[0,T], \tag{3.5}$$

where

$$u_n(t) = -\int_0^t e^{-\lambda_n s}\left(\int_{\partial\Omega}\psi_n(\sigma)f(\sigma,t-s)d\sigma\right)ds,\ n\in\mathbb{N}.$$

PROOF. Since $\{\varphi_n,\ n\in\mathbb{N}\}$ is an orthonormal basis of $L^2(\Omega)$ and $u\in C^1([0,T], L^2(\Omega))$ by Proposition 3.2, we have

$$\forall s\in[0,T],\ u(\cdot,s) = \sum_{n\in\mathbb{N}} u_n(s)\varphi_n \text{ in } L^2(\Omega),$$

with

$$s\mapsto u_n(s) = \int_\Omega u(x,s)\varphi_n(x)dx \in C^1([0,T].$$

Moreover, for all $s\in[0,T]$, we get that

$$\begin{aligned} u_n'(s) &= \int_\Omega \partial_t u(x,s)\varphi_n(x)dx \\ &= -\int_\Omega(-\Delta+q(x))u(x,t)\varphi_n(x)dx \\ &= -\int_\Omega u(x,s)A_q\varphi_n(x)dx + \int_{\partial\Omega}\partial_\nu u(\sigma,s)\varphi_n(\sigma)d\sigma - \int_{\partial\Omega}u(\sigma,s)\partial_\nu\varphi_n(\sigma)d\sigma, \end{aligned}$$

upon applying the Green formula. As a consequence, we have

$$\begin{aligned} u_n'(s) &= -\lambda_n\int_\Omega u(x,s)\varphi_n(x)dx - \int_{\partial\Omega}f(\sigma,s)\psi_n(\sigma)d\sigma \\ &= -\lambda_n u_n(s) - \int_{\partial\Omega}f(\sigma,s)\psi_n(\sigma)d\sigma, \end{aligned}$$

and hence $e^{-\lambda_n s}\frac{d}{ds}(e^{\lambda_n s}u_n(s)) = u_n'(s)+\lambda_n u_n(s) = -\int_{\partial\Omega}f(\sigma,s)\psi_n(\sigma)d\sigma$. Thus, taking into account that $u_n(0)=0$, we get for every $t\in[0,T]$, that

$$e^{\lambda_n t}u_n(t) = -\int_0^t e^{\lambda_n s}\left(\int_{\partial\Omega}f(\sigma,t)\psi_n(\sigma)d\sigma\right)ds,$$

upon integrating over $[0,t]$. $\square$

In general the series in (3.5) converges in $L^2(\Omega)$ but the convergence can be upgraded to $H^2(\Omega)$ by assuming that the function $f\in\mathscr{H}$ satisfies the following condition

$$\exists\varepsilon\in(0,T_0),\ \forall(\sigma,t)\in\partial\Omega\times[T_0-\varepsilon,T_0],\ f(\sigma,t)=0, \tag{3.6}$$

for some fixed $T_0\in(0,T)$. Indeed, in this case we deduce from the identity $u(\cdot,T_0)=f(\cdot,T_0)=0$ on $\partial\Omega$ that $u(\cdot,T_0)\in H^2(\Omega)\cap H_0^1(\Omega)=\mathrm{dom}(A_q)$. As a consequence we have $\sum_{n=1}^{+\infty} \lambda_n |u_n(T_0)|^2 < \infty$ and the series in (3.5) converges in $H^2(\Omega)$ for $t = T_0$:

$$u(\cdot, T_0) = \sum_{n=1}^{+\infty} u_n(T_0)\varphi_n \text{ in } H^2(\Omega).$$

Thus, by continuity of the trace operator $g \mapsto (\partial_\nu g)_{|\partial\Omega}$ from $H^2(\Omega)$ to $H^{1/2}(\partial\Omega)$, we obtain that

$$\begin{aligned} \partial_\nu u(\cdot, T_0) &= \sum_{n=1}^{+\infty} u_n(T_0)\psi_n \\ &= -\sum_{n=1}^{+\infty} \left(\int_\varepsilon^{T_0} e^{-\lambda_n s} \left(\int_{\partial\Omega} f(\sigma, T_0 - s)\psi_n(\sigma) d\sigma \right) ds \right) \psi_n \text{ in } H^{1/2}(\partial\Omega). \end{aligned} \tag{3.7}$$

Having proved (3.7), we may now establish the:

LEMMA 3.9. *For all $f \in \mathscr{H}$ fulfilling (3.6), it holds true that*

$$\Lambda_q f = -\int_\varepsilon^{T_0} \sum_{n=1}^{+\infty} e^{-\lambda_n s} \left(\int_{\partial\Omega} f(\sigma, T_0 - s)\psi_n(\sigma) d\sigma \right) \psi_n ds \text{ in } H^{1/2}(\partial\Omega).$$

PROOF. In light of (3.7), we may write that

$$u_n(T_0) = -\int_\varepsilon^{T_0} e^{-\lambda_n s}\gamma_n(s) ds \text{ with } \gamma_n(s) := \int_{\partial\Omega} f(\sigma, T_0 - s)\psi_n(\sigma) d\sigma,\ s \in [\varepsilon, T_0]. \tag{3.8}$$

As $|\gamma_n(s)| \leqslant \|f(\cdot, T_0 - s)\|_{L^2(\partial\Omega)} \|\psi_n\|_{L^2(\partial\Omega)}$, we get from (1.7) that

$$\begin{aligned} &\sum_{n=n_M}^{+\infty} e^{-\lambda_n s} |\gamma_n(s)| \, \|\psi_n\|_{H^{1/2}(\partial\Omega)} \\ &\qquad \leqslant c^2 \left(\sum_{n=1}^{+\infty} \lambda_n^2 e^{-\lambda_n \varepsilon} \right) \|f(\cdot, T_0 - s)\|_{H^{3/2}(\partial\Omega)},\ s \in [\varepsilon, T_0]. \end{aligned}$$

Since $\sum_{n=1}^{+\infty} \lambda_n^2 e^{-\lambda_n \varepsilon} < +\infty$ by (3.3), and $s \mapsto \|f(\cdot, T_0 - s)\|_{H^{3/2}(\partial\Omega)} \in L^1(\varepsilon, T_0)$, the Lebesgue dominated convergence theorem then yields

$$\sum_{n=1}^{+\infty} \left(\int_\varepsilon^{T_0} e^{-\lambda_n s}\gamma_n(s) ds \right) \psi_n = \int_\varepsilon^{T_0} \sum_{n=1}^{+\infty} e^{-\lambda_n s}\gamma_n(s)\psi_n ds,$$

where the convergence of the series is considered in $H^{1/2}(\partial\Omega)$. □

Armed with Lemma 3.9, we turn now to proving Theorem 3.7.

3.3.2. *Proof of Theorem* 3.7. We split the proof into four steps.

First step: Set up. For $f \in \mathscr{H}_{\text{in}}$ obeying the condition (3.6), we apply Lemma 3.9 and get

$$\Lambda_j(f) = -\int_\varepsilon^{T_0} \sum_{n=1}^{+\infty} e^{-\lambda_{j,n} s} \left(\int_{\Gamma_{\text{in}}} \psi_{j,n}(\sigma) f(\sigma, T_0 - s) d\sigma \right) \psi_{j,n} ds \text{ in } H^{1/2}(\Gamma_{\text{out}}),\ j = 1, 2.$$

This and the assumption $\Lambda_1(f) = \Lambda_2(f)$ for all $f \in \mathscr{H}_{\text{in}}$, yield for a.e. $\sigma' \in \Gamma_{\text{out}}$ that

(3.9)
$$\int_\varepsilon^{T_0} \sum_{n=1}^{+\infty} \left(\int_{\Gamma_{\text{in}}} \left(e^{-\lambda_{1,n}s}\psi_{1,n}(\sigma)\psi_{1,n}(\sigma') - e^{-\lambda_{2,n}s}\psi_{2,n}(\sigma)\psi_{2,n}(\sigma') \right) f(\sigma, T_0 - s) d\sigma \right) ds = 0.$$

Let $g \in H^{3/2}(\partial\Omega)$ be such that supp $g \subset \Gamma_{\text{in}}$, fix $\varepsilon \in (0, T_0)$ and pick $h \in C_0(0, T_0 - \varepsilon)$, the set of compactly supported continuous functions in $(0, T_0 - \varepsilon)$. Then, the function

$$f(\sigma, t) := \begin{cases} g(\sigma)h(t) & \text{if } (\sigma, t) \in \partial\Omega \times (0, T_0 - \varepsilon) \\ 0 & \text{if } (\sigma, t) \in \partial\Omega \times [T_0 - \varepsilon, T_0), \end{cases}$$

lies in $\mathscr{H}_{\text{in}}$ and verifies (3.6). Thus, by applying (3.9) with f expressed in this form, we find for a.e. $\sigma' \in \Gamma_{\text{out}}$ that

$$\int_\varepsilon^{T_0} \sum_{n=1}^{+\infty} \left(\int_{\Gamma_{\text{in}}} \left(e^{-\lambda_{1,n}s}\psi_{1,n}(\sigma)\psi_{1,n}(\sigma') - e^{-\lambda_{2,n}s}\psi_{2,n}(\sigma)\psi_{2,n}(\sigma') \right) g(\sigma) d\sigma \right) h(T_0 - s) ds = 0.$$

Since h is arbitrary in $C_0(0, T_0 - \varepsilon)$, then $s \mapsto h(T_0 - s)$ is arbitrary in $C_0(\varepsilon, T_0)$, so the above identity yields

$$\sum_{n=1}^{+\infty} \int_{\Gamma_{\text{in}}} \left(e^{-\lambda_{1,n}s}\psi_{1,n}(\sigma)\psi_{1,n}(\sigma') - e^{-\lambda_{2,n}s}\psi_{2,n}(\sigma)\psi_{2,n}(\sigma') \right) g(\sigma) d\sigma = 0, \ s \in (\varepsilon, T_0), \ \sigma' \in \Gamma_{\text{out}}, \tag{3.10}$$

by density of $C_0(\varepsilon, T_0)$ in $L^1(\varepsilon, T_0)$. This follows from the fact that

$$s \mapsto \sum_{n=1}^{+\infty} e^{-\lambda_{j,n}s} \left(\int_{\Gamma_{\text{in}}} \psi_{j,n}(\sigma) g(\sigma) d\sigma \right) \psi_{j,n} \in L^\infty((\varepsilon, T_0), H^{1/2}(\Gamma_{\text{out}})),$$

arising from the estimate

$$\begin{aligned} \sum_{n=n_M}^{+\infty} e^{-\lambda_{j,n}s} \left| \int_{\Gamma_{\text{in}}} \psi_{j,n}(\sigma) g(\sigma) d\sigma \right| \|\psi_{j,n}\|_{H^{1/2}(\Gamma_{\text{out}})} & \\ \leqslant \|g\|_{L^2(\Gamma_{\text{in}})} \sum_{n=n_M}^{+\infty} e^{-\lambda_{j,n}\varepsilon} \|\psi_{j,n}\|^2_{H^{1/2}(\partial\Omega)} & \\ \leqslant c^2 \|g\|_{H^{3/2}(\partial\Omega)} \left(\sum_{n=n_M}^{+\infty} \lambda_{j,n}^2 e^{-\lambda_{j,n}\varepsilon} \right) & \\ < \infty, & \end{aligned}$$

which is derived for all $s \in (\varepsilon, T_0)$ from (1.7) and (3.3). Now, since ε is arbitrary in $(0, T_0)$, we infer from (3.10) that

$$\sum_{n=1}^{+\infty} \int_{\Gamma_{\text{in}}} \left(e^{-\lambda_{1,n} s} \psi_{1,n}(\sigma) \psi_{1,n}(\sigma') - e^{-\lambda_{2,n} s} \psi_{2,n}(\sigma) \psi_{2,n}(\sigma') \right) g(\sigma) d\sigma = 0, \quad \sigma' \in \Gamma_{\text{out}},\ s \in (0, T_0). \tag{3.11}$$

Second step: Analytic continuation. For $j = 1, 2$, we set

$$F_j(\sigma', s) = \sum_{n=1}^{+\infty} e^{-\lambda_{j,n} s} \left(\int_{\Gamma_{\text{in}}} \psi_{j,n}(\sigma) g(\sigma) d\sigma \right) \psi_{j,n}(\sigma'),\ (\sigma', s) \in \Gamma_{\text{out}} \times (0, +\infty), \tag{3.12}$$

and we establish the:

LEMMA 3.10. *The $H^{1/2}(\Gamma_{\text{out}})$-function $s \mapsto F_j(\cdot, s)$, $j = 1, 2$, is analytic in $(0, +\infty)$.*

PROOF. For K, a compact subset of $(0, +\infty)$, we set $\varepsilon := \inf\{s,\ s \in K\} > 0$. Then, with reference to (1.7), we have

$$\begin{aligned} e^{-\lambda_{j,n} s} \left| \int_{\Gamma_{\text{in}}} \psi_{j,n}(\sigma) g(\sigma) d\sigma \right| \|\psi_{j,n}\|_{H^{1/2}(\partial\Omega)} & \\ &\leqslant e^{-\lambda_{j,n}\varepsilon} \|\psi_{j,n}\|_{L^2(\Gamma_{\text{in}})} \|g\|_{L^2(\Gamma_{\text{in}})} \|\psi_{j,n}\|_{H^{1/2}(\partial\Omega)} \\ &\leqslant e^{-\lambda_{j,n}\varepsilon} \|\psi_{j,n}\|^2_{H^{1/2}(\partial\Omega)} \|g\|_{H^{3/2}(\partial\Omega)} \\ &\leqslant c^2 \|g\|_{\mathscr{H}} e^{-\lambda_{j,n}\varepsilon} \lambda_{j,n}^2,\ s \in K,\ n \in \mathbb{N}, \end{aligned}$$

for $j = 1, 2$, where c is the constant appearing in (3.3). Therefore, the series

$$\sum_{n=1}^{+\infty} e^{-\lambda_{j,n} s} \left(\int_{\Gamma_{\text{in}}} \psi_{j,n}(\sigma) g(\sigma) d\sigma \right) \psi_{j,n}$$

converges in $H^{1/2}(\partial\Omega)$, uniformly for $s \in K$. As a consequence, the mapping $s \mapsto F_j(\cdot, s)$ is analytic in K, since this is obviously the case for each function $s \mapsto e^{-\lambda_{j,n} s}(\int_{\Gamma_{\text{in}}} \psi_{j,n}(\sigma) g(\sigma) d\sigma) \psi_{j,n}$ with $n \in \mathbb{N}$. Finally, K being arbitrary in $(0, +\infty)$, we end up getting from this that F_j is analytic in $(0, +\infty)$. □

Now, putting (3.11)-(3.12) together, we infer from Lemma 3.10 that:

$$F_1(\cdot, s) = F_2(\cdot, s) \text{ in } H^{1/2}(\Gamma_{\text{out}}),\ s \in (0, +\infty). \tag{3.13}$$

Next, for each $s \in (0, +\infty)$ fixed, we have

$$\sum_{n=n_M}^{+\infty} \lambda_{j,n} e^{-\lambda_{j,n} s} \|\psi_{j,n}\|_{L^2(\Gamma_{\text{in}})} \leqslant c \sum_{n=n_M}^{+\infty} \lambda_{j,n}^2 e^{-\lambda_{j,n} s} < \infty,$$

by (1.7) and (3.3), whence $\sigma \mapsto \sum_{n=1}^{+\infty} e^{-\lambda_{j,n} s} |\psi_{j,n}(\sigma)| \|\psi_{j,n}\|_{H^{1/2}(\Gamma_{\text{out}})} \in L^2(\Gamma_{\text{in}})$. Therefore, it holds true that $\sigma \mapsto \sum_{n=1}^{+\infty} e^{-\lambda_{j,n} s} |\psi_{j,n}(\sigma) g(\sigma)| \|\psi_{j,n}\|_{H^{1/2}(\Gamma_{\text{out}})} \in L^1(\Gamma_{\text{in}})$ for all $s \in (0, +\infty)$. Thus, by applying the Lebesgue dominated convergence theorem, we get that for every $s \in (0, +\infty)$,

$$F_j(\cdot, s) = \sum_{n=1}^{+\infty} e^{-\lambda_{j,n} s} \left(\int_{\Gamma_{\text{in}}} \psi_{j,n}(\sigma) g(\sigma) d\sigma \right) \psi_{j,n}$$
$$= \int_{\Gamma_{\text{in}}} \left(\sum_{n=1}^{+\infty} e^{-\lambda_{j,n} s} \psi_{j,n}(\sigma) \psi_{j,n} \right) g(\sigma) d\sigma,$$

the convergence of the series being taken in the sense of $H^{1/2}(\Gamma_{\text{out}})$. This and (3.13) yield for a.e. $\sigma' \in \Gamma_{\text{out}}$ and all $s \in (0, +\infty)$, that

$$\int_{\Gamma_{\text{in}}} \left(\sum_{n=1}^{+\infty} e^{-\lambda_{1,n} s} \psi_{1,n}(\sigma) \psi_{1,n}(\sigma') \right) g(\sigma) d\sigma = \int_{\Gamma_{\text{in}}} \left(\sum_{n=1}^{+\infty} e^{-\lambda_{2,n} s} \psi_{2,n}(\sigma) \psi_{2,n}(\sigma') \right) g(\sigma) d\sigma. \tag{3.14}$$

Moreover, since $\sigma \mapsto \sum_{n=1}^{+\infty} e^{-\lambda_{j,n} s} \psi_{j,n}(\sigma) \psi_{j,n} \in L^2(\Gamma_{\text{in}}, H^{1/2}(\Gamma_{\text{out}}))$, $j = 1, 2$, and since g is arbitrary in $H^{3/2}(\Gamma_{\text{in}})$, we deduce from (3.14) and the density of $H^{3/2}(\Gamma_{\text{in}})$ in $L^2(\Gamma_{\text{in}})$, that for all $s \in (0, +\infty)$, the identity

$$\sum_{n=1}^{+\infty} e^{-\lambda_{1,n} s} \psi_{1,n}(\sigma) \psi_{1,n}(\sigma') = \sum_{n=1}^{+\infty} e^{-\lambda_{2,n} s} \psi_{2,n}(\sigma) \psi_{2,n}(\sigma'), \tag{3.15}$$

holds in $L^2(\Gamma_{\text{in}}, H^{1/2}(\Gamma_{\text{out}}))$, and consequently in $L^2(\Gamma_{\text{in}} \times \Gamma_{\text{out}})$.

Third step: Generalized Dirichlet series. Let $\{\lambda'_{j,n},\ n \in \mathbb{N}\}$ be the sequence of strictly increasing eigenvalues of $A_j = A_{q_j}$, $j = 1, 2$. For each $n \in \mathbb{N}$, we denote by $m_{j,n}$ the geometric multiplicity[9] of the eigenvalue $\lambda'_{j,n}$ and we introduce a family $\{\varphi_{j,n,i},\ i = 1, \ldots, m_{j,n}\}$ of eigenfunctions of A_j, which satisfy

$$A_j \varphi_{j,n,i} = \lambda'_{j,n} \varphi_{j,n,i},\ i = 1, \ldots, m_{j,n},$$

and form a $L^2(\Omega)$-orthonormal basis of the eigenspace $\ker(A_j - \lambda'_{j,n})$. Next, we put

$$\theta_{j,n}(\sigma, \sigma') := \sum_{i=1}^{m_{j,n}} \psi_{j,n,i}(\sigma) \psi_{j,n,i}(\sigma'),\ (\sigma, \sigma') \in \Gamma_{\text{in}} \times \Gamma_{\text{out}}, \tag{3.16}$$

where $\psi_{j,n,i} := \partial_\nu \varphi_{j,n,i}$. For every fixed $s \in (0, +\infty)$, it is clear from (1.7) and (3.3) that both series appearing in (3.15) are absolutely convergent in $L^2(\Gamma_{\text{in}} \times \Gamma_{\text{out}})$, so we infer from (3.16) that

$$\sum_{n=1}^{+\infty} e^{-\lambda'_{1,n} s} \theta_{1,n}(\sigma, \sigma') = \sum_{n=1}^{+\infty} e^{-\lambda'_{2,n} s} \theta_{2,n}(\sigma, \sigma'),\ s \in (0, +\infty),\ (\sigma, \sigma') \in \Gamma_{\text{in}} \times \Gamma_{\text{out}}. \tag{3.17}$$

Moreover, each function $\theta_{j,n}$, for $j = 1, 2$ and $n \in \mathbb{N}$, being not identically zero in $\Gamma_{\text{in}} \times \Gamma_{\text{out}}$ according to Lemma 3.6, it follows from (3.17) and the standard theory of Dirichlet series, that

$$\lambda'_{1,n} = \lambda'_{2,n} \text{ and } \theta_{1,n} = \theta_{2,n} \text{ on } \Gamma_{\text{in}} \times \Gamma_{\text{out}},\ n \in \mathbb{N}. \tag{3.18}$$

[9] That is to say that $m_{j,n}$ is the dimension of the linear subspace $\ker(A_j - \lambda'_{j,n})$ in $L^2(\Omega)$.

Fourth step: End of the proof. We are left with the task of showing that $m_{1,n} = m_{2,n}$ for all $n \in \mathbb{N}$, and that the eigenfunctions $\varphi_{2,n}$ can be chosen in such a way that

$$\psi_{1,n,i} = \psi_{2,n,i} \text{ on } \Gamma_{\text{in}} \cup \Gamma_{\text{out}},\ i = 1, \ldots, m_n,$$

where we have set $m_n := m_{1,n} = m_{2,n}$. Prior to doing so, we recall that for all non empty open subset $\Gamma \subset \partial\Omega$, the dimension of the subspace spanned by $\{(\psi_{j,n,i})_{|\Gamma},\ i = 1, \ldots, m_{j,n}\}$ in $L^2(\Gamma)$, is equal to $m_{j,n}$, i.e. that

$$m_{j,n} = \dim\{\psi_{j,n,i},\ i = 1, \ldots, m_{j,n}\},\ j = 1, 2.$$

a) We start by establishing that $m_{1,n} = m_{2,n} = m_n$ and that there exists $M_n \in \mathcal{O}_{m_n}(\mathbb{R})$, the set of orthogonal matrices of size m_n, such that we have

$$\Psi_{2,n} = M_n \Psi_{1,n}, \tag{3.19}$$

with $\Psi_{j,n} := (\psi_{j,n,1}, \ldots, \psi_{j,n,m_{j,n}})^T$, $j = 1, 2$,

To this end, we notice that the set $\Gamma_{n,1} := \{\sigma \in \Gamma_{\text{in}} \cap \Gamma_{\text{out}},\ \psi_{1,n,1}(\sigma) \neq 0\}$ has positive Lebesgue measure, since $\psi_{1,n,1}$ is not identically zero in $\Gamma_{\text{in}} \cap \Gamma_{\text{out}}$. Similarly, the functions $\psi_{1,n,1}$ and $\psi_{1,n,2}$ being linearly independent in $L^2(\Gamma_{\text{in}} \cap \Gamma_{\text{out}})$, the Lebesgue measure of the set

$$\Gamma_{n,2} := \left\{ (\sigma_1, \sigma_2) \in (\Gamma_{\text{in}} \cap \Gamma_{\text{out}})^2,\ \det \begin{pmatrix} \psi_{1,n,1}(\sigma_1) & \psi_{1,n,2}(\sigma_1) \\ \psi_{1,n,1}(\sigma_2) & \psi_{1,n,2}(\sigma_2) \end{pmatrix} \neq 0 \right\}$$

is positive[10]. Thus, by induction on i, we can build a subset $\Gamma_{n,m_{1,n}} \subset (\Gamma_{\text{in}} \cap \Gamma_{\text{out}})^{m_{1,n}}$, with positive Lebesgue measure, such that following matrix

$$P_{1,n}(\boldsymbol{\sigma}) := \begin{pmatrix} \psi_{1,n,1}(\sigma_1) & \ldots & \psi_{1,n,m_{1,n}}(\sigma_1) \\ \vdots & & \vdots \\ \psi_{1,n,1}(\sigma_{m_{1,n}}) & \ldots & \psi_{1,n,m_{1,n}}(\sigma_{m_{1,n}}) \end{pmatrix}$$

is invertible for a.e. $\boldsymbol{\sigma} := (\sigma_1, \ldots, \sigma_{m_{1,n}}) \in \Gamma_{n,m_{1,n}}$.

Next, with reference to (3.18), we get upon applying (3.16) with $\sigma' = \sigma_j$ for $j = 1, \ldots, m_{1,n}$, that

$$\sum_{i=1}^{m_{1,n}} \psi_{1,n,i}(\sigma_j)\psi_{1,n,i}(\sigma)$$
$$= \sum_{i=1}^{m_{2,n}} \psi_{2,n,i}(\sigma_j)\psi_{2,n,i}(\sigma),\ \sigma \in \Gamma_{\text{in}},\ \boldsymbol{\sigma} = (\sigma_1, \ldots, \sigma_{m_{1,n}}) \in \Gamma_{n,m_{1,n}}.$$

This can be equivalently rewritten as $P_{1,n}(\boldsymbol{\sigma})\Psi_{1,n}(\sigma) = P_{2,n}(\boldsymbol{\sigma})\Psi_{2,n}(\sigma)$ for a.e. $\sigma \in \Gamma_{\text{in}}$, where $P_{2,n}(\boldsymbol{\sigma})$ is the following $m_{1,n} \times m_{2,n}$ matrix:

$$P_{2,n}(\boldsymbol{\sigma}) := \begin{pmatrix} \psi_{2,n,1}(\sigma_1) & \ldots & \psi_{2,n,m_{2,n}}(\sigma_1) \\ \vdots & & \vdots \\ \psi_{2,n,1}(\sigma_{m_{1,n}}) & \ldots & \psi_{2,n,m_{2,n}}(\sigma_{m_{1,n}}) \end{pmatrix}.$$

[10]Otherwise, we would have $\psi_{1,n,2}(\sigma_1)\psi_{1,n,1} - \psi_{1,n,1}(\sigma_1)\psi_{1,n,2} = 0$ in $L^2(\Gamma_{\text{in}} \cap \Gamma_{\text{out}})$ for a.e. $\sigma_1 \in \Gamma_{\text{in}} \cap \Gamma_{\text{out}}$, and hence $\psi_{1,n,1}(\sigma_1) = 0$ since $\psi_{1,n,1}$ and $\psi_{1,n,2}$ are linearly independent in $L^2(\Gamma)$, which is a contradiction with the fact that $\Gamma_{n,1}$ has non-zero Lebesgue measure.

Therefore, putting $M_n(\boldsymbol{\sigma}) := P_{1,n}(\boldsymbol{\sigma})^{-1}P_{2,n}(\boldsymbol{\sigma})$ for a.e. $\boldsymbol{\sigma} \in \Gamma_{n,m_{1,n}}$, we get that $\Psi_{1,n}(\sigma) = M_n(\boldsymbol{\sigma})\Psi_{2,n}(\sigma)$ for a.e. $\sigma \in \Gamma_{\text{in}}$. Further, taking $\sigma = \sigma_j$ in (3.17), we get in the same way as before that $\Psi_{1,n}(\sigma') = M_n(\boldsymbol{\sigma})\Psi_{2,n}(\sigma')$ for a.e. $\sigma' \in \Gamma_{\text{out}}$. As a consequence, we have

$$\Psi_{1,n}(\sigma) = M_n(\boldsymbol{\sigma})\Psi_{2,n}(\sigma),\ \sigma \in \Gamma_{\text{in}} \cup \Gamma_{\text{out}},\ \boldsymbol{\sigma} \in \Gamma_{n,m_{1,n}}. \tag{3.20}$$

Since $\dim\{\psi_{j,n,i},\ i = 1,\dots,m_{j,n}\} = m_{j,n}$ in $L^2(\Gamma_{\text{in}} \cup \Gamma_{\text{out}})$, $j = 1,2$, we infer from (3.20) that $m_{1,n} \leqslant m_{2,n}$. Moreover, as $j = 1$ and $j = 2$ play symmetric roles here, we have $m_{2,n} \leqslant m_{1,n}$, so we end up getting that $m_{1,n} = m_{2,n}$.

It remains to show that $M_n(\boldsymbol{\sigma}) \in \mathcal{O}_{m_n}(\mathbb{R})$ for a.e. $\boldsymbol{\sigma} \in \Gamma_{n,m_n}$. This can be done by plugging each of the two following equalities $\Psi_{1,n}(\sigma) = M_n(\boldsymbol{\sigma})\Psi_{2,n}(\sigma)$ for a.e. $\sigma \in \Gamma_{\text{in}}$ and $\Psi_{1,n}(\sigma') = M_n(\boldsymbol{\sigma})\Psi_{2,n}(\sigma')$ for a.e. $\sigma' \in \Gamma_{\text{out}}$, in (3.17). We obtain that $M_n(\boldsymbol{\sigma})\psi_{2,n}(\sigma) \cdot M_n(\boldsymbol{\sigma})\psi_{2,n}(\sigma') = \psi_{2,n}(\sigma) \cdot \psi_{2,n}(\sigma')$, where the symbol $\cdot$ stands for the Euclidian scalar product in $\mathbb{R}^{m_n}$. Therefore, we have $(M_n(\boldsymbol{\sigma})^T M_n(\boldsymbol{\sigma}) - I_{m_n})\psi_{2,n}(\sigma)\cdot\psi_{2,n}(\sigma') = 0$ for a.e. $(\sigma,\sigma') \in \Gamma_{\text{in}}\times\Gamma_{\text{out}}$, where I_{m_n} denotes the identity matrix of size m_n. The family $\{\psi_{2,n,i},\ i = 1,\dots,m_n\}$, being linearly independent in $L^2(\Gamma_{\text{out}})$, this entails that $(M_n(\boldsymbol{\sigma})^T M_n(\boldsymbol{\sigma}) - I_{m_n})\psi_{2,n}(\sigma) = 0$ for a.e. $\sigma \in \Gamma_{\text{in}}$. Similarly, using that $\{\psi_{2,n,i},\ i = 1,\dots,m_n\}$ is linearly independent in $L^2(\Gamma_{\text{in}})$, we get that $M_n(\boldsymbol{\sigma})^T M_n(\boldsymbol{\sigma}) - I_{m_n} = 0$, which establishes that $M_n(\boldsymbol{\sigma}) \in \mathcal{O}_{m_n}(\mathbb{R})$.

b) We turn now to showing that $\psi_{1,n,i} = \psi_{2,n,i}$ on $\partial\Omega$ for all $i \in \{1,\dots,m_n\}$, up to some appropriate choice of the eigenfunctions $\varphi_{2,n,i}$. To do that, we write $\varphi_{2,n} = (\varphi_{2,n,1},\dots,\varphi_{2,n,m_n})^T$ and we consider

$$\varphi'_{2,n} = (\varphi'_{2,n,1},\dots,\varphi'_{2,n,m_n})^T := M_n(\boldsymbol{\sigma})^T \varphi_{2,n},$$

where $\boldsymbol{\sigma}$ is arbitrary in Γ_{n,m_n}.

Writing M_n instead of $M_n(\boldsymbol{\sigma})$ in the sequel, we have for all $(i,k) \in \{1,\dots,m_n\}^2$ that

$$\langle \varphi'_{2,n,i}, \varphi'_{2,n,k}\rangle_{L^2(\Omega)} = \sum_{r,s=1}^{m_n} (M_n^T)_{ir}(M_n^T)_{ks}\langle \varphi_{2,n,r}, \varphi_{2,n,s}\rangle_{L^2(\Omega)} = \sum_{r,s=1}^{m_n} (M_n)_{ri}(M_n)_{sk}\delta_{rs},$$

where δ denotes the Kronecker symbol[11]. Thus it holds true for all $(i,k) \in \{1,\dots,m_n\}^2$ that $\langle \varphi'_{2,n,i}, \varphi'_{2,n,k}\rangle_{L^2(\Omega)} = \sum_{r=1}^{m_n}(M_n)_{ri}(M_n)_{rk} = \sum_{r=1}^{m_n}(M_n^T)_{ir}(M_n)_{rk}$ $= (M_n^T M_n)_{ik} = \delta_{ik}$. Consequently, the family $\{\varphi'_{2,n,i},\ i = 1,\dots,m_n\}$ is orthonormal in $L^2(\Omega)$. Moreover, for all $i \in \{1,\dots,m_n\}$ and for a.e. $\sigma \in \partial\Omega$, we get upon writing $\nu(\sigma) = (\nu_1(\sigma),\dots,\nu_d(\sigma))^T$, that

$$\psi'_{2,n,i}(\sigma) := \nabla\varphi'_{2,n,i}(\sigma) \cdot \nu(\sigma) = \sum_{\ell=1}^{d} \partial_\ell\varphi'_{2,n,i}(\sigma)\nu_\ell(\sigma)$$

---

[11] That is to say that

$$\delta_{rs} := \begin{cases} 1 & \text{if } r = s \\ 0 & \text{otherwise.} \end{cases}$$

reads

$$\begin{aligned}\psi'_{2,n,i}(\sigma) &= \sum_{\ell=1}^{d} \partial_\ell \left(\sum_{r=1}^{m_n} (M_n^T)_{ir} \varphi_{2,n,r}(\sigma) \right) \nu_\ell(\sigma) \\ &= \sum_{r=1}^{m_n} (M_n)_{ri} \left(\sum_{\ell=1}^{d} \partial_\ell \varphi_{2,n,r}(\sigma) \nu_\ell(\sigma) \right) \\ &= \sum_{r=1}^{m_n} (M_n)_{ri} \psi_{2,n,r}(\sigma).\end{aligned}$$

Therefore, we have $\Psi'_{2,n} = M_n^T \Psi_{2,n}$ and hence $\Psi'_{2,n} = \Psi_{1,n}$ on $\partial\Omega$, by virtue of (3.19). This terminates the proof of Theorem 3.7.

References

[1] G. Alessandrini and J. Sylvester, *Stability for a multidimensional inverse spectral theorem*, Comm. Partial Differential Equations **15** (1990), no. 5, 711–736, DOI 10.1080/03605309908820705. MR1070844

[2] M. Bellassoued, M. Choulli, and M. Yamamoto, *Stability estimate for an inverse wave equation and a multidimensional Borg-Levinson theorem*, J. Differential Equations **247** (2009), no. 2, 465–494, DOI 10.1016/j.jde.2009.03.024. MR2523687

[3] M. I. Belishev, *Recent progress in the boundary control method*, Inverse Problems **23** (2007), no. 5, R1–R67, DOI 10.1088/0266-5611/23/5/R01. MR2353313

[4] G. Borg, *Eine Umkehrung der Sturm-Liouvilleschen Eigenwertaufgabe. Bestimmung der Differentialgleichung durch die Eigenwerte* (German), Acta Math. **78** (1946), 1–96, DOI 10.1007/BF02421600. MR15185

[5] B. Canuto and O. Kavian, *Determining coefficients in a class of heat equations via boundary measurements*, SIAM J. Math. Anal. **32** (2001), no. 5, 963–986, DOI 10.1137/S003614109936525X. MR1828313

[6] B. Canuto and O. Kavian, *Determining two coefficients in elliptic operators via boundary spectral data: a uniqueness result* (English, with English and Italian summaries), Boll. Unione Mat. Ital. Sez. B Artic. Ric. Mat. (8) **7** (2004), no. 1, 207–230. MR2044267

[7] R. Carlson, *An inverse spectral problem for Sturm-Liouville operators with discontinuous coefficients*, Proc. Amer. Math. Soc. **120** (1994), no. 2, 475–484, DOI 10.2307/2159884. MR1197532

[8] M. Choulli, *Une introduction aux problèmes inverses elliptiques et paraboliques* (French), Mathématiques & Applications (Berlin) [Mathematics & Applications], vol. 65, Springer-Verlag, Berlin, 2009. MR2554831

[9] M. Choulli and P. Stefanov, *Stability for the multi-dimensional Borg-Levinson theorem with partial spectral data*, Comm. Partial Differential Equations **38** (2013), no. 3, 455–476, DOI 10.1080/03605302.2012.747538. MR3019446

[10] I. M. Gel′fand and B. M. Levitan, *On the determination of a differential equation from its spectral function* (Russian), Izvestiya Akad. Nauk SSSR. Ser. Mat. **15** (1951), 309–360. MR0045281

[11] H. Isozaki, *Some remarks on the multi-dimensional Borg-Levinson theorem*, J. Math. Kyoto Univ. **31** (1991), no. 3, 743–753, DOI 10.1215/kjm/1250519727. MR1127097

[12] A. Katchalov, Y. Kurylev, M. Lassas, and N. Mandache, *Equivalence of time-domain inverse problems and boundary spectral problems*, Inverse Problems **20** (2004), no. 2, 419–436, DOI 10.1088/0266-5611/20/2/007. MR2065431

[13] O. Kavian, Y. Kian, and E. Soccorsi, *Uniqueness and stability results for an inverse spectral problem in a periodic waveguide*, J. Math. Pures Appl. (9) **104** (2015), no. 6, 1160–1189, DOI 10.1016/j.matpur.2015.09.002. MR3420415

[14] Y. Kian, *A multidimensional Borg-Levinson theorem for magnetic Schrödinger operators with partial spectral data*, J. Spectr. Theory **8** (2018), no. 1, 235–269, DOI 10.4171/JST/195. MR3762132

[15] Y. Kian, L. Oksanen, E. Soccorsi, and M. Yamamoto, *Global uniqueness in an inverse problem for time fractional diffusion equations*, J. Differential Equations **264** (2018), no. 2, 1146–1170, DOI 10.1016/j.jde.2017.09.032. MR3720840

[16] N. Levinson, *The inverse Sturm-Liouville problem*, Mat. Tidsskr. B **1949** (1949), 25–30. MR32067

[17] A. Nachman, J. Sylvester, and G. Uhlmann, *An n-dimensional Borg-Levinson theorem*, Comm. Math. Phys. **115** (1988), no. 4, 595–605. MR933457

[18] R. G. Novikov, *A multidimensional inverse spectral problem for the equation $-\Delta\psi + (v(x) - Eu(x))\psi = 0$* (Russian), Funktsional. Anal. i Prilozhen. **22** (1988), no. 4, 11–22, 96, DOI 10.1007/BF01077418; English transl., Funct. Anal. Appl. **22** (1988), no. 4, 263–272 (1989). MR976992

[19] L. Päivärinta and V. Serov, *An n-dimensional Borg-Levinson theorem for singular potentials*, Adv. in Appl. Math. **29** (2002), no. 4, 509–520, DOI 10.1016/S0196-8858(02)00027-1. MR1943362

[20] V. Pohjola, *Multidimensional Borg-Levinson theorems for unbounded potentials*, Asymptot. Anal. **110** (2018), no. 3-4, 203–226, DOI 10.3233/asy-181484. MR3885496

[21] A. G. Ramm, *Inverse problems*, Mathematical and Analytical Techniques with Applications to Engineering, Springer, New York, 2005. Mathematical and analytical techniques with applications to engineering; With a foreword by Alan Jeffrey. MR2838778

[22] M. Reed and B. Simon, *Methods of modern mathematical physics. IV. Analysis of operators*, Academic Press [Harcourt Brace Jovanovich, Publishers], New York-London, 1978. MR0493421

Aix-Marseille Univ, Université de Toulon, CNRS, CPT, Marseille, France
Email address: eric.soccorsi@univ-amu.fr

Contemporary Mathematics
Volume **757**, 2020
https://doi.org/10.1090/conm/757/15249

Simultaneous and indirect control of waves: Some recent developments and open problems

Louis Tebou

ABSTRACT. The control of a single wave equation has known some tremendous amount of activity since the pioneering works of Russell in the sixties and seventies, and especially after the publication of Lions' monographs on controllability. However, the control of multi-component systems of wave equations can be arguably said to be at an early stage. In these notes, I discuss some of my recent works in this area, focusing on the notion of simultaneous control (using the same control scheme in all components of the system at hand), and indirect control (using a single control for a system consisting of two components). I also discuss some problems that I deem important to investigate.

CONTENTS

1. Simultaneous stabilization

The notion of simultaneous control was introduced by Russell in the framework of simultaneous controllability of a system of uncoupled wave equations in his study of the boundary controllability of Maxwell equations in rectangular domains [**82**]. To solve that problem, he transformed it into a controllability problem for a system of two uncoupled wave equations, one having the Dirichlet boundary conditions while the other one had the Neumann boundary conditions. That simultaneous controllability result was later generalized by Lions [**59**] to a larger class of domains, and to uncoupled plate equations. As for the simultaneous internal controllability of uncoupled wave equations (two or more equations), Haraux initiated that work in [**35**], where he established some unique continuation results leading to approximate controllability in all space dimensions. He also proved an exact controllability result in one space dimension where the control region was an arbitrary open subinterval of the interval under consideration, and he proved another one in

2010 *Mathematics Subject Classification.* Primary 93D15.

higher space dimensions where the control region was the whole domain under consideration. The one-dimensional controllability result of Haraux was generalized to all space dimensions and under the Bardos-Lebeau-Rauch geometric control condition by Tebou [**85, 88**]. In particular Haraux derived stabilization results from his one dimensional observability estimates in [**36**], while the present author derived stabilization in [**85**]. Simultaneous stabilization means that we want to analyze the stability of a multi-component system by using the same damping mechanism in all the components of the system. In particular, we are concerned with the case where the damping is localized in a nonempty open subset of the domain under consideration.

1.1. Lamé systems with localized damping. Let Ω be a bounded domain in $\mathbb{R}^N$ with smooth boundary Γ. Let $T > 0$.

Consider the system of uncoupled wave equations

$$\begin{aligned}&u_{jtt} - a_j\Delta u_j = 0 \text{ in } Q\\ &u_j = 0 \text{ on } \Gamma \times (0,T)\\ &u_j(x,0) = u_j^0(x), \quad u_{jt}(x,0) = u_j^1(x) \text{ in } \Omega, \quad j = 1,\ 2, \ldots,\ q,\end{aligned}$$

where $(u_j^0, u_j^1) \in H_0^1(\Omega) \times L^2(\Omega)$, and a_j are positive constants, for each j.

In 1988, Haraux showed for an arbitrary nonempty open set ω, the following unique continuation, and observability inequality, respectively:

- If $\sum_{j=1}^q u_j(x,t) = 0$ in $\omega \times (0,T)$ then $u_j^0 = 0, \quad u_j^1 = 0$ in $\Omega, \quad \forall j.$ provided that $a_j \neq a_k$ for all $j,\ k$ with $j \neq k$.
- If $N = 1$ and T is large enough, or $\omega = \Omega$, then there exists $C > 0$: for all j and all $(u_j^0, u_j^1) \in L^2(\Omega) \times H^{-1}(\Omega)$

$$\sum_{j=1}^q \{||u_j^0||^2_{L^2(\Omega)} + ||u_j^1||^2_{H^{-1}(\Omega)}\} \leq C\int_0^T \int_\omega |\sum_{j=1}^q u_j(x,t)|^2\, dxdt$$

provided that $a_j \neq a_k$ for all $j,\ k$ with $j \neq k$.

Later on, the observability result result of Haraux was improved by the present author under the following geometric control condition:

(GCC): Bardos-Lebeau-Rauch (1992): ω is an admissible control region in time T if every ray of geometric optics enters ω in a time less than T.

THEOREM 1.1. *([**85, 88**]) Let T_0 denote the best controllability time for a single wave equation with unit speed of propagation. Suppose that $T > T_0 \max\{a_j^{-\frac{1}{2}}; j = 1,\ 2,\ \ldots,\ q\}$ and (ω, T) satisfies (GCC). There exists a constant $C > 0$ such that for all $(u_j^0, u_j^1) \in H_0^1(\Omega) \times L^2(\Omega)$, $j = 1,\ 2, \ldots,\ q$:*

$$\sum_{j=1}^q \{||u_j^0||^2_{H_0^1(\Omega)} + ||u_j^1||^2_{L^2(\Omega)}\} \leq C\int_0^T \int_\omega |\sum_{j=1}^q u_{jt}(x,t)|^2\, dxdt,$$

with $C = C(\Omega, \omega, T, (a_j)_j, q)$, if and only if $a_j \neq a_k$ for all $j,\ k$ with $j \neq k$.

Now, let $(y_j^0, y_j^1)_j \in \left(\left[H_0^1(\Omega)\right]^N \times \left[L^2(\Omega)\right]^N\right)^q$, let d be a nonnegative function in $L^\infty(\Omega)$, and onsider the damped elastodynamic system

$$\begin{aligned}
&y_{jtt} - \mu_j \Delta y_j - (\mu_j + \lambda_j)\nabla \mathrm{div}(y_j) + d\sum_{k=1}^{q} y_{kt} = 0 \text{ in } \Omega \times (0,\infty)\\
&y_j = 0 \text{ on } \Gamma \times (0,\infty)\\
&y_j(x,0) = y_j^0(x), \quad y_{jt}(x,0) = y_j^1(x), \text{ in } \Omega,\\
&j = 1,\ 2, ...,\ q,
\end{aligned}$$

where, for each j, μ_j and λ_j are the Lamé constants.

The total energy is given, for all $t \geq 0$, by

$$2E(t) = \sum_{j=1}^{q} \int_\Omega \{|y_{jt}(x,t)|^2 + \mu_j |\nabla y_j(x,t)|^2 + (\mu_j + \lambda_j)|\mathrm{div}(y_j(x,t))|^2\}\, dx$$

The energy E is a nonincreasing function of the time variable as

$$\frac{dE}{dt}(t) = -\int_\Omega d(x) \left|\sum_{k=1}^{q} y_{kt}(x,t)\right|^2 dx.$$

The questions that we would like to address are the following:

- **Question 1:** Does the energy E decay to zero as the time variable goes to infinity?
- **Question 2:** Under which conditions is the Lamé system exponentially stable?

To answer those questions, we are going to reformulate the dynamic Lamé systems as an abstract evolution system. To this end, introduce the Hilbert space $\mathcal{H} = \left(\left[H_0^1(\Omega)\right]^N \times \left[L^2(\Omega)\right]^N\right)^q$ over the field $\mathbb{C}$ of complex numbers, equipped with the norm

$$||Z||_{\mathcal{H}}^2 = \sum_{j=1}^{q} \int_\Omega \{|v_j(x)|^2 + \mu_j |\nabla u_j(x)|^2 + (\mu_j + \lambda_j)|\mathrm{div}(u(x))|^2\}\, dx,$$

$\forall Z = ((u_j, v_j)_j) \in \mathcal{H}$.

Set $Z_j = (y_j, y_{j,t})$. The Lamé system may be recast as the first order abstract evolution equation

$$\dot{Z}_j = \mathcal{A}_j Z_j, \quad Z_j(0) = (y_j^0, y_j^1),\ j = 1,\ 2,\ ...,\ q,$$

where the dot denotes differentiation with respect to time, and the unbounded operator $\mathcal{A}_j$ is given by

$$\mathcal{A}_j \begin{pmatrix} u_j \\ v_j \end{pmatrix} = \begin{pmatrix} v_j \\ \mu_j \Delta u_j + (\mu_j + \lambda_j)\nabla \mathrm{div} u_j - d\sum_{\ell=1}^{q} v_\ell \end{pmatrix}$$

with

$$D(\mathcal{A}_j) = \left\{(u_j, v_j) \in [H_0^1(\Omega)]^N \times [H_0^1(\Omega)]^N; \mu_j \Delta u_j + (\mu_j + \lambda_j)\nabla \mathrm{div} u_j \in [L^2(\Omega)]^N\right\}.$$

It can be checked that one has (assuming for instance that Γ is C^2)

$$D(\mathcal{A}_j) = [H^2(\Omega) \cap H_0^1(\Omega)]^N \times [H_0^1(\Omega)]^N.$$

Thus, for each j, the operator $\mathcal{A}_j$ has a compact resolvent. Consequently the spectrum of $\mathcal{A}_j$ is discrete for each j. With the help of Lumer-Phillips Theorem, ([**71**], p. 14), one can show that the operator $\mathcal{A}$ defined by $\mathcal{A}Z = (\mathcal{A}_j Z_j)_j$ is the infinitesimal generator of a C_0-semigroup of contractions on $\mathcal{H}$. Indeed, $D(\mathcal{A})$ is dense in $\mathcal{H}$, $\mathcal{A}$ is dissipative

$$\Re(\mathcal{A}Z, Z) = -\int_\Omega d(x)|\sum_{j=1}^{q} v_j(x)|^2\,dx \le 0,\ \forall Z \in D(\mathcal{A}),$$

and (denoting by $\mathcal{I}$ the identity operator on $\mathcal{H}$):

$$R(\mathcal{I} - \mathcal{A}) = \mathcal{H},\ \text{by Lax-Milgram Lemma.}$$

Our first stability result reads:

THEOREM 1.2 (Strong stability). *Let ω be a nonempty open subset of Ω. Suppose that d is positive in ω. The elastodynamic system is strongly stable:*

$$\lim_{t\to\infty} E(t) = 0$$

if and only if the propagation speeds are pairwise distinct:

$$\mu_j \neq \mu_k \text{ and } \lambda_j + 2\mu_j \neq \lambda_k + 2\mu_k,\quad \forall j,k,\ j \neq k.$$

PROOF SKETCH. We may apply Benchimol [**13**] or Arendt-Batty [**8**] strong stability criterion; it suffices to show that $\mathcal{A}$ has no purely imaginary eigenvalue. One easily checks that $0 \in \rho(\mathcal{A})$. Now, let λ be a nonzero real number and let $Z = (u,v) \in D(\mathcal{A})$ with

$$(*)\qquad \mathcal{A}Z = i\lambda Z.$$

We shall show that $Z = (0,0)$. It follows from $(*)$:

$$d(x)\sum_{j=1}^{q} u_j = 0 \text{ in } \Omega, \text{ and so, } -\lambda^2 u_j - \mu_j \Delta u_j - (\mu_j + \lambda_j)\nabla \text{div} u_j = 0 \text{ in } \Omega.$$

Therefore, setting $\varphi_j = \text{div}(u_j)$ and $\ell_j = 1/(\lambda_j + 2\mu_j)$, it follows

$$\sum_{j=1}^{q} u_j = 0 \text{ in } \omega, \text{ and } -\lambda^2 \ell_j \varphi_j - \Delta\varphi_j = 0 \text{ in } \omega.$$

Using elementary algebra, one derives from the last two equations

$$\sum_{j=1}^{q} \ell_j^k \varphi_j = 0 \text{ in } \omega,\ k = 0,\ 1,\ ...,\ q-1.$$

The determinant of that linear system is a Vandermonde determinant and is given by

$$D_q = \prod_{1\le j<k\le q} (\ell_k - \ell_j).$$

One checks that $D_q \neq 0$ if and only if $\lambda_j + 2\mu_j \neq \lambda_k + 2\mu_k$ for all $j,\ k$ with $j \neq k$. In this case, $\varphi_j = 0$ in ω for each j.

$$D_q = \prod_{1\le j<k\le q} (\ell_k - \ell_j).$$

Consequently,

$$-\lambda^2 u_j - \mu_j \Delta u_j = 0 \text{ in } \omega.$$

Repeating the same arguments as above, we find, (setting $m_j = 1/\mu_j$):

$$\sum_{j=1}^{q} m_j^k u_j = 0 \text{ in } \omega,\ k = 0,\ 1,\ ...,\ q-1.$$

As earlier, one derives $u_j = 0$ in ω for each j if and only if $\mu_j \neq \mu_k$ for all $j,\ k$ with $j \neq k$.

The Imanuvilov-Yamamoto Carleman estimate for the static Lamé system [**44**], then yields $u_j = 0$ in Ω for each j. Hence $Z = (0,0)$. □

Now, are going to state and prove our exponential stabilization result. For the simplicity of exposition, we choose the damping region to be a neighborhood of the whole boundary.

THEOREM 1.3 (Exponential stability). *Let* $(y_j^0, y_j^1)_j \in \left(\left[H_0^1(\Omega)\right]^N \times \left[L^2(\Omega)\right]^N\right)^q$. *Suppose*

$$\mu_j \neq \mu_k,\ \lambda_j + 2\mu_j \neq \lambda_k + 2\mu_k,\ \textit{and}\ \lambda_j\mu_k = \lambda_k\mu_j,\quad \forall j,k,\ j \neq k.$$

Assume that ω *is a neighborhood of* $\partial\Omega$, *and suppose that the damping is effective in* ω:

$$\exists d_0 > 0 : d(x) \geq d_0 \ a.e.\ \omega.$$

There exist positive constants M *and* κ, *independent of the initial data, such that the following energy decay estimate holds:*

$$E(t) \leq Me^{-\kappa t}E(0),\ \textit{for all}\ t \geq 0.$$

PROOF SKETCH. Thanks to the semigroup exponential stability criterion of Huang [**41**] or Pruss [**72**], and given that we already have strong stability, the exponential decay estimate will follow from the resolvent estimate

$$\exists C_0 > 0 : ||(ib\mathcal{I} - \mathcal{A})^{-1}||_{\mathcal{L}}(\mathcal{H} \leq C_0,\quad \forall b \in \mathbb{R}.$$

To this end, let $U \in \mathcal{H}$, and let b be a real number. Since the range of $ib\mathcal{I} - \mathcal{A}$ is $\mathcal{H}$, there exists $Z \in D(\mathcal{A})$ such that

$$ibZ - \mathcal{A}Z = U. \tag{1.1}$$

We shall prove

$$||Z||_{\mathcal{H}} \leq C_0||U||_{\mathcal{H}}, \tag{1.2}$$

where, here and in the sequel, C_0 is a generic positive constant that may eventually depend on Ω, ω, and d, and the other parameters of the system, but not on b.

To prove the estimate (1.2), first, we note that if $Z = (u,v)$, and $U = (f,g)$, then the equation $ibZ - \mathcal{A}Z = U$ may be recast as $(1 \leq j \leq q)$:

$$\begin{aligned} &ibu_j - v_j = f_j \\ &ibv_j - \mu_j\Delta u_j - (\lambda_j + \mu_j)\nabla \text{div} u_j + d\sum_{k=1}^{q} v_j = g_j. \end{aligned} \tag{1.3}$$

The proof of the inequality requires several steps which can be summed up as: first we shall use the multipliers technique to prove

$$||Z||_{\mathcal{H}}^2 \leq C_0||U||_{\mathcal{H}}^2 + C_0 b^2 \sum_{j=1}^{q} \int_{\Omega} \zeta^2 |u_j|^2\, dx.$$

Then using elementary algebra, we shall derive

$$||Z||^2_{\mathcal{H}} \leq C_0||U||^2_{\mathcal{H}} - 2C_0 b^2 \Re \sum_{1\leq j<k\leq q} \int_\Omega \zeta^2 u_j \bar{u}_k \, dx.$$

Finally, we will use appropriate multipliers and the parameter constraints to derive the claimed estimate.

Taking the inner product with Z on both sides of (1.1), then taking the real parts, we immediately derive

$$\int_\Omega d(x) \left| \sum_{k=1}^q v_k(x) \right|^2 dx \leq ||U||_{\mathcal{H}} ||Z||_{\mathcal{H}}. \tag{1.4}$$

It now follows from the first equation in (1.3):

$$\begin{aligned} b^2 \int_\Omega d(x) \left| \sum_{k=1}^q u_k(x) \right|^2 dx &\leq 2\int_\Omega d(x) \left| \sum_{k=1}^q v_k(x) \right|^2 dx + 2\int_\Omega d(x) \left| \sum_{k=1}^q f_k(x) \right|^2 dx \\ &\leq 2||U||_{\mathcal{H}}||Z||_{\mathcal{H}} + C_0||U||^2_{\mathcal{H}}. \end{aligned} \tag{1.5}$$

Let $\alpha > 0$ and β be real constants. Multiply the second equation in (1.3) by $\beta \bar{u}_j$, integrate on Ω, take the sum over j, and take real parts to find

$$\begin{aligned} &\beta \Re \int_\Omega \sum_{j=1}^q g_j \bar{u}_j \, dx \\ &= \beta \Re \int_\Omega (\sum_{j=1}^q \{ibv_j - \mu_j \Delta u_j - (\lambda_j + \mu_j)\nabla \mathrm{div} u_j + d\sum_{k=1}^q v_k)\bar{u}_j\} \, dx. \end{aligned} \tag{1.6}$$

From which, one derives (Green's formula, Cauchy Schwarz inequality, and first equation in (1.3)):

$$\begin{aligned} &\beta \sum_{j=1}^q (\mu_j |\nabla u_j|_2^2 + (\lambda_j + \mu_j)|\mathrm{div} u_j|_2^2 - |v_j|_2^2) \\ &\leq C_0 \left(||U||_{\mathcal{H}}||Z||_{\mathcal{H}} + ||U||^{\frac{1}{2}}_{\mathcal{H}}||Z||^{\frac{3}{2}}_{\mathcal{H}} \right). \end{aligned} \tag{1.7}$$

Now, multiply the second equation in (1.3) by $2\alpha x \nabla \bar{u}_j$, take real parts and and the sum over j to find

$$\begin{aligned} &2\alpha \sum_{j=1}^q \Re \int_\Omega \{ibv_j - \mu_j \Delta u_j - (\mu_j + \lambda_j)\nabla \mathrm{div} u_j\} \cdot (x\nabla \bar{u}_j) \, dx \\ &= 2\alpha \sum_{j=1}^q \Re \int_\Omega \{g_j - d\sum_{\ell=1}^q v_\ell\} \cdot (x\nabla \bar{u}_j) \, dx. \end{aligned} \tag{1.8}$$

Thanks to the Cauchy-Schwarz inequality, one derives from the right hand side of (1.8) the estimate

$$\begin{aligned} \left| 2\alpha \sum_{j=1}^q \Re \int_\Omega \{g_j - d\sum_{\ell=1}^q v_\ell\} \cdot (x\nabla \bar{u}_j) \, dx \right| &\leq C_0 \sum_{j=1}^q \left\{ |g_j|_2 |\nabla u_j|_2 + |d\sum_{\ell=1}^q v_\ell|_2 |\nabla u_j|_2 \right\} \\ &\leq C_0 \left(||U||_{\mathcal{H}}||Z||_{\mathcal{H}} + ||U||^{\frac{1}{2}}_{\mathcal{H}}||Z||^{\frac{3}{2}}_{\mathcal{H}} \right). \end{aligned} \tag{1.9}$$

Using Green's formula in each of the terms in the left hand side of (1.8), we find respectively:

(1.10)

$$
\begin{aligned}
2\alpha\Re\sum_{j=1}^{q} ib\int_\Omega v_j\cdot(x\nabla\bar u_j)\,dx &= -2\alpha\Re\sum_{j=1}^{q}\int_\Omega v_j\cdot(x\nabla(\bar v_j+\bar f_j))\,dx\\
&= -\alpha\Re\sum_{j=1}^{q}\int_\Omega x\cdot\nabla|v_j|^2\,dx - 2\alpha\Re\sum_{j=1}^{q}\int_\Omega v_j\cdot(x\nabla\bar f_j)\,dx\\
&= N\alpha\sum_{j=1}^{q}|v_j|_2^2 - 2\alpha\Re\sum_{j=1}^{q}\int_\Omega v_j\cdot(x\nabla\bar f_j)\,dx,
\end{aligned}
$$

$$
\begin{aligned}
&-2\alpha\textstyle\sum_{j=1}^{q}\mu_j\Re\displaystyle\int_\Omega \Delta u_j\cdot(x\nabla\bar u_j)\,dx\\
&= 2\alpha\Re\sum_{j=1}^{q}\mu_j\int_\Omega \nabla u_j\nabla(x\nabla\bar u_j)\,dx - 2\alpha\mu_j\sum_{j=1}^{q}\mu_j\int_\Gamma(x\cdot\nu)|\partial_\nu u_j|^2\,d\gamma\\
&= 2\alpha\sum_{j=1}^{q}\mu_j\int_\Omega|\nabla u_j|^2 + \alpha\sum_{j=1}^{q}\mu_j\int_\Omega x\cdot\nabla(|\nabla u_j|^2))\,dx\\
&-2\alpha\sum_{j=1}^{q}\mu_j\int_\Gamma(x\cdot\nu)|\partial_\nu u_j|^2\,d\gamma\\
&= (2-N)\alpha\sum_{j=1}^{q}\mu_j\int_\Omega|\nabla u_j|^2\,dx - \alpha\sum_{j=1}^{q}\mu_j\int_\Gamma(x\cdot\nu)|\partial_\nu u_j|^2\,d\gamma,
\end{aligned}
\tag{1.11}
$$

and

(1.12)

$$
\begin{aligned}
&-2\alpha\sum_{j=1}^{q}(\mu_j+\lambda_j)\Re\int_\Omega \nabla\mathrm{div}u_j\cdot(x\nabla\bar u_j)\,dx\\
&= 2\alpha\sum_{j=1}^{q}(\mu_j+\lambda_j)\Re\int_\Omega \mathrm{div}u_j\mathrm{div}(x\nabla\bar u_j)\,dx - 2\alpha\sum_{j=1}^{q}(\mu_j+\lambda_j)\int_\Gamma(x\cdot\nu)|\partial_\nu u_j\cdot\nu|^2\,d\gamma\\
&= 2\alpha\sum_{j=1}^{q}(\mu_j+\lambda_j)|\mathrm{div}u_j|_2^2 + \alpha\sum_{j=1}^{q}(\mu_j+\lambda_j)\int_\Omega x\cdot\nabla(|\mathrm{div}u_j|^2)\,dx\\
&-2\alpha\sum_{j=1}^{q}(\mu_j+\lambda_j)\int_\Gamma(x\cdot\nu)|\partial_\nu u_j\cdot\nu|^2\,d\gamma\\
&= (2-N)\alpha\sum_{j=1}^{q}(\mu_j+\lambda_j)|\mathrm{div}u_j|_2^2 - \alpha\sum_{j=1}^{q}(\mu_j+\lambda_j)\int_\Gamma(x\cdot\nu)|\partial_\nu u_j\cdot\nu|^2\,d\gamma.
\end{aligned}
$$

It readily follows from (1.8)-(1.12)

$$
\begin{aligned}
&\sum_{j=1}^{q}\left(\alpha N|v_j|_2^2 - (N-2)\alpha\left(\mu_j|\nabla u_j|_2^2 + (\lambda_j+\mu_j)|\mathrm{div}u_j|_2^2\right)\right)\\
&\le C_0\left(||U||_{\mathcal H}||Z||_{\mathcal H} + ||U||_{\mathcal H}^{\frac12}||Z||_{\mathcal H}^{\frac32}\right)\\
&+C_0\sum_{j=1}^{q}\int_{\partial\Omega}(\mu_j|\partial_\nu u_j|^2 + (\lambda_j+\mu_j)|\partial_\nu u_j\cdot\nu|^2)\,d\Gamma.
\end{aligned}
\tag{1.13}
$$

Gathering the inequalities (1.7) and (1.13), and choosing α and β with $\alpha(N-2) < \beta < \alpha N$, it follows:

$$\begin{aligned}
&\sum_{j=1}^{q} \left(|v_j|_2^2 + \left(\mu_j|\nabla u_j|_2^2 + (\lambda_j+\mu_j)|\mathrm{div} u_j|_2^2\right)\right) \\
&\le C_0\left(||U||_{\mathcal{H}}||Z||_{\mathcal{H}} + ||U||_{\mathcal{H}}^{\frac{1}{2}}||Z||_{\mathcal{H}}^{\frac{3}{2}}\right) \\
&+C_0\sum_{j=1}^{q}\int_{\partial\Omega}\left(\mu_j|\partial_\nu u_j|^2 + (\lambda_j+\mu_j)|\partial_\nu u_j\cdot\nu|^2\right)d\Gamma.
\end{aligned} \tag{1.14}$$

To complete the proof, we have to get rid of the boundary integral term in (1.13). For this purpose, let ω_1 be another neighborhood of Γ that is strongly contained in ω. Let $h \in \left[C^1(\bar{\Omega})\right]^N$ be a vector field with

$$h = 0 \text{ in } \Omega\setminus\omega_1, \quad h = \nu \text{ on } \partial\Omega.$$

Now, multiply the second equation in (1.3) by $2h\nabla\bar{u}_j$, and proceed as above to derive:

$$\begin{aligned}
&C_0\sum_{j=1}^{q}\int_{\partial\Omega}\left(\mu_j|\partial_\nu u_j|^2 + (\lambda_j+\mu_j)|\partial_\nu u_j\cdot\nu|^2\right)d\Gamma. \\
&\le C_0\sum_{j=1}^{q}\int_{\omega_1}\left\{|v_j|^2 + \left(\mu_j\nabla u_j|^2 + (\lambda_j+\mu_j)|\mathrm{div} u_j|^2\right)\right\}dx \\
&+C_0\left(||U||_{\mathcal{H}}||Z||_{\mathcal{H}} + ||U||_{\mathcal{H}}^{\frac{1}{2}}||Z||_{\mathcal{H}}^{\frac{3}{2}}\right).
\end{aligned}$$

We shall now absorb the localized energy. To this end, let $\eta \in C^1(\bar{\Omega})$ be a nonnegative function with

$$0 \le \eta \le 1 \text{ in } \Omega, \quad \eta = 1 \text{ in } \omega_1 \text{ and } \eta = 0 \text{ in } \Omega\setminus\omega_2,$$

where ω_2 is another neighborhood of Γ with $\omega_1 \Subset \omega_2 \Subset \omega$.

Multiplying the second equation in (1.3) by $\eta^2\bar{u}_j$, taking real parts and using Green's formula, we derive, (assuming $|b| > 1$, and using the first equation in (1.3)):

$$\begin{aligned}
&C_0\sum_{j=1}^{q}\int_{\omega_1}\left\{|v_j|^2 + \left(\mu_j\nabla u_j|^2 + (\lambda_j+\mu_j)|\mathrm{div} u_j|^2\right)\right\}dx \\
&\le C_0 b^2\sum_{j=1}^{q}\int_{\omega_2}|u_j|^2\,dx \\
&+C_0\left(||U||_{\mathcal{H}}||Z||_{\mathcal{H}} + ||U||_{\mathcal{H}}^{\frac{1}{2}}||Z||_{\mathcal{H}}^{\frac{3}{2}} + ||U||_{\mathcal{H}}^2\right).
\end{aligned}$$

Now, let $\zeta \in C^1(\bar{\Omega})$ be a nonnegative function with

$$0 \le \zeta \le 1 \text{ in } \Omega, \quad \zeta = 1 \text{ in } \omega_2 \text{ and } \zeta = 0 \text{ in } \Omega\setminus\omega.$$

We have

$$\begin{aligned}
C_0 b^2\sum_{j=1}^{q}\int_{\omega_2}|u_j|^2\,dx &\le C_0 b^2\int_\Omega \zeta^2\left|\sum_{j=1}^{q}u_j\right|^2 dx \\
&\quad -2C_0 b^2\Re\sum_{1\le j<k\le q}\int_\Omega \zeta^2 u_j\bar{u}_k\,dx.
\end{aligned}$$

Combining those estimates, and reporting the resulting estimate in (1.14), we find

$$
\begin{aligned}
b^2 \sum_{j=1}^{q} |u_j|_2^2 + ||Z||_{\mathcal{H}}^2 \quad & \leq C_0 b^2 \left| \Re \sum_{1 \leq j < k \leq q} \int_\Omega \zeta^2 u_j \bar{u}_k \, dx \right| \\
& + C_0 \left(||U||_{\mathcal{H}} ||Z||_{\mathcal{H}} + ||U||_{\mathcal{H}}^{\frac{1}{2}} ||Z||_{\mathcal{H}}^{\frac{3}{2}} + ||U||_{\mathcal{H}}^2 \right).
\end{aligned} \tag{1.15}
$$

It remains now to get rid of the integral term in (1.15); this is where the parameter constraints come into play.

Multiply the equation $-b^2 u_j - \mu_j \Delta u_j - (\lambda_j + \mu_j) \nabla \mathrm{div} u_j + d \sum_{\ell=1}^{q} v_\ell = g_j + ibf_j$ by $\mu_k \zeta^2 \bar{u}_k$, and the equation $-b^2 u_k - \mu_k \Delta u_k - (\lambda_k + \mu_k) \nabla \mathrm{div} u_k + d \sum_{\ell=1}^{q} v_\ell = g_k + ibf_k$ by $\mu_j \zeta^2 \bar{u}_j$, then taking real parts and using Green's formula, we get

$$
\begin{aligned}
& -b^2 \mu_k \Re \int_\Omega \zeta^2 u_j \bar{u}_k \, dx + \mu_j \mu_k \Re \int_\Omega \zeta^2 \nabla u_j \nabla \bar{u}_k \, dx \\
& + 2\mu_j \mu_k \Re \int_\Omega \zeta \bar{u}_k \cdot (\nabla \zeta \nabla u_j) + (\lambda_j + \mu_j) \mu_k \Re \int_\Omega \zeta^2 \mathrm{div} u_j \mathrm{div} \bar{u}_k \, dx \\
& + 2(\lambda_j + \mu_j) \mu_k \Re \int_\Omega \zeta (\mathrm{div} u_j) \bar{u}_k \cdot \nabla \zeta \, dx \\
& = \mu_k \int_\Omega \zeta^2 \left(g_j + ibf_j - d \sum_{\ell=1}^{q} v_\ell \right) \bar{u}_k,
\end{aligned} \tag{1.16}
$$

and

$$
\begin{aligned}
& -b^2 \mu_j \Re \int_\Omega \zeta^2 \bar{u}_j u_k \, dx + \mu_j \mu_k \Re \int_\Omega \zeta^2 \nabla \bar{u}_j \nabla u_k \, dx \\
& + 2\mu_j \mu_k \Re \int_\Omega \zeta \bar{u}_j \cdot (\nabla \zeta \nabla u_k) + (\lambda_k + \mu_k) \mu_j \Re \int_\Omega \zeta^2 \mathrm{div} \bar{u}_j \mathrm{div} u_k \, dx \\
& + 2(\lambda_k + \mu_k) \mu_j \Re \int_\Omega \zeta (\mathrm{div} u_k) \bar{u}_j \cdot \nabla \zeta \, dx \\
& = \mu_j \int_\Omega \zeta^2 \left(g_k + ibf_k - d \sum_{\ell=1}^{q} v_\ell \right) \bar{u}_j.
\end{aligned} \tag{1.17}
$$

Subtracting (1.17) from (1.16), and taking the sum over the indices j and k, we derive

$$
\begin{aligned}
& -b^2 \sum_{1 \leq j < k \leq q} \Re \int_\Omega \zeta^2 \bar{u}_j u_k \, dx \\
& = \sum_{1 \leq j < k \leq q} \frac{2\mu_j \mu_k}{\mu_k - \mu_j} \Re \int_\Omega \zeta (\bar{u}_j \cdot (\nabla \zeta \nabla u_k) - \bar{u}_k \cdot (\nabla \zeta \nabla u_j)) \, dx \\
& + \sum_{1 \leq j < k \leq q} \frac{2(\lambda_k + \mu_k)\mu_j}{\mu_k - \mu_j} \Re \int_\Omega \zeta (\bar{u}_j \mathrm{div} u_k) - \bar{u}_k \mathrm{div} u_j) \cdot \nabla \zeta \, dx \\
& + \sum_{1 \leq j < k \leq q} \frac{1}{\mu_k - \mu_j} \Re \int_\Omega \zeta^2 (g_k + ibf_k - d \sum_{\ell=1}^{q} v_\ell) \bar{u}_j \, dx \\
& - \sum_{1 \leq j < k \leq q} \frac{1}{\mu_k - \mu_j} \Re \int_\Omega \zeta^2 (g_j + ibf_j - d \sum_{\ell=1}^{q} v_\ell) \bar{u}_k \, dx.
\end{aligned} \tag{1.18}
$$

Applying Cauchy-Schwarz inequality to each term in the right hand side of (1.18) then yields

$$\left|b^2 \sum_{1\leq j<k\leq q} \Re \int_\Omega \zeta^2 \bar{u}_j u_k\, dx\right| \leq C_0 \sum_{j=1}^{q} |u_j|_2^2$$
$$+C_0 \left(||U||_{\mathcal{H}}||Z||_{\mathcal{H}} + ||U||_{\mathcal{H}}^{\frac{1}{2}}||Z||_{\mathcal{H}}^{\frac{3}{2}} + ||U||_{\mathcal{H}}^2\right).$$

Hence

$$b^2 \sum_{j=1}^{q} |u_j|_2^2 + ||Z||_{\mathcal{H}}^2 \quad \leq C_0 \sum_{j=1}^{q} |u_j|_2^2$$
$$+C_0 \left(||U||_{\mathcal{H}}||Z||_{\mathcal{H}} + ||U||_{\mathcal{H}}^{\frac{1}{2}}||Z||_{\mathcal{H}}^{\frac{3}{2}} + ||U||_{\mathcal{H}}^2\right).$$

Choosing $|b|$ large, enough, and using Young inequality, we get the claimed inequality for numbers b with $|b| > b_0$ for some $b_0 > 0$. The inequality for all real numbers b follows from the continuity of the resolvent for $|b| \leq b_0$. □

COROLLARY 1 (An observability estimate). *Let $T > 0$. Let ω be a neighborhood of the boundary of Ω.*

Consider the uncoupled elastodynamic system

$$\begin{array}{l} y_{jtt} - \mu_j \Delta y_j - (\mu_j + \lambda_j)\nabla div(y_j) = 0 \text{ in } \Omega \times (0,T) \\ y_j = 0 \text{ on } \Gamma \times (0,T) \\ y_j(x,0) = y_j^0(x), \quad y_{jt}(x,0) = y_j^1(x), \text{ in } \Omega, \\ j = 1,\ 2, \ldots,\ q. \end{array}$$

There exists $T_0 > 0$ such that for any $T > T_0$, there exists $C > 0$:

$$E(0) \leq \int_0^T \int_\omega |\sum_{j=1}^{q} y_{jt}(x,t)|^2\, dxdt,$$

provided that

$$\mu_j \neq \mu_k,\ \lambda_j + 2\mu_j \neq \lambda_k + 2\mu_k,\ \text{and } \lambda_j \mu_k = \lambda_k \mu_j, \quad \forall j, k,\ j \neq k.$$

The corollary is a consequence of a result of Haraux [**36**] establishing the equivalence between exponential stability and observability for second order evolution equations.

REMARK 1.1. The exponential stability just established for the Lamé systems remains valid when the damping region is allowed to satisfy the more general geometric constraint of Liu [**62**], given by, (see [**62**, Theorem 4.1], or [**66**, pp.93-95]):

(GC). There exist open sets $\Omega_j \subset \Omega$ with piecewise smooth boundary $\partial\Omega_j$, and points $x_0^j \in \mathbb{R}^N$, $j = 1,\ 2, \ldots, J$, such that $\Omega_i \cap \Omega_j = \emptyset$, for any $1 \leq i < j \leq J$, and:

$$\Omega \cap \mathcal{N}_\delta \left[\left(\bigcup_{j=1}^{J} \Gamma_j\right) \bigcup \left(\Omega \setminus \bigcup_{j=1}^{J} \Omega_j\right)\right] \subset \omega,$$

for some $\delta > 0$, where

$$\begin{array}{l} \mathcal{N}_\delta(S) = \bigcup_{x\in S}\{y \in \mathbb{R}^N; |x-y| < \delta\}, \text{ for } S \subset \mathbb{R}^N, \\ \Gamma_j = \left\{x \in \partial\Omega_j; m^j(x)\cdot \nu^j(x) > 0\right\}, \end{array}$$

ν^j being the unit normal vector pointing into the exterior of Ω_j, and $m^j(x) = x - x_0^j$.

A typical example of control region ω is now provided in 2D.

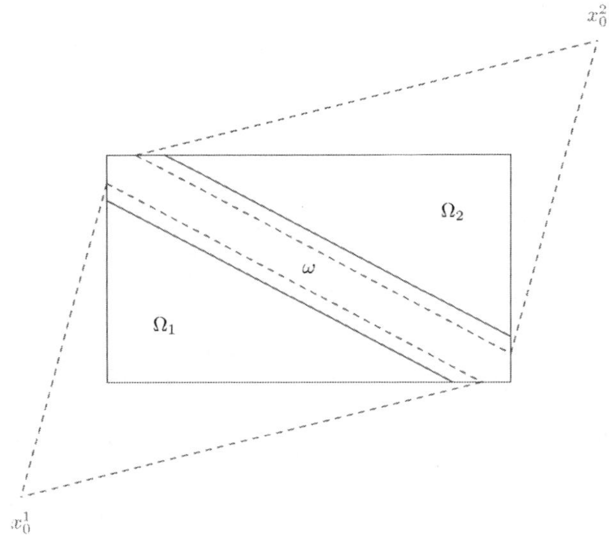

FIGURE 1. Depiction of a subset ω satisfying GC when Ω is a rectangular region; the boundary of ω is solid, while the portion of the boundary of Ω_j, $(j = 1, 2)$, inside Ω is dashed.

REMARK 1.2 ([**90**]). We note that the geometric condition just stated is a generalization of the geometric condition found in Lions' book on controllability [**59**], where the control region is a neighborhood of the portion of the boundary $\Gamma(x_0) = \{x \in \partial\Omega; (x - x_0) \cdot \nu(x) \geq 0\}$, for some $x_0 \in \mathbb{R}^N$. The point x_0 may be viewed as the position of an observer, and $\Gamma(x_0)$ is the portion of the boundary that is invisible to the observer. That control region naturally arises when one uses a multiplier built using the radial vector field $m(x) = x - x_0$. The idea of Liu in [**62**] was to build a larger class of control regions by using, instead of one observer, finitely many observers. So the geometric constraint (GC) provides a much larger class of admissible control regions. One cannot obtain in the above figure ω as a neighborhood of a diagonal by using a single observer; instead, two observers are needed. The use of a single observer would require ω to be a neighborhood of two adjacent sides. We note that, when $J = 1$, (GC) reduces to the geometric constraint found in Lions' book. We emphasize that the constraint (GC) is a sufficient condition, not a necessary one. In particular, in the one dimensional setting, the damping support may be chosen to be an arbitrary nonempty open subinterval of the domain under consideration.

1.2. A system of plate-wave equations. Consider the following damped system

$$
\begin{aligned}
&y_{tt} - \Delta y + a(x)(y_t + z_t) = 0 \text{ in } \Omega \times (0, \infty) \\
&(1 - \gamma\Delta)z_{tt} + \Delta^2 z + a(x)(y_t + z_t) = 0 \text{ in } \Omega \times (0, \infty) \\
&y = 0, \quad z = 0, \quad \partial_\nu z = 0 \text{ on } \Gamma \times (0, \infty) \\
&y(0) = y^0, \quad y_t(0) = y^1, \quad z(0) = z^0, \quad z_t(0) = z^1 \text{ in } \Omega.
\end{aligned}
\tag{1.19}
$$

where the function a is nonnegative and bounded measurable in Ω, the initial data y^0, y^1, z^0, and z^1 lie in appropriate Hilbert spaces.

Set $V_0 = L^2(\Omega)$ and for $\gamma > 0$, let V_γ be the Sobolev space $H_0^1(\Omega)$ over the field $\mathbb{C}$ of complex numbers equipped with the norm

$$||u||_{V_\gamma}^2 = |B_\gamma^{\frac{1}{2}} u|_2^2 = \int_\Omega \{|u|^2 + \gamma|\nabla u|^2\}\, dx,$$

where $B_\gamma u = u - \gamma \Delta u$.

Introduce the Hilbert space over the field $\mathbb{C}$ of complex numbers

$$\mathcal{H}_\gamma = (H_0^1(\Omega) \times L^2(\Omega)) \times (H_0^2(\Omega) \times V_\gamma)$$

equipped with the norm

$$||Z||_\gamma^2 = \int_\Omega \{|\nabla u|^2 + |v|^2 + |\Delta w|^2 + |B_\gamma^{\frac{1}{2}} z|^2\}\, dx, \quad \forall Z = (u, v, w, z) \in \mathcal{H}_\gamma.$$

If we set $Z = \begin{pmatrix} y \\ y_t \\ z \\ z_t \end{pmatrix}$, then (1.19) may be recast as:

$$\begin{aligned} &Z' - \mathcal{A}_\gamma Z = 0 \text{ in } (0, \infty), \\ &Z(0) = \begin{pmatrix} y^0 \\ y^1 \\ z^0 \\ z^1 \end{pmatrix}, \end{aligned} \tag{1.20}$$

where the unbounded operator $\mathcal{A}_\gamma$ is given by:

$$\mathcal{A}_\gamma = \begin{pmatrix} 0 & I & 0 & 0 \\ \Delta & -aI & 0 & -aI \\ 0 & 0 & 0 & I \\ 0 & -B_\gamma^{-1}(aI) & -B_\gamma^{-1}\Delta^2 & -B_\gamma^{-1}(aI) \end{pmatrix}$$

with

$$\begin{aligned} &D(\mathcal{A}_\gamma) = \Big\{(u, v, w, z) \in H_0^1(\Omega) \times H_0^1(\Omega) \times H_0^2(\Omega) \times H_0^2(\Omega); \\ &-\Delta u + a(v + z) \in L^2(\Omega) \& \Delta^2 w + a(v + z) \in B_\gamma(V_\gamma)\Big\} = \tilde{W}_1 \times \tilde{V}_1 \times W_\gamma \times \tilde{V}_2, \end{aligned}$$

where $W_0 = H^4(\Omega) \cap H_0^2(\Omega)$ for $\gamma = 0$, and

$$\tilde{W}_1 = H^2(\Omega) \cap H_0^1(\Omega), \quad \tilde{V}_1 = H_0^1(\Omega), \quad \tilde{V}_2 = H_0^2(\Omega), \quad W_\gamma \subset H^3(\Omega) \cap H_0^2(\Omega).$$

Our strong stability result reads:

THEOREM 1.4. [**87**] *Let ω be an arbitrary nonvoid open set contained in Ω. Suppose that the damping coefficient a is positive in ω. The linear operator $\mathcal{A}_\gamma$ generates a C_0 semigroup of contractions $(S_\gamma(t))_{t\geq 0}$ on $\mathcal{H}_\gamma$. Further assume that either $\gamma \geq 1$, or else ($0 \leq \gamma < 1$ and meas$(\partial\omega \cap \partial\Omega) > 0$). Then the semigroup $(S_\gamma(t))_{t\geq 0}$ is strongly stable:*

$$\lim_{t\to\infty} ||S_\gamma(t)Z^0||_\gamma = 0, \quad \forall Z^0 \in \mathcal{H}_\gamma.$$

REMARK 1.3. Given that the operator $\mathcal{A}_\gamma$ has a compact resolvent, the proof of the strong stability of the semigroup amounts to showing that $\mathcal{A}_\gamma$ has no purely imaginary eigenvalues [**8**, **13**].

On the other hand, one can check, using a compact perturbation argument [**31**, **92**], that for $\gamma > 0$, the semigroup $(S_\gamma(t))_{t\geq 0}$ is not exponentially stable. So, our exponential stability result involves the case $\gamma = 0$.

THEOREM 1.5. [**87**] *Set $\gamma = 0$. Let ω be an open set contained in Ω which satisfies the Liu geometric constraint* [**62**]. *Suppose that the damping coefficient a further satisfies*

$$\exists a_0 > 0 : a(x) \geq a_0 \ a.e. \ \omega.$$

Then the semigroup $(S_0(t))_{t\geq 0}$ is exponentially stable, namely, there exist positive constants M and λ such that

$$||S_0(t)Z^0||_0 \leq Me^{-\lambda t}||Z^0||_0, \quad \forall t \geq 0, \quad \forall Z^0 \in \mathcal{H}_0.$$

As a consequence of that exponential stability result, Proposition 1 in [**36**], which establishes an equivalence between observability and stabilization for second order evolution equations with bounded damping operators, we have

COROLLARY 2. *Let $T > 0$ be a real number. Let (u, v) solve the uncoupled system*

$$\begin{aligned} &u_{tt} - \Delta u = 0 \ in \ Q := \Omega \times (0,T) \\ &v_{tt} + \Delta^2 v = 0 \ in \ Q \\ &u = 0, \quad v = 0, \quad \partial_\nu v = 0 \ on \ \Gamma \times (0,T) \\ &(u(0), u_t(0)) = (u^0, u^1) \in H_0^1(\Omega) \times L^2(\Omega) \\ &(v(0), v_t(0)) = (v^0, v^1) \in H_0^2(\Omega) \times L^2(\Omega). \end{aligned} \tag{1.21}$$

Set

$$E_{u,v}(0) = \int_\Omega \{|u^1(x)|^2 + |\nabla u^0(x)|^2 + |v^1(x)|^2 + |\Delta v^0(x)|^2\} \, dx.$$

Suppose that the set ω satisfies the geometric constraint (GC). There exists a time T_0 such that for every $T > T_0$, there exists a positive constant C independent of the initial data such that the following observability estimate

$$E_{u,v}(0) \leq C \int_0^T \int_\omega |u_t(x,t) + v_t(x,t)|^2 \, dxdt,$$

holds.

1.3. A Timoshenko beam. Let $L > 0$, and set $\Omega = (0, L)$, and $\omega = (l_1, l_2)$ with $0 \leq l_1 < l_2 \leq L$.

Consider the damped Timoshenko system:

$$\begin{cases} \rho_1 y_{tt} - k(y_x + z)_x + a(x)(y_t + z_t) = 0 \text{ in } (0,L) \times (0,\infty) \\ \rho_2 z_{tt} - \sigma z_{xx} + k(y_x + z) + a(x)(y_t + z_t) = 0 \text{ in } (0,L) \times (0,\infty), \end{cases}$$

with the boundary conditions:

(**DD**) $y(0,t) = 0, \quad y(L,t) = 0, \quad z(0,t) = 0, \quad z(L,t) = 0$, or else

(**DN**) $y(0,t) = 0, \quad y(L,t) = 0, \quad z_x(0,t) = 0, \quad z_x(L,t) = 0, \quad t > 0$ and the initial conditions:

$$y(x,0) = y^0(x), \quad y_t(x,0) = y^1(x), \quad z(x,0) = z^0(x), \quad z_t(x,0) = z^1(x), \quad x \in \Omega.$$

The damping coefficient a is a nonnegative bounded measurable function, which is positive in ω only.

REMARK 1.4. The Timoshenko system describes the motion of a beam when the effects of rotatory inertia are accounted for; $y=$ transverse displacement, and $z=$ shear angle displacement.

The type of damping used here naturally arises in the modeling of mixtures of elastic materials [**23**], or the modelling of certain chemical reactions involving multiple components [**96**].

The fact that this damping is degenerate makes the stabilization problem more challenging to analyze.

Introduce the energy

$$E(t) = \frac{1}{2}\int_\Omega \{\rho_1|y_t(x,t)|^2 + k|y_x(x,t)+z(x,t)|^2\}\,dx$$
$$+\tfrac{1}{2}\int_\Omega \{\rho_2|z_t(x,t)|^2 + \sigma|z_x(x,t)|^2\}\,dx, \quad \forall t\geq 0.$$

The energy E is a nonincreasing function of the time variable t as we have for every $t\geq 0$, (hereafter, $'$ denotes differentiation with respect to time)

$$E'(t) = -\int_\Omega a(x)|y_t(x,t)+z_t(x,t)|^2\,dx.$$

Our main purpose in this section is to answer the following questions:

- Does the energy $E(t)$ decay to zero as the time variable t goes to infinity?
- If so, how fast? And if not, why?

To answer those question, we are going to reformulate our system as an abstract evolution system. To this end, set $Z = \begin{pmatrix} y \\ y' \\ z \\ z' \end{pmatrix}$. Our initial system may then be recast as:

$$Z' - \mathcal{A}Z = 0 \text{ in } (0,\infty), \qquad Z(0) = \begin{pmatrix} y^0 \\ y^1 \\ z^0 \\ z^1 \end{pmatrix}$$

where the unbounded operator $\mathcal{A}$ is given by

$$\mathcal{A} = \begin{pmatrix} 0 & I & 0 & 0 \\ \frac{k}{\rho_1}\partial_x^2 & -\frac{a}{\rho_1}I & \frac{k}{\rho_1}\partial_x & -\frac{a}{\rho_1}I \\ 0 & 0 & 0 & I \\ -\frac{k}{\rho_2}\partial_x & -\frac{a}{\rho_2}I & \frac{\sigma}{\rho_2}\partial_x^2 - \frac{k}{\rho_2}I & -\frac{a}{\rho_2}I \end{pmatrix}$$

In the (**DD**) case:

$$D(\mathcal{A}) = \Big\{(u,v,w,z) \in \big(H_0^1(\Omega)\times H_0^1(\Omega)\big)^2; k(u_x+w)_x - a(v+z) \in L^2(\Omega),$$
$$\text{and } \sigma w_{xx} - k(u_x+w) - a(v+z) \in L^2(\Omega)\Big\}$$

and, in the (**DN**) case:

$$D(\mathcal{A}) = \Big\{(u,v,w,z) \in (H_0^1(\Omega))^2 \times V^2; k(u_x+w)_x - a(v+z) \in L^2(\Omega),$$
$$\text{and } \sigma w_{xx} - k(u_x+w) - a(v+z) \in L^2(\Omega)\Big\}$$

$$V = \Big\{u \in H^1(\Omega); \int_\Omega u(x)\,dx = 0\Big\}.$$

One checks that in the (**DD**) case, one has $D(\mathcal{A}) = \big((H^2(\Omega)\cap H_0^1(\Omega))\times H_0^1(\Omega)\big)^2$
In the (**DN**) case, $D(\mathcal{A}) = (H^2(\Omega)\cap H_0^1(\Omega))\times H_0^1(\Omega)\times (H^2(\Omega)\cap V)\times V$.

Thus, in either case, the operator $\mathcal{A}$ has a compact resolvent. Consequently the spectrum of $\mathcal{A}$ is discrete.

Introduce the Hilbert spaces over the field $\mathbb{C}$ of complex numbers $\mathcal{H}_1 = \left(H_0^1(\Omega) \times L^2(\Omega)\right)^2$ and $\mathcal{H}_2 = H_0^1(\Omega) \times L^2(\Omega) \times V \times L^2(\Omega)$, equipped with the norm

$$||Z||^2_{\mathcal{H}_i} = \int_\Omega \{\rho_1|v|^2 + k|u_x + w|^2 + \rho_2|z|^2 + \sigma|w_x|^2\}\, dx,$$

$$\forall Z = (u, v, w, z) \in \mathcal{H}_i, \quad i = 1,\ 2.$$

Our strong stability result here reads:

THEOREM 1.6 (Strong stability, [**89**]). *Suppose that ω is an arbitrary nonempty open interval in Ω. Let the damping coefficient a be positive in ω. In either of the (**DD**) or (**DN**) case, the associated operator $\mathcal{A}$ generates a C_0 semigroup of contractions $(S_i(t))_{t\geq 0}$ on the corresponding Hilbert space $\mathcal{H}_i$, (i=1, 2), which is strongly stable:*

$$\lim_{t\to\infty} ||S_i(t)Z^0||_{\mathcal{H}} = 0, \quad \forall Z^0 \in \mathcal{H}_i,$$

if and only if $\partial\omega \cap \partial\Omega \neq \emptyset$.

PROOF SKETCH.
- Semigroup generation: pretty straightforward thanks to Lumer-Phillips Theorem.
- Strong stability: Thanks to a Benchimol result [**13**], it suffices to show that $\mathcal{A}$ has no imaginary eigenvalue.

One easily checks that zero is not an eigenvalue of $\mathcal{A}$. Now, let b be a nonzero real number, and let $Z = (u, v, w, z) \in D(\mathcal{A})$ such that $\mathcal{A}Z = ibZ$. We shall prove that $Z = (0, 0, 0, 0)$ if and only if $\partial\omega \cap \partial\Omega \neq \emptyset$. Note that $\mathcal{A}Z = ibZ$ may be recast as:

$$\begin{aligned} &-b^2u - \hat{k}(u_x + w)_x + ib\hat{a}(u + w) = 0 \text{ in } (0, L)\\ &-b^2w - \hat{\sigma}w_{xx} + \check{k}(u_x + w) + ib\check{a}(u + w) = 0 \text{ in } (0, L). \end{aligned}$$

One easily checks that $u = -w$ in ω. The heart of the matter now is to check under which condition, one has $u \equiv 0$ in ω; indeed, if we can prove that $u \equiv 0$ in ω, then $w \equiv 0$ in ω, and basic uniqueness results may then be invoked to conclude that $Z = (0, 0, 0, 0)$. Therefore, it remains to find under which condition $u \equiv 0$ in ω. Since $u = -w$ in ω, our system reduces to:

$$\begin{aligned} &-b^2u - \hat{k}(u_x - u)_x = 0 \text{ in } \omega\\ &b^2u + \hat{\sigma}u_{xx} + \check{k}(u_x - u) = 0 \text{ in } \omega. \end{aligned}$$

Adding both equations, one gets rid of b, thereby obtaining

$$(\hat{\sigma} - \hat{k})u_{xx} + (\hat{k} + \check{k})u_x - \check{k}u = 0 \text{ in } \omega.$$

The characteristic equation for this differential equation is given by:

$$(\hat{\sigma} - \hat{k})r^2 + (\hat{k} + \check{k})r - \check{k} = 0.$$

Using the latter equation, one shows that $u \equiv 0$ in ω when $\partial\omega \cap \partial\Omega \neq \emptyset$, and that the operator $\mathcal{A}$ does have imaginary eigenvalues if $\partial\omega \cap \partial\Omega = \emptyset$. Hence the associated semigroup is strongly stable if and only if $\partial\omega \cap \partial\Omega \neq \emptyset$. □

We now state our exponential stability result:

THEOREM 1.7 (Exponential stability, [**89**]). *Suppose that ω is an arbitrary nonempty open interval in Ω with $\partial\omega \cap \partial\Omega \neq \emptyset$. Let the damping coefficient a satisfy*

$$a(x) \geq a_0 > 0, \text{ a.e. in } \omega.$$

For each $i = 1,\ 2$, the semigroup $(S_i(t))_{t\geq 0}$ is exponentially stable, `viz.`, *there exist positive constants M and λ with*

$$||S_i(t)Z^0||_{\mathcal{H}_i} \leq M\exp(-\lambda t)||Z^0||_{\mathcal{H}_i}, \quad \forall Z^0 \in \mathcal{H}_i.$$

PROOF: MAIN IDEAS. Set $\omega = (l_1, L)$. We focus on the (**DD**) case, and set $\mathcal{H} = \mathcal{H}_1$. Thanks to the exponential stability criterion of Huang [**41**] or Pruss, [**72**], it suffices to show:

(1) $\mathrm{i}\mathbb{R} \subset \rho(\mathcal{A})$
(2) $\sup\{||(ib-\mathcal{A})^{-1}||_{\mathcal{L}(\mathcal{H})};\ b \in \mathbb{R}\} < \infty$

The first point follows from the proof of the strong stability theorem.
Proving the second point amounts to showing:

$$\exists C_0 > 0 : \forall U \in \mathcal{H}, \text{ one has } ||(ib-\mathcal{A})^{-1}U||_{\mathcal{H}} \leq C_0||U||_{\mathcal{H}}, \quad \forall b \in \mathbb{R}.$$

Now, let $b \in \mathbb{R}$, $U = (f, g, h, l) \in \mathcal{H}$, and let $Z = (u, v, w, z) \in D(\mathcal{A})$ such that

$$(ib - \mathcal{A})Z = U.$$

We shall prove:

$$||Z||_{\mathcal{H}} \leq C_0||U||_{\mathcal{H}}.$$

Our initial estimate:

$$\int_\Omega a(x)|v(x)+z(x)|^2\,dx = \Re(U,Z) \leq ||U||_{\mathcal{H}}||Z||_{\mathcal{H}}.$$

Recast the U-Z equation as:

$$\begin{aligned}
&ibu - v = f\\
&ibv - \hat{k}(u_x+w)_x + \hat{a}(v+z) = g\\
&ibw - z = h\\
&ibz - \hat{\sigma}w_{xx} + \check{k}(u_x+w) + \check{a}(v+z) = l.
\end{aligned}$$

Using appropriate multipliers, we arrive at (for large enough $|b|$)

$$\begin{aligned}
&\check{k}b^2|u|_2^2 + \check{k}\hat{k}|u_x+w|_2^2 + \hat{\sigma}b^2|w|_2^2 + \hat{\sigma}^2|w_x|_2^2\\
&\leq C_0 b^2\int_\Omega \eta^2(|u|^2+|w|^2)\,dx + C_0(||U||_{\mathcal{H}}||Z||_{\mathcal{H}} + ||U||_{\mathcal{H}}^{\frac{1}{2}}||Z||_{\mathcal{H}}^{\frac{3}{2}} + ||U||_{\mathcal{H}}^2)\\
&\leq C_0 b^2\int_\Omega \eta^2|u+w|^2\,dx - 2C_0b^2\Re\int_\Omega \eta^2 u\bar{w}\,dx\\
&+C_0(||U||_{\mathcal{H}}||Z||_{\mathcal{H}} + ||U||_{\mathcal{H}}^{\frac{1}{2}}||Z||_{\mathcal{H}}^{\frac{3}{2}} + ||U||_{\mathcal{H}}^2).
\end{aligned}$$

Resorting once more to adequate multipliers, we find

$$-b^2(\hat{\sigma}-\hat{k})\Re\int_\Omega \eta^2 u\bar{w}\,dx = \hat{\sigma}\Re\int_\Omega((\hat{g}+ibf)\eta^2 - 2\hat{k}\eta_x\eta(u_x+w))\bar{w}\,dx$$

$$-\hat{k}\Re\int_\Omega((\hat{l}+ibh-\check{k}(u_x+w))\eta^2\bar{u} - 2\hat{\sigma}\eta_x\eta w_x\bar{u} + \hat{\sigma}\bar{w}_x w)\,dx.$$

If $\hat{k} \neq \hat{\sigma}$, one then gets the estimate:

$$\check{k}b^2|u|_2^2 + \check{k}\hat{k}|u_x + w|_2^2 + \hat{\sigma}b^2|w|_2^2 + \hat{\sigma}^2|w_x|_2^2$$
$$\leq C_0(||U||_{\mathcal{H}}||Z||_{\mathcal{H}} + ||U||_{\mathcal{H}}^{\frac{1}{2}}||Z||_{\mathcal{H}}^{\frac{3}{2}} + ||U||_{\mathcal{H}}^2).$$

If $\hat{k} = \hat{\sigma}$, we need a different approach. Using the multiplier $\eta^2\bar{w}$, we derive

$$-b^2\Re\int_\Omega \eta^2 u\bar{w}\,dx = -\hat{k}\Re\int_\Omega \eta^2 u_x\bar{w}_x\,dx$$

$$-\hat{k}\Re\int_\Omega (\eta^2 w_x + 2\eta_x\eta(u_x + w))\bar{w}\,dx + \Re\int_\Omega (\hat{g} + ibf)\eta^2\bar{w}\,dx.$$

Introduce $\varphi = u + w$, which solves:

$$-b^2\varphi - \hat{k}\varphi_{xx} = \hat{g} + \hat{l} + ib(f + h) + \hat{k}w_x - \check{k}(u_x + w) \text{ in } \Omega.$$

Using that equation, one checks that

$$\int_\Omega \eta^2|\varphi_x|^2\,dx \leq C_0(||U||_{\mathcal{H}}||Z||_{\mathcal{H}} + ||U||_{\mathcal{H}}^2 + ||U||_{\mathcal{H}}^{\frac{1}{2}}||Z||_{\mathcal{H}}^{\frac{3}{2}}).$$

With the help of the multiplier $\eta^2\bar{w}_x$, we derive from the same equation:

$$\begin{aligned}
&\hat{k}\int_\Omega \eta^2|w_x|^2\,dx - \check{k}\Re\int_\Omega \eta^2 u_x\bar{w}_x\,dx\\
&= -\Re\int_\Omega (\eta^2 h_x + 2\eta_x\eta(ib\bar{w} + \bar{h}))ib\varphi\,dx - \hat{k}\Re\varphi_x(L)\overline{w_x(L)}\\
&+\Re\int_\Omega \eta^2\varphi_x(\check{k}(\bar{u}_x + \bar{w}) - \bar{\hat{l}})\,dx\\
&-\Re\int_\Omega \eta^2(\hat{g} + \hat{l} - \check{k}w)\bar{w}_x - ib(\eta^2(f + h))_x\bar{w}\,dx\\
&\leq C_0(||U||_{\mathcal{H}}||Z||_{\mathcal{H}} + ||U||_{\mathcal{H}}^{\frac{1}{2}}||Z||_{\mathcal{H}}^{\frac{3}{2}} + ||U||_{\mathcal{H}}^2 + ||U||_{\mathcal{H}}^{\frac{1}{4}}||Z||_{\mathcal{H}}^{\frac{7}{4}} + ||U||_{\mathcal{H}}^{\frac{1}{8}}||Z||_{\mathcal{H}}^{\frac{15}{8}})\\
&+C_0|w|_2^2.
\end{aligned}$$

On the other hand, the estimate of φ_x leads to:

$$\int_\Omega \eta^2|w_x|^2\,dx + \Re\int_\Omega \eta^2 u_x\bar{w}_x\,dx = \Re\int_\Omega \eta^2\varphi_x\bar{w}_x\,dx$$
$$\leq C_0(||U||_{\mathcal{H}}||Z||_{\mathcal{H}} + ||U||_{\mathcal{H}}^{\frac{1}{2}}||Z||_{\mathcal{H}}^{\frac{3}{2}} + ||U||_{\mathcal{H}}^{\frac{1}{4}}||Z||_{\mathcal{H}}^{\frac{7}{4}}).$$

It now follows from the last two inequalities:

$$\begin{aligned}
&\left|\Re\int_\Omega \eta^2 u_x\bar{w}_x\,dx\right|\\
&\leq C_0(||U||_{\mathcal{H}}||Z||_{\mathcal{H}} + ||U||_{\mathcal{H}}^{\frac{1}{2}}||Z||_{\mathcal{H}}^{\frac{3}{2}} + ||U||_{\mathcal{H}}^2 + ||U||_{\mathcal{H}}^{\frac{1}{4}}||Z||_{\mathcal{H}}^{\frac{7}{4}} + ||U||_{\mathcal{H}}^{\frac{1}{8}}||Z||_{\mathcal{H}}^{\frac{15}{8}})\\
&+C_0|w|_2^2.
\end{aligned}$$

Hence

$$\begin{aligned}
&\check{k}b^2|u|_2^2 + \check{k}\hat{k}|u_x + w|_2^2 + \hat{\sigma}b^2|w|_2^2 + \hat{\sigma}^2|w_x|_2^2\\
&\leq C_0(||U||_{\mathcal{H}}||Z||_{\mathcal{H}} + ||U||_{\mathcal{H}}^{\frac{1}{2}}||Z||_{\mathcal{H}}^{\frac{3}{2}} + ||U||_{\mathcal{H}}^2 + ||U||_{\mathcal{H}}^{\frac{1}{4}}||Z||_{\mathcal{H}}^{\frac{7}{4}} + ||U||_{\mathcal{H}}^{\frac{1}{8}}||Z||_{\mathcal{H}}^{\frac{15}{8}}).
\end{aligned}$$

□

2. Indirect controllability

This section is devoted to the notion of indirect controllability; here we have two coupled systems, one is controlled and the other one is uncontrolled. The action of the control is transmitted from the controlled system to the uncontrolled one by the coupling.

2.1. Hyperbolic equations with internal coupling. The controllability of wave equations has a history of more than four decades now. It started in the late sixties, and developed timidly through the early eighties thanks to works by Russell [**32, 75–81**], and Chen [**19–21**] By the mid-eighties, Lions introduced his now famous Hilbert uniqueness method aka H.U.M [**58–60**]. This method reduces the controllability problem for a given distributed system to obtaining an inverse or observability inequality for the associated adjoint system. This tool revolutionized the field of control theory for distributed systems, and made it very attractive to mathematicians of all stripes; many of whom did not hesitate to pitch in. These new interests produced an astounding number of results, so that the space allotted to this contribution would not be enough to quote them all. We will limit ourselves to listing some of the techniques developed through the years by different authors to tackle controllability problems involving hyperbolic equations:

- The Fourier method (this includes moments, harmonic and nonharmonic analysis techniques); this method is well-suited for one-dimensional problems or problems in special (rectangular, spherical) domains, e.g. [**9, 10, 47, 73, 75, 81**],
- the characteristics method, e.g. [**17, 29, 81**],
- the multipliers method: it was introduced in control theory by Russell [**80**], and this was later exemplified in many works, e.g. [**1, 5, 6, 39, 45, 46, 48, 49, 59, 60, 62, 66, 70, 95, 101, 102**],
- microlocal analysis method: this method was developed by Bardos-Lebeau-Rauch [**11**] to tackle control issues for hyperbolic equations. They gave a general sufficient and almost necessary geometric condition for exact controllability:

  (GCC): *The set ω is an admissible region for exact controllability in time T if every ray of geometric optics enters $\omega \times (0,T)$ in a time less than T.*

  This method describes the optimal control region for exact controllability of hyperbolic equations with C^∞ coefficients in general C^∞ domains. The C^∞ requirements in [**11**] for both the domain and coefficients was later relaxed in [**15**] to C^3 for the domain and C^2 for the coefficients. This technique has been used in many other works.
- Carleman estimates method: The Carleman estimate technique for global controllability was initiated by Fursikov and Imanuvilov [**29**] in the framework of parabolic equations. Its counterpart for hyperbolic equations was devised by Zhang [**99, 100**] for the ordinary wave equation. Subsequently, Fu-Yong-Zhang [**27**] and Duyckaerts-Zhang-Zuazua [**24**] adapted it to general hyperbolic equations. It is fair to state here that the first precise Carleman estimate for the wave equation was established by Ruiz [**74**] in the context of unique continuation for weak solutions of the wave equation

with a potential. It is worth noting that the introduction of precise Carleman estimates made it possible to study control problems for semilinear parabolic and hyperbolic equations with superlinear growth, especially in the higher space dimensions setting. Other Carleman estimates were developed by Imanuvilov [**42**], Lasiecka-Triggiani [**51**], Lasiecka-Triggiani-Zhang [**55**], Triggiani [**93**] and Triggiani-Yao [**94**].

- The Riemannian geometry approach: This method was developed by Yao [**98**] for hyperbolic equations with C^∞ coefficients and was subsequently relaxed to account for C^1 coefficients and energy level first order terms by Lasiecka-Triggiani-Yao [**53, 54**].

As those developments in the linear framework were happening, attempts were being made to study the controllability of semilinear wave equations. The first controllability result for a semilinear wave equation was produced by Fattorini[**25**]; that result was an approximate controllability result in the one-dimensional space setting. Later on, that result was extended to higher space dimensions by Chewning [**22**] and Russell [**81**]. The main ingredient in their proofs is the implicit function method as introduced by Markus in the framework of the controllability of nonlinear finite dimensional systems [**65**]. However, the implicit function method leads to local controllability results only. The works of Zuazua [**101**–**103**], produced in the late eighties and early nineties were the first to prove global or exact controllability results for semilinear wave equations using the Schauder or Leray-Schauder fixed point theorems; in particular superlinear nonlinearities are involved in [**103**], which deals with the one-dimensional space setting. The results in [**101, 102**] were later improved and generalized by Lasiecka-Triggiani [**50, 52**], Fursikov-Imanuvilov [**29**], and Zhang [[**99**]. The exact controllability result in [**103**] was later improved by Cannarsa-Komornik-Loreti [**17**] by accounting for more general nonlinearities, but still in the one-dimensional space framework. The first global controllability result in higher space dimensions involving a superlinear nonlinearity is due to Li-Zhang [**57**]; the nonlinearity f satisfies the growth assumption:

$$\lim_{|s|\to\infty} \frac{f(s)}{s\sqrt{\log(|s|)}} = 0.$$

Subsequently, that result was improved by Duyckaerts-Zhang-Zuazua [**24**] to account for nonlinearities f satisfying:

$$\lim_{|s|\to\infty} \frac{f(s)}{s(\log(|s|))^r} = 0, \quad 0 \le r < 3/2.$$

All the results mentioned so far involve a single wave equation.

As for coupled wave equations that are indirectly controlled, the literature is not as rich; most of the works involving coupled wave equations require that the coupling parameter be small in magnitude e.g. [**1, 6**], with a controllability time that blows up as the coupling parameter tends to zero. That smallness requirement is due to the multiplier technique tool utilized. The present author studied the controllability of two coupled wave equations using the Carleman estimates tool, and was able to establish controllability results without any smallness constraint and with the same controllability time as for a single wave equation, [**84, 86**]. We now present some of our finding for coupled hyperbolic equations.

Let the coefficients matrix $(b_{ij})_{i,j}$, satisfy:

$$b_{ij} \in C^1(\bar{\Omega}); \quad b_{ij} = b_{ji}, \quad \forall i, \ j = 1, 2, ..., N,$$
$$\exists a_0 > 0 : b_{ij}(x) z_i z_j \geq a_0 z_i z_i, \quad \forall (x, z) \in \bar{\Omega} \times \mathbb{R}^N.$$

The Einstein summation convention on repeated indices is used throughout.

The functions a, b, c, d lie in $L^\infty(0, T; L^s(\Omega))$, $s \geq \max(2, N)$ for $N \neq 2$, and $s > 2$ for $N = 2$. and the functions k_{ij}, l_{ij} lie in $W_0^{1,\infty}(0, T; L^s(\Omega))$.

Consider the controllability problems: Given (z^0, z^1) and (w^0, w^1), and $\varepsilon > 0$, find a control h such that if (z, w) solves the system

$$\begin{cases} z_{tt} - \partial_i(b_{ij}(x)\partial_j z) + az + cw - \operatorname{div}(k_{11} z) - (l_{11} z)_t \\ -\operatorname{div}(k_{21} w) - (l_{21} w)_t = h 1_\omega \text{ in } Q \\ \\ w_{tt} - \partial_i(b_{ij}(x)\partial_j w) + bz + dw - \operatorname{div}(k_{12} z) - (l_{12} z)_t \\ -\operatorname{div}(k_{22} w) - (l_{22} w)_t = 0 \text{ in } Q \\ \\ z = 0, \quad w = 0 \text{ on } \Sigma = \partial\Omega \times (0, T) \\ \\ z(0) = z^0; \quad z_t(0) = z^1 \quad w(0) = w^0; \quad w_t(0) = w^1 \text{ in } \Omega, \end{cases}$$

then (exact controllability)

$$z(T) = 0, \quad z_t(T) = 0, \quad w(T) = 0, \quad w_t(T) = 0 \text{ in } \Omega,$$

or else (approximate controllability)

$$||z(T)||_1 + ||z_t(T)||_2 \leq \varepsilon, \quad ||w(T)||_1 + ||w_t(T)||_2 \leq \varepsilon.$$

REMARK 2.1. Before going on, some important facts worth mentioning:

- For exact controllability, T and ω must be large enough.
- For approximate controllability, only T must be large enough.
- Lions' HUM reduces exact controllability to an inverse (observability) estimate for the adjoint system.

Consider the coupled (adjoint) system

$$\begin{cases} u_{tt} - \partial_i(b_{ij}(x)\partial_j u) + au + bv + k_{11} \cdot \nabla u + l_{11} u_t \\ +k_{12} \cdot \nabla v + l_{12} v_t = 0 \text{ in } Q \\ \\ v_{tt} - \partial_i(b_{ij}(x)\partial_j v) + cu + dv + k_{21} \cdot \nabla u + l_{21} u_t \\ +k_{22} \cdot \nabla v + l_{22} v_t = 0 \text{ in } Q \\ \\ u = 0, \quad v = 0 \text{ on } \Sigma = \partial\Omega \times (0, T) \\ \\ u(0) = u^0; \quad u_t(0) = u^1 \quad v(0) = v^0; \quad v_t(0) = v^1 \text{ in } \Omega. \end{cases}$$

The coupled system is well-posed in $H_0^1(\Omega) \times L^2(\Omega) \times L^2(\Omega) \times H^{-1}(\Omega)$. Introduce the energies:

$$E_u(t) = \frac{1}{2} \int_\Omega \{|u_t(x, t)|^2 + (b_{ij}(x)\partial_j u(x, t)\partial_i u(x, t)\} \, dx,$$

$$\widehat{E}_u(t) = \frac{1}{2} \left(||u(., t)||^2_{L^2(\Omega)} + ||u_t(., t)||^2_{H^{-1}(\Omega)} \right).$$

For each $t \in [0,T]$, set

$$E(t) = E_u(t) + \widehat{E}_v(t), \quad \widehat{E}(t) = \widehat{E}_u(t) + \widehat{E}_v(t).$$

Introduce a function $q \in C^2(\bar{\Omega})$ satisfying for some $m_0 \geq 4$:

i) $(2b_{il}(b_{kj}q_{x_k})_{x_l} - b_{ij,x_l}b_{kl}q_{x_k})\, z_i z_j \geq m_0 b_{ij} z_i z_j, \quad \forall (x,z) \in \bar{\Omega} \times \mathbb{R}^N.$
ii) $\min\left\{|\nabla q(x)|; x \in \bar{\Omega}\right\} > 0.$
iii) $\frac{1}{4} b_{ij}(x) q_{x_i}(x) q_{x_j}(x) \geq R_1^2 \geq R_0^2 > 0, \quad \forall x \in \bar{\Omega},$

where $R_0 = \min\left\{\sqrt{q(x)}; x \in \bar{\Omega}\right\}$, and $R_1 = \max\left\{\sqrt{q(x)}; x \in \bar{\Omega}\right\}$. Let ν be the unit normal pointing into the exterior of Ω, and set

$$\Gamma_0 = \left\{x \in \partial\Omega; b_{ij}\nu_i q_{x_j}(x) > 0\right\}.$$

Our first observability result reads:

THEOREM 2.1. [**86**]*Let ω and $\mathcal{O}$ be neighborhoods of Γ_0. Let $r \in \mathcal{D}(0,T)$ be an appropriate cutoff function. Assume that a, c, $d \in L^\infty(0,T; L^s(\Omega))$, with $s > 2$ for $N \in \{1, 2\}$ and $s \geq N$ for $N \geq 3$. Let $b \in L^\infty(Q)$, and let $k_{ij} \in (W_0^{1,s}(Q) \cap L^\infty(Q))^N$, $l_{ij} \in W_0^{1,s}(Q) \cap L^\infty(Q)$, $i,j = 1,2$. Suppose that $k_{12} \equiv 0$, $l_{12} \equiv 0$, supp$(k_{22}) \subset \omega_0 \times (0,T)$, and supp$(l_{22}) \subset \omega_0 \times (0,T)$, where ω_0 is another neighborhood of Γ_0 whose closure $\bar{\omega}_0$ is contained in $\mathcal{O} \cap \omega$. Suppose that there exists $b_0 > 0$ such that $b(x,t) \geq b_0$ for almost every (x,t) in $\mathcal{O} \times (0,T)$.*

For every $T > 2R_1$, there exists a positive constant C such that for all $(u^0, u^1) \in H_0^1(\Omega) \times L^2(\Omega)$, and $(v^0, v^1) \in L^2(\Omega) \times H^{-1}(\Omega)$, one has the observability estimate:

$$E(0) \leq C \int_0^T \int_\omega (r^2|u_t|^2 + |u|^2)\, dxdt$$

for the corresponding solution pair (u,v) of the adjoint system.

REMARK 2.2. Some comments are in order.

(1) No smallness assumption is made on the zeroth order couplings.
(2) The controllability time is the same as for a single wave equation.
(3) The restrictions $k_{12} \equiv 0$ and $l_{12} \equiv 0$ are for well-posedness purposes.
(4) The support constraints on k_{22} and l_{22} are used in the proof of the observability estimate to absorb some unwanted terms, but they may be replaced with smallness constraints instead.
(5) Once Theorem 1 is proven, the Hilbert uniqueness method (H.U.M) of Lions may be used to show that the optimal control for the controllability problem is given by $h = (r^2\hat{u}_t)_t - \hat{u}$ where $(\hat{u}, \hat{v})$ denotes an appropriate solution of the adjoint system.

PROOF OF THEOREM 2.1: KEY ELEMENTS.

- Energy estimates show

$$E(0) \leq C \int_{Q_0} \{|u_t|^2 + |\nabla u|^2 + |v|^2\}\, dxdt,$$

where Q_0 is an appropriate subset of Q.

- Fu-Yong-Zhang Carleman estimate shows

$$\int_{Q_0} (|u_t|^2 + |\nabla u|^2 + |v|^2)\, dxdt \leq Ce^{-\mu\lambda}E(0) + C\int_0^T r^2 \int_{\omega_0} |v|^2\, dxdt$$
$$+C\int_0^T \int_\omega (r^2|u_t|^2 + |u|^2)\, dxdt$$

where $\lambda > 0$ is large enough, and $\mu > 0$ is fixed.

- Use a localizing argument to absorb $C\displaystyle\int_0^T r^2 \int_{\omega_0} |v|^2\, dxdt$. □

Before stating our next observability result, we recall the following Carleman estimate for two coupled systems, [**86**]:

LEMMA 2.1. *Let b_{ij} be given as above. Assume that Γ_0 is as in Theorem 1, and ω is any neighborhood of Γ_0. Then there exists $\lambda_0 > 1$ and a positive constant $C = C(\Omega, T)$, such that for all a, b, c, $d \in L^\infty(0,T; L^s(\Omega))$, with $s > 2$ for $N = 2$ and $s \geq \max(2, N)$ for $N \neq 2$, all $\lambda \geq \lambda_0$ and all u, $v \in C([0,T]; L^2(\Omega))$ satisfying $u(x,0) = u(x,T) = v(x,0) = v(x,T) = 0$ for $x \in \Omega$, $\mathcal{P}u \in H^{-1}(Q)$, $\mathcal{P}v \in H^{-1}(Q)$, and $(u, \mathcal{P}\eta) = \langle \mathcal{P}u, \eta\rangle_{H^{-1}(Q),H_0^1(Q)}$, $(v, \mathcal{P}\eta) = \langle \mathcal{P}v, \eta\rangle_{H^{-1}(Q),H_0^1(Q)}$, $\forall\eta \in H_0^1(Q)$ with $\mathcal{P}\eta \in L^2(Q)$,one has:*

$$\begin{aligned}
&\lambda \textstyle\int_Q e^{2\lambda\varphi}(u^2+v^2)\, dxdt\\
&\leq C\left(||e^{\lambda\varphi}(\mathcal{P}u + au + bv)||^2_{H^{-1}(Q)} + ||e^{\lambda\varphi}(\mathcal{P}v + cu + dv)||^2_{H^{-1}(Q)}\right)\\
&+\tfrac{C}{\lambda^{(2-2N/s)}}\Big(||e^{\lambda\varphi}(au + bv)||^2_{L^2(0,T;H^{-N/s}(\Omega))}\\
&+||e^{\lambda\varphi}(cu + dv)||^2_{L^2(0,T;H^{-N/s}(\Omega))}\Big)\\
&+C\lambda^2||e^{\lambda\varphi}u||^2_{L^2(0,T;L^2(\omega))} + C\lambda^2||e^{\lambda\varphi}v||^2_{L^2(0,T;L^2(\omega))}.
\end{aligned}$$

Set

$$\delta = ||a||_{\infty,s} + ||b||_{\infty,s} + ||c||_{\infty,s} + ||d||_{\infty,s} + \sum_{i,j=1}^{2} ||\mathrm{div}(k_{ij})||_{\infty,s}$$
$$+\sum_{i,j=1}^{2} ||l_{ij,t}||_{\infty,s}$$

$$\delta_0 = \sum_{i,j=1}^{2} ||k_{ij}||_\infty + \sum_{i,j=1}^{2} ||l_{ij}||_\infty$$

where $||.||_{\infty,s} = ||.||_{L^\infty(0,T;L^s(\Omega))}$, and $||.||_\infty = ||.||_{L^\infty(Q)}$.

THEOREM 2.2. [**86**] *Let ω, $\mathcal{O}$, a, d and s be as in Theorem 2.1, and suppose that $b \in L^\infty(0,T; L^s(\Omega))$, $c \in L^\infty(Q)$, and there exists $b_0 > 0$ such that $b(x,t) \geq b_0$ for almost every (x,t) in $\mathcal{O} \times (0,T)$. Let $k_{ij} \in (W_0^{1,s}(Q) \cap L^\infty(Q))^N$, $l_{ij} \in W_0^{1,s}(Q) \cap L^\infty(Q)$, $i, j = 1, 2$. Suppose that $k_{21} \equiv 0$, $l_{21} \equiv 0$, supp$(k_{ij}) \subset \omega_0 \times (0,T)$, and supp$(l_{ij}) \subset \omega_0 \times (0,T)$.*

For every $T > 2R_1$, there exists a positive constant $C_0 = C_0(\Omega, \omega, \mathcal{O}, T, N, s)$ such that for all $(u^0, u^1) \in L^2(\Omega) \times H^{-1}(\Omega)$, and $(v^0, v^1) \in H_0^1(\Omega) \times L^2(\Omega)$, one *has the observability estimate:*

$$\widehat{E}(0)^2 \leq e^{C_0(1+\delta_0+\delta^{\frac{2s}{3s-2N}})} \left(\int_0^T \int_\omega |u|^2 \, dxdt \right) (\widehat{E}_u(0) + E_v(0))$$

for all solution pair (u, v) of the adjoint system.

PROOF OF THEOREM 2.2: MAIN IDEAS.

STEP 1. Prove the energy estimates

- $$\widehat{E}(t) \leq \left[\exp C_0(1 + \delta_0 + \delta^{\frac{N+s}{2s}})|t - \tau| \right] \widehat{E}(\tau), \qquad \forall \tau, t \in [0, T],$$
- $$\int_{T_0}^{T_0'} h\widehat{E}(t) \, dt \leq C_0(1 + \delta + \delta_0) \int_{Q_0} \{|u|^2 + |v|^2\} \, dxdt,$$

where h is an appropriate cut-off function.

STEP 2. Derive from Step 1

$$\widehat{E}(0) \leq e^{C_0(1+\delta_0+\delta^{\frac{N+s}{2s}})} \int_{Q_0} \{|u|^2 + |v|^2\} \, dxdt.$$

STEP 3. Duyckaerts-Zhang-Zuazua Carleman estimate [**24**], yields

$$\int_{Q_0} (|u|^2 + |v|^2) \, dxdt \leq e^{-C_0\lambda} \widehat{E}(0) + e^{C_0\lambda} \int_0^T \int_\omega |u|^2 \, dxdt$$
$$+ e^{C_0\lambda} \int_0^T r^2 \int_{\omega_0} |v|^2 \, dxdt,$$

for some constants $C_0 = C_0(\Omega, T, N, s, \omega) > 0$, and for all $\lambda \geq C_0(1 + \delta_0 + \delta^{\frac{2s}{3s-2N}})$.

STEP 4. Use a localizing argument to absorb $e^{C_0\lambda} \int_0^T r^2 \int_{\omega_0} |v|^2 \, dxdt$.

Let a, b, c, $d \in L^s(\Omega)$, with s as in Theorem 2.1. Assume now $l_{ij} \equiv 0$, and $k_{ij} \equiv 0$, $i, j = 1, 2$. Let ω, $\mathcal{O}$, be as in Theorem 2.1, and suppose that there exists $b_0 > 0$ such that $b(x) \geq b_0$ for almost every x in $\mathcal{O}$.

Further assume that either:

$$a \geq 0, \quad d \geq 0, \quad 2a - |b + c| \geq 0, \text{ and } 2d - |b + c| \geq 0, \text{ a.e. } x \in \Omega$$

or else

$$a \geq 0, \quad d \geq 0, \text{ a.e. } x \in \Omega, \quad 1 - C_s^2 |b + c|_s > 0, \text{ and } \lambda_0^2 - |b + c|_s > 0,$$

where λ_0^2 is the first eigenvalue of the operator $-\partial_i(b_{ij}(x)\partial_j)$ under Dirichlet boundary conditions, and C_s denotes the best constant in the Sobolev inequality:

$$||w||^2_{\frac{2s}{s-2}} \leq C_s^2 \int_\Omega b_{ij}(x)\partial_j w(x)\partial_i w(x) \, dx, \quad \forall w \in H_0^1(\Omega).$$

□

THEOREM 2.3. *Assume the hypotheses just stated. For every $T > 2R_1$, there exists a positive constant $C_0 = C_0(\Omega, \omega, \mathcal{O}, T, N, s)$ such that for all $(u^0, u^1) \in H_0^1(\Omega) \times L^2(\Omega)$, and $(v^0, v^1) \in (H^2(\Omega) \cap H_0^1(\Omega)) \times H_0^1(\Omega)$, one has the observability estimate:*

$$(E_u(0) + E_v(0))^2 \leq e^{C_0(1+\delta^{\frac{2s}{3s-2N}})} \left(\int_0^T \int_\omega |u_t|^2 \, dxdt \right) (E_u(0) + \check{E}_v(0))$$

for all solution pair (u, v) of the adjoint system, and where $2\check{E}_v(0) = ||v^0||^2_{H^2(\Omega)} + ||v^1||^2_{H_0^1(\Omega)}$.

SKETCH OF THE PROOF OF THEOREM 2.3. For this proof, we shall use Theorem 2.2, and the following result

LEMMA 2.2. *Let a, b, c, and d be given as in Theorem 2.3. Thenthere exists a positive constant $C_0 = C_0(\Omega, b + c)$ such that*

$$\begin{aligned} &|| - \partial_i(b_{ij}(x)\partial_j u) + au + bv||^2_{H^{-1}(\Omega)} + || - \partial_i(b_{ij}(x)\partial_j v) + cu + dv||^2_{H^{-1}(\Omega)} \\ &\geq C_0 \int_\Omega \{b_{ij}(x)\partial_j u \partial_i u + b_{ij}(x)\partial_j v \partial_i v\} \, dx, \quad \forall u, v \in H_0^1(\Omega). \end{aligned}$$

Set $\hat{w} = u_t$ and $\hat{z} = v_t$. Then these functions solve the system

$$\begin{cases} \hat{w}_{tt} - \partial_i(b_{ij}(x)\partial_j \hat{w}) + a\hat{w} + b\hat{z} = 0 \text{ in } Q \\ \hat{z}_{tt} - \partial_i(b_{ij}(x)\partial_j \hat{z}) + c\hat{w} + d\hat{z} = 0 \text{ in } Q \\ \hat{w} = 0, \quad \hat{z} = 0 \text{ on } \Sigma = \partial\Omega \times (0, T) \\ \hat{w}(0) = u^1 \in L^2(\Omega); \quad \hat{w}_t(0) = \partial_i(b_{ij}(x)\partial_j u^0) - au^0 - bv^0 \in H^{-1}(\Omega) \\ \hat{z}(0) = v^1 \in H_0^1(\Omega); \quad \hat{z}_t(0) = \partial_i(b_{ij}(x)\partial_j v^0) - cu^0 - dv^0 \in L^2(\Omega). \end{cases}$$

Introduce the following energy associated with that system

$$\widehat{E}_{\hat{w},\hat{z}}(t) = \widehat{E}_{\hat{w}}(t) + \widehat{E}_{\hat{z}}(t) \quad \forall t \in [0, T].$$

Thanks to Theorem 2.2, one has:

$$\widehat{E}_{\hat{w},\hat{z}}(0)^2 \leq e^{C_0(1+\delta^{\frac{2s}{3s-2N}})} \left(\int_0^T \int_\omega |\hat{w}|^2 \, dxdt \right) (\widehat{E}_{\hat{w}}(0) + E_{\hat{z}}(0)).$$

Some elementary calculations show that

$$\widehat{E}_{\hat{w}}(0) + E_{\hat{z}}(0) \leq C_0(E_u(0) + \check{E}_v(0)),$$

while Lemma 2 yields

$$\widehat{E}_{\hat{w},\hat{z}}(0) \geq C_0(E_u(0) + E_v(0)).$$

Hence

$$(E_u(0) + E_v(0))^2 \leq e^{C_0(1+\delta^{\frac{2s}{3s-2N}})} \left(\int_0^T \int_\omega |u_t|^2 \, dxdt \right) (E_u(0) + \check{E}_v(0)).$$

□

THEOREM 2.4. *Suppose that the hypotheses of Theorem 2.3 hold. Forevery $T > 2R_1$, there exists a positive constant $C = C(\Omega, \omega, \mathcal{O}, T, N, s, a, b, c, d)$ such that for all $(u^0, u^1) \in (H^2(\Omega) \cap H_0^1(\Omega)) \times H_0^1(\Omega)$, and $(v^0, v^1) \in H_0^1(\Omega) \times L^2(\Omega)$, one has the observability estimate:*

$$\check{E}_u(0) + E_v(0) \leq C \int_0^T \int_\omega \{|u_t|^2 + |u_{tt}|^2\} \, dxdt.$$

3. Indirect stabilization

3.1. Mindlin-Timoshenko plate.

$$\begin{aligned}
&\rho_1 y_{tt} - k\mathrm{div}(\nabla y + z) = 0 \text{ in } \Omega \times (0,\infty)\\
&\rho_2 z_{tt} - \mu\Delta z - (\lambda+\mu)\nabla\mathrm{div} z + k(\nabla y + z) + a z_t = 0 \text{ in } \Omega \times (0,\infty)\\
&y = 0, \quad z = 0 \text{ on } \partial\Omega \times (0,\infty)\\
&y(.,0) = y^0 \in H_0^1(\Omega), \quad y_t(.,0) = y^1 \in L^2(\Omega),\\
&z(.,0) = z^0 \in [H_0^1(\Omega)]^N, \quad z_t(.,0) = z^1 \in [L^2(\Omega)]^N.
\end{aligned}$$

In the one-dimensional setting, the system , known as the Timoshenko beam equations, describes the motion of a beam when the effects of rotatory inertia are accounted for; the transverse displacement is represented by y while z denotes the shear angle displacement.

In 2D, that system is known as the Mindlin-Timoshenko plate equations, where y represents the vertical deflection and z stands for the rotation angles of a filament.

The constants ρ_1, ρ_2, k, and μ are physical constants and are all positive. In particular, the constants λ and μ are the Lamé constants with $\lambda + \mu > 0$.

Although the damping is distributed everywhere in the domain under consideration, the stabilization problem at hand is not easy, since there is no damping in the vertical deflection equation. We are counting on the coupling to effectively transmit the damping from the rotation angles of a filament to the vertical deflection, so as to produce an exponentially stable system. We already know from earlier works [**26**, **69**] that the system at hand is polynomially stable. We also know that under certain boundary conditions, excluding the Dirichlet boundary conditions being used here, exponential stability fails in 2D, [**26**]. However, it is known in the one-dimensional setting that this system is exponentially stable if and only if $\frac{k}{\rho_1} = \frac{2\mu+\lambda}{\rho_2}$. This exponential stability of the Timoshenko beam holds under various boundary conditions including the Dirichlet ones being used in this work, e.g. [**2**, **97**]. It is then natural to wonder whether such an exponential stability result extends to higher dimensions, and if so, under what conditions. Our purpose in this note is to investigate that outstanding issue with the hope of improving and generalizing to higher dimensions those known stability results. We note that the condition $\frac{k}{\rho_1} = \frac{2\mu+\lambda}{\rho_2}$ is unnecessary for the exponential stability of the Timoshenko beam when the vertical deflection equation is also damped e.g. [**7**, **18**, **69**, **85**].

It is well-known that the indirectly damped Timoshenko beam, ($N = 1$), is exponentially stable if and only if

$$(*1) \qquad \frac{k}{\rho_1} = \frac{2\mu+\lambda}{\rho_2}.$$

QUESTIONS. Is the Mindlin-Timoshenko system exponentially stable under $(*1)$?

What happens when $(*1)$ fails?

To study the stabilization problem at hand, we are going to recast the plate system as an abstract evolution system. To this end, setting $Z = \begin{pmatrix} y \\ y' \\ z \\ z' \end{pmatrix}$, the

Mindlin-Timoshenko system may then be recast as:

$$Z' - \mathcal{A}Z = 0 \text{ in } (0,\infty), \qquad Z(0) = \begin{pmatrix} y^0 \\ y^1 \\ z^0 \\ z^1 \end{pmatrix}$$

where the unbounded operator $\mathcal{A}$ is given by

$$\mathcal{A} = \begin{pmatrix} 0 & I & 0 & 0 \\ \frac{k}{\rho_1}\Delta & 0 & \frac{k}{\rho_1}\mathrm{div} & 0 \\ 0 & 0 & 0 & I \\ -\frac{k}{\rho_2}\nabla & 0 & \frac{\mu}{\rho_2}\Delta + \frac{\lambda+\mu}{\rho_2}\nabla\mathrm{div} - \frac{k}{\rho_2}I & -\frac{a}{\rho_2}I \end{pmatrix}$$

with,

$$\begin{aligned} D(\mathcal{A}) = \Big\{&(u,v,w,z) \in \left(H_0^1(\Omega)\right)^2 \times \left([H_0^1(\Omega)]^N\right)^2; \\ &k\mathrm{div}(\nabla u + w) \in L^2(\Omega), \\ &\text{and } \mu\Delta w + (\lambda+\mu)\nabla\mathrm{div}w - k(\nabla u + w) - az \in [L^2(\Omega)]^N \Big\} \end{aligned}$$

It can be checked that one has (assuming for instance that Γ is C^2)

$$D(\mathcal{A}) = (H^2(\Omega) \cap H_0^1(\Omega)) \times H_0^1(\Omega) \times ([H^2(\Omega)]^N \cap [H_0^1(\Omega)]^N) \times [H_0^1(\Omega)]^N.$$

Thus, the operator $\mathcal{A}$ has a compact resolvent. Consequently the spectrum of $\mathcal{A}$ is discrete.

Introduce the Hilbert space over the field $\mathbb{C}$ of complex numbers $\mathcal{H} = H_0^1(\Omega) \times L^2(\Omega) \times [H_0^1(\Omega)]^N \times [L^2(\Omega)]^N$, equipped with the norm

$$||Z||_{\mathcal{H}}^2 = \rho_1|v|_2^2 + k|\nabla u + w|_2^2 + \rho_2|z|_2^2 + \mu|\nabla w|_2^2 + (\lambda+\mu)|\mathrm{div}w|_2^2,$$

$$\forall Z = (u,v,w,z) \in \mathcal{H}.$$

THEOREM 3.1 (Strong stability, [**91**]). *Suppose that ω is an arbitrary nonempty open set in Ω. Let the damping coefficient a be positive in ω. The operator $\mathcal{A}$ generates a C_0 semigroup of contractions $(S(t))_{t\geq 0}$ on the Hilbert space $\mathcal{H}$, which is strongly stable:*

$$\lim_{t\to\infty} ||S(t)Z^0||_{\mathcal{H}} = 0, \quad \forall Z^0 \in \mathcal{H}.$$

PROOF SKETCH. First, we shall prove that the unbounded operator $\mathcal{A}$ generates a C_0 semigroup of contractions $(S(t))_{t\geq 0}$, then we shall show that $i\mathbb{R} \subset \rho(\mathcal{A})$.

We have:

- $\overline{D(\mathcal{A})} = \mathcal{H}$ and the operator $\mathcal{A}$ is dissipative, as:

$$\Re(\mathcal{A}Z, Z) = -\int_\Omega a(x)|z(x)|^2\,dx \leq 0, \quad \forall Z = (u,v,w,z) \in \mathcal{D}(\mathcal{A}).$$

- $\mathcal{I} - \mathcal{A}$ is onto, by Lax-Milgram Lemma, ($\mathcal{I}$ denotes the identity operator).

To prove strong stability, it suffices, thanks to a Benchimol strong stability criterion, for linear operators with compact resolvent [**13**], to show that $\mathcal{A}$ has no imaginary eigenvalue. One easily checks that zero is not an eigenvalue of $\mathcal{A}$. Now, let b be a nonzero real number, and let $Z = (u,v,w,z) \in D(\mathcal{A})$ such that $\mathcal{A}Z = ibZ$. We shall prove that $Z = (0,0,0,0)$. Note that $\mathcal{A}Z = ibZ$ may be recast as:

$$\begin{aligned} &-b^2u - \hat{k}\mathrm{div}(\nabla u + w) = 0 \text{ in } \Omega \\ &-b^2w - \hat{\mu}\Delta w - (\hat{\mu} + \hat{\lambda})\nabla\mathrm{div}w + \check{k}(\nabla u + w) + ib\check{a}w = 0 \text{ in } \Omega. \end{aligned}$$

One easily checks that $w = 0$ in ω, by multiplying the first equation by $\bar{u}$, the second by $\bar{w}$, integrating by parts and taking the imaginary parts. Using the second equation, it follows that $\nabla u = 0$ in ω; so, using the first equation, one derives that $u = 0$ in ω as b is nonzero. Now, thanks to a Carleman estimate of imanuvilov-Puel for elliptic equations [**43**], there exist positive constants C, $\tau_0 \geq 1$ and $s_0 \geq 1$ such that, for every $\tau \geq \tau_0$ and every $s \geq s_0$, the component u satisfies:

$$
\begin{aligned}
&s\tau^2 \int_\Omega \{|\nabla u|^2 + s^2\tau^2\varphi^2|u|^2\}e^{2s\varphi}\,dx \\
&\leq C \int_\Omega \{b^4|u|^2 + |\mathrm{div} w|^2\}e^{2s\varphi}\,dx \\
&+Cs\tau^2 \int_\omega \{|\nabla u|^2 + s^2\tau^2\varphi^2|u|^2\}e^{2s\varphi}\,dx, \quad \forall b \in \mathbb{R},
\end{aligned}
$$

where φ is an appropriate weight function, and C is independent of s, τ, and b. Similarly, according to a Carleman estimate of Imanuvilov-Yamamoto for the static Lamé system [**44**], there exist positive constants C, $\tau_1 \geq 1$ and $s_1 \geq 1$ such that, for every $\tau \geq \tau_1$ and every $s \geq s_1$, the component w satisfies:

$$
\begin{aligned}
&\tau^2 \int_\Omega \{|\nabla w|^2 + s^2\tau^2\varphi^2|w|^2\}e^{2s\varphi}\,dx \\
&\leq C \int_\Omega \{(b^4+1)|w|^2 + |\nabla u|^2\}e^{2s\varphi}\,dx \\
&+C\tau^2 \int_\omega \{|\nabla w|^2 + s^2\tau^2\varphi^2|w|^2\}e^{2s\varphi}\,dx, \quad \forall b \in \mathbb{R},
\end{aligned}
$$

where φ is an appropriate weight function, and C is independent of s, τ and b.

Combining those two estimates, noticing that $|\mathrm{div} w(x)| \leq |\nabla w(x)|$ for almost every x in Ω, and choosing s and τ large enough, one can use the left hand side to absorb the first integrals in the right hand side. Next, using the fact that $u = 0$ and $w = 0$ in $\emptyset$, one derives that $u = 0$ and $w = 0$ in Ω; hence $Z = (0,0,0,0)$. □

THEOREM 3.2 (exponential stability, [**91**]). *Suppose that the damping coefficient a satisfies:*

$$\exists a_0 > 0 : a(x) \geq a_0, \qquad a.e.\ x \in \Omega.$$

Assume that $\frac{k}{\rho_1} = \frac{2\mu+\lambda}{\rho_2}$. *The semigroup* $(S(t))_{t\geq 0}$ *is exponentially stable,* viz., *there exist positive constants* M *and* ζ *with:*

$$||S(t)Z^0||_{\mathcal{H}} \leq M\exp(-\zeta t)||Z^0||_{\mathcal{H}}, \quad \forall Z^0 \in \mathcal{H}.$$

PROOF OF THEOREM 3.2: MAIN IDEAS. We shall use the frequency domain approach, which amounts to showing the two facts :

(1) $i\mathbb{R} \subset \rho(\mathcal{A})$, and
(2) $\sup\{||(ib - \mathcal{A})^{-1}||_{\mathcal{L}(\mathcal{H})};\ b \in \mathbb{R}\} < \infty$.

Thanks to the proof of Theorem 3.1, we already have the first point. It remains to prove the second point. For this purpose, it suffices to show that there exists $C_0 > 0$ such that for every $U \in \mathcal{H}$, one has:

$$||(ib - \mathcal{A})^{-1}U||_{\mathcal{H}} \leq C_0||U||_{\mathcal{H}}, \quad \forall b \in \mathbb{R},$$

where hereafter, C_0 denotes a generic positive constant that may eventually depend on Ω, ω, and the other parameters of the system, but never on b.

Let $b \in \mathbb{R}$, $U = (f,g,h,l) \in \mathcal{H}$, and let $Z = (u,v,w,z) \in D(\mathcal{A})$ such that

$$(ib - \mathcal{A})Z = U.$$

We shall prove $||Z||_{\mathcal{H}} \leq C_0||U||_{\mathcal{H}}$. To this end, multiply both sides of the equation by Z, then take the real part of the inner product in $\mathcal{H}$ to derive:

$$\int_\Omega a(x)|z(x)|^2\,dx = \Re(U,Z) \leq ||U||_{\mathcal{H}}||Z||_{\mathcal{H}}, \text{ or } |z|_2^2 \leq C_0||U||_{\mathcal{H}}||Z||_{\mathcal{H}}.$$

Equation $(ib - \mathcal{A})Z = U$ may be recast as:

$$\begin{cases} ibu - v = f \\ ibv - \hat{k}\mathrm{div}(\nabla u + w) = g \\ ibw - z = h \\ ibz - \hat{\mu}\Delta w - (\hat{\lambda} + \hat{\mu})\nabla \mathrm{div} w + \check{k}(\nabla u + w) + \breve{a}z = l. \end{cases}$$

Thanks to the estimate on z, we derive from the third equation

$$b^2 \int_\Omega |w(x)|^2\,dx \leq C_0||U||_{\mathcal{H}}||Z||_{\mathcal{H}} + C_0||U||_{\mathcal{H}}^2.$$

Multiplying the last equation in by $\bar{w}$, taking the real parts, and integrating by parts over Ω, one finds

$$\begin{aligned} &\Re ib \int_\Omega z \cdot \bar{w}\,dx + \int_\Omega \{\hat{\mu}|\nabla w|^2 + (\hat{\lambda}+\hat{\mu})|\mathrm{div} w|^2\}\,dx \\ &= \Re \int_\Omega \{(l - \breve{a}z)\cdot \bar{w}\,dx - \check{k}\Re \int_\Omega (\nabla u + w)\cdot \bar{w}\,dx. \end{aligned}$$

With Cauchy-Schwarz inequality, one readily derives, for every b with $|b| > 1$:

$$\begin{aligned} &|z|_2^2 + \int_\Omega \{\hat{\mu}|\nabla w|^2 + (\hat{\lambda}+\hat{\mu})|\mathrm{div} w|^2\}\,dx \\ &\leq C_0(||U||_{\mathcal{H}}||Z||_{\mathcal{H}} + |b|^{-1}||U||_{\mathcal{H}}^{\frac{1}{2}}||Z||_{\mathcal{H}}^{\frac{3}{2}}) \end{aligned}$$

At this stage, we note that the last inequality provides a good estimate for w; so, in order to complete the proof of Theorem 3.2, it remains to get estimates for u and v. This is the crucial point in the proof. First, we will estimate $|v|_2$ in terms of $|\nabla u + w|_2$, then we will estimate $|\nabla u + w|_2$. In particular, the factor $|b|^{-1}$ will be very helpful in the proof of the polynomial decay estimate given by the next theorem, leading to a better decay estimate; it can be dropped in the rest of the proof of Theorem 3.2 for $|b| > 1$.

Estimating $|v|_2$. Taking the conjugate of the equation $ibu - v = f$, then multiplying the new equation by $-v$ and integrating over Ω, we find

$$ib \int_\Omega \bar{u}v\,dx + |v|_2^2 = -\int_\Omega \bar{f}v\,dx.$$

Next, multiplying the equation $ibv - \hat{k}\mathrm{div}(\nabla u + w) = g$ by $-\bar{u}$ and using Green's formula, we find

$$\begin{aligned} -ib\int_\Omega \bar{u}v\,dx &= \hat{k}\int_\Omega (\nabla u + w)\cdot\nabla\bar{u} - \int_\Omega \bar{u}g\,dx \\ &= \hat{k}\int_\Omega |\nabla u + w|^2\,dx - \hat{k}\int_\Omega (\nabla u + w)\cdot\bar{w}\,dx - \int_\Omega \bar{u}g\,dx. \end{aligned}$$

Now, thanks to Cauchy-Schwarz and Poincaré inequalities, one gets

$$\left|\int_\Omega (\bar{f}v + g\bar{u})\,dx\right| \leq |f|_2|v|_2 + |g|_2|u|_2 \leq C_0||U||_{\mathcal{H}}||Z||_{\mathcal{H}}.$$

It then follows

$$|v|_2^2 \leq C_0(||U||_{\mathcal{H}}||Z||_{\mathcal{H}} + ||U||_{\mathcal{H}}^{\frac{1}{2}}||Z||_{\mathcal{H}}^{\frac{3}{2}}) + \hat{k}|\nabla u + w|_2^2.$$

Using the speeds equality constraint, one finds:

$$\check{k}\int_\Omega |\nabla u + w|^2\,dx \leq C_0(||U||_{\mathcal{H}}||Z||_{\mathcal{H}} + ||U||_{\mathcal{H}}^{\frac{1}{2}}||Z||_{\mathcal{H}}^{\frac{3}{2}}) + \hat{k}\Re\int_\Gamma \varphi\partial_\nu\bar{u}\,d\gamma.$$

Using appropriate first order multipliers, one gets rid of the boundary integral, and the claimed estimate follows. Applying the continuity of the resolvent and Huang or Pruss exponential stability criterion, one obtains the desired exponential stability of the semigroup. □

THEOREM 3.3 (Polynomial stability, [**91**]). *Suppose that the damping coefficient a is as in Theorem 3.2. Assume that $\frac{k}{\rho_1} \neq \frac{2\mu+\lambda}{\rho_2}$. The semigroup $(S(t))_{t\geq 0}$ is polynomially stable,* viz.*, there exists a positive constant M such that:*

$$||S(t)Z^0||_{\mathcal{H}} \leq M\frac{||Z^0||_{D(\mathcal{A})}}{(1+t)^{\frac{1}{2}}}, \quad \forall Z^0 \in D(\mathcal{A}).$$

3.2. Kirchhoff plate-wave. This part is a Joint work with Ahmed Hajej (U. Cergy-Pontoise, France) and Zayd Hajjej (U. Gabes, Tunisia), J. Math. Anal. Appl. 2019

Consider the following weakly coupled system of Kirchhoff plate and wave equations:

$$\begin{cases} u_{tt} - \gamma\Delta u_{tt} + \Delta^2 u + \alpha v = 0 & \text{in} \quad \Omega\times(0,\infty)\\ v_{tt} - \Delta v + v_t + \alpha u = 0 & \text{in} \quad \Omega\times(0,\infty)\\ u = \partial_\nu u = 0 & \text{on} \quad \Gamma_0\times(0,\infty)\\ \Delta u + (1-\mu)B_1 u = 0 & \text{on} \quad \Gamma_1\times(0,\infty)\\ \partial_\nu\Delta u - \gamma\partial_\nu u_{tt} + (1-\mu)B_2 u = 0 & \text{on} \quad \Gamma_1\times(0,\infty)\\ v = 0 & \text{on} \quad \Gamma\times(0,\infty)\\ u(0) = u^0 \in V, \quad u_t(0) = u^1 \in H_0^1(\Omega), & \\ v(0) = v^0 \in H_0^1(\Omega), \quad v_t(0) = v^1 \in L^2(\Omega). & \end{cases}$$

Undamped Kirchhoff plate/ damped wave Ω is an open set of $\mathbb{R}^2$ with regular boundary $\Gamma = \partial\Omega = \Gamma_0 \cup \Gamma_1$ such that $\overline{\Gamma}_0 \cap \overline{\Gamma}_1 = \emptyset$,

The constant $\gamma > 0$ is the rotational inertia of the plate and the constant $0 < \mu < \frac{1}{2}$ is the Poisson coefficient

The boundary operators B_1, B_2 are defined by

$$B_1 u = 2\nu_1\nu_2 u_{xy} - \nu_1^2 u_{yy} - \nu_2^2 u_{xx},$$

$$B_2 u = \partial_\tau\left((\nu_1^2 - \nu_2^2)u_{xy} + \nu_1\nu_2(u_{yy} - u_{xx})\right),$$

where $\nu = (\nu_1, \nu_2)$ is the unit outer normal vector to Γ and $\tau = (-\nu_2, \nu_1)$ is a unit tangent vector. Introduce the energy, (setting $P_\gamma u = u - \gamma\Delta u$)

$$E(t) = \frac{1}{2}\int_\Omega \{|P_\gamma^{\frac{1}{2}} u_t|^2 + |\Delta u|^2 + |v_t|^2 + |\nabla v|^2 + 2\alpha uv\}(x,t)\,dx.$$

We have:

$$E(t) \leq \frac{C_\alpha}{(t+1)^{\frac{1}{3}}}\left(||u^0||^2_{H^3(\Omega)} + ||u^1||^2_{H^2(\Omega)} + ||v^0||^2_{H^2(\Omega)} + ||v^1||^2_{H^1_0(\Omega)}\right).$$

To prove that energy estimate, we relied on the frequency domain method, interpolation inequalities, a good choice of functional inequalities, and the Borichev-Tomilov criterion for polynomial stability of semigroups [**14**].

A Damped Kirchhoff plate/ undamped wave

$$\begin{cases} u_{tt} - \gamma\Delta u_{tt} + \Delta^2 u + \alpha v + u_t = 0 & \text{in} \quad \Omega \times (0,\infty) \\ v_{tt} - \Delta v + \alpha u = 0 & \text{in} \quad \Omega \times (0,\infty) \\ u = \partial_\nu u = 0 & \text{on} \quad \Gamma_0 \times (0,\infty) \\ \Delta u + (1-\mu)B_1 u = 0 & \text{on} \quad \Gamma_1 \times (0,\infty) \\ \partial_\nu \Delta u - \gamma \partial_\nu u_{tt} + (1-\mu)B_2 u = 0 & \text{on} \quad \Gamma_1 \times (0,\infty) \\ v = 0 & \text{on} \quad \Gamma \times (0,\infty) \\ u(0) = u^0 \in V, \quad u_t(0) = u^1 \in H_0^1(\Omega), & \\ v(0) = v^0 \in H_0^1(\Omega), \quad v_t(0) = v^1 \in L^2(\Omega). & \end{cases}$$

Introduce the energy

$$E(t) = \frac{1}{2}\int_\Omega \{|P_\gamma u_t|^2 + |\Delta u|^2 + |v_t|^2 + |\nabla v|^2 + 2\alpha uv\}(x,t)\,dx.$$

Using the frequency domain method, we show:

$$E(t) \le \frac{C_\alpha}{(t+1)^{\frac{1}{4}}}\left(||u^0||^2_{H^3(\Omega)} + ||u^1||^2_{H^2(\Omega)} + ||v^0||^2_{H^2(\Omega)} + ||v^1||^2_{H_0^1(\Omega)}\right).$$

4. Some open problems

4.1. Simultaneous control. Consider the system of uncoupled wave equations

$$\begin{aligned} &u_{jtt} - a_j \Delta u_j = 0 \text{ in } Q \\ &u_j = 0 \text{ on } \Gamma \times (0,T) \\ &u_j(x,0) = u_j^0(x), \quad u_{jt}(x,0) = u_j^1(x) \text{ in } \Omega, \quad j = 1,\ 2, ...,\ q, \end{aligned}$$

where $(u_j^0, u_j^1) \in H_0^1(\Omega) \times L^2(\Omega)$ for each j.

1988: Haraux (1988) shows for arbitrary nonempty open set ω:
If $\sum_{j=1}^q u_j(x,t) = 0$ in $\omega \times (0,T)$ then $u_j^0 = 0, \quad u_j^1 = 0$ in $\Omega, \quad \forall j$.
provided that $a_j \neq a_k$ for all $j,\ k$ with $j \neq k$.

QUESTIONS. Can we quantify that unique continuation result?

For instance, for a single wave equation, we have the weak observability inequality due to Bellassoued [**12**]:

$$\begin{aligned} &||u^0||^2_{L^2(\Omega)} + ||u^1||^2_{H^{-1}(\Omega)} \\ &\le \frac{C\Big\{||u^0||^2_{H_0^1(\Omega)} + ||u^1||^2_{L^2(\Omega)}\Big\}}{\lambda} \\ &+ Ce^{\mu\lambda}\int_0^T r(t)^2 \int_{\omega_0} |u(x,t)|^2\,dxdt. \end{aligned}$$

For the uncoupled system, assuming $a_j \neq a_k$ for all j, k with $j \neq k$, do we have:

$$\sum_{j=1}^{q}\left\{||u_j^0||^2_{L^2(\Omega)} + ||u_j^1||^2_{H^{-1}(\Omega)}\right\}$$
$$\leq \frac{C\sum_{j=1}^{q}\left\{||u_j^0||^2_{H_0^1(\Omega)} + ||u_j^1||^2_{L^2(\Omega)}\right\}}{\lambda}$$
$$+Ce^{\mu\lambda}\int_0^T r(t)^2\int_{\omega_0}|\sum_{j=1}^{q}u_j(x,t)|^2\,dxdt?$$

Introduce a function $q \in C^2(\bar{\Omega})$ satisfying:

i) ii) $\min\left\{|\nabla q(x)|; x \in \bar{\Omega}\right\} > 0$.

iii) $\frac{1}{4}\min\{a_j; j = 1,\ 2, ...,\ q\}|\nabla q(x)|^2 \geq R_1^2 \geq R_0^2 > 0, \quad \forall x \in \bar{\Omega},$

where $R_0 = \min\left\{\sqrt{q(x)}; x \in \bar{\Omega}\right\}$, and $R_1 = \max\left\{\sqrt{q(x)}; x \in \bar{\Omega}\right\}$. Let ν be the unit normal pointing into the exterior of Ω, and set

$$\Gamma_0 = \{x \in \partial\Omega; \nu \cdot \nabla q(x) > 0\}.$$

Let $T > 2R_1$. Choose a constant $\mu \in (0,1)$ such that

$$(2R_1/T)^2 < \mu < 2R_1/T.$$

Set $\varphi(x,t) = d(x) - \mu(t - T/2)^2$.

Define a differential operator $\mathcal{P}$ by $\mathcal{P}u = u_{tt} - (b_{ij}(x)u_{x_j})_{x_i}$.

For a single wave equation, we have the Carleman estimates:

LEMMA 4.1 (Carleman estimates). *Assume that Γ_0 is defined as above and ω is a neighborhood of Γ_0.*

i) [**27**, Fu-Yong-Zhang] *Then there exists $\lambda_0 > 1$ and a positive constant $C = C(\Omega,T)$, such that for all $\lambda \geq \lambda_0$ and any $u \in H_0^1(Q)$ with $\mathcal{P}u \in L^2(Q)$, it holds:*

$$\lambda\int_Q e^{2\lambda\varphi}(\lambda^2|u|^2 + |u_t|^2 + |\nabla u|^2)\,dxdt \leq C||e^{\lambda\varphi}(\mathcal{P}u)||^2_{L^2(Q)}$$
$$+C\lambda^2\int_0^T\int_\omega e^{2\lambda\varphi}(\lambda^2|u|^2 + |u_t|^2)\,dxdt.$$

ii) [**24**, Duyckaerts-Zhang-Zuazua] *Let $V \in L^\infty(0,T;L^m(\Omega))$ with $m \in [N,+\infty)$. Then there exist $\lambda_0 > 1$ and a positive constant $C = C(\Omega,T)$, such that for all $\lambda \geq \lambda_0$ and any $u \in C([0,T];L^2(\Omega))$ satisfying $u(x,0) = u(x,T) = 0$ for $x \in \Omega$, $\mathcal{P}u \in H^{-1}(Q)$, and*

$$(u,\mathcal{P}\eta) = \langle\mathcal{P}u,\eta\rangle_{H^{-1}(Q),H_0^1(Q)}, \quad \forall\eta \in H_0^1(Q) \text{ with } \mathcal{P}\eta \in L^2(Q),$$

it holds:

$$\lambda||e^{\lambda\varphi}u||^2_{L^2(Q)} \leq C\Big(||e^{\lambda\varphi}(\mathcal{P}u - Vu)||^2_{H^{-1}(Q)}$$
$$+\tfrac{1}{\lambda^{(2-2N/m)}}||e^{\lambda\varphi}Vu||^2_{L^2(0,T;H^{-N/m}(\Omega))} + \lambda^2||e^{\lambda\varphi}u||^2_{L^2(0,T;L^2(\omega))}\Big).$$

For the uncoupled system, do we have, assuming $a_j \neq a_k$ for all j, k with $j \neq k$:

There exist $\lambda_0 > 1$ and a positive constant $C = C(\Omega, T)$, such that for all $\lambda \geq \lambda_0$ and any $u \in \left[H_0^1(Q)\right]^q$ with $\mathcal{P}u = (\mathcal{P}_j u_j)_j \in \left[L^2(Q)\right]^q$, it holds (with $\mathcal{P}_j u_j = u_{j,tt} - a_j \Delta u_j$):

$$\lambda \sum_{j=1}^{q} \int_Q e^{2\lambda\varphi}(\lambda^2|u_j|^2 + |u_{jt}|^2 + |\nabla u_j|^2)\,dxdt \leq C \sum_{j=1}^{q} ||e^{\lambda\varphi}(\mathcal{P}_j u_j)||^2_{L^2(Q)}$$
$$+C\lambda^2 \int_0^T \int_\omega e^{2\lambda\varphi}\left(\lambda^2|\sum_{j=1}^{q} u_j|^2 + |\sum_{j=1}^{q} u_{jt}|^2\right) dxdt?$$

4.2. Wave equations: Simultaneous boundary damping. Consider the system of damped wave equations (to fix ideas, we assume the damping on the whole boundary):

$$\begin{aligned}
&w_{jtt} - a_j \Delta w_j = 0 \text{ in } Q\\
&a_j \partial_\nu w_j + c w_j + d \sum_{j=1}^{q} w_{jt} = 0 \text{ on } \Gamma \times (0,T)\\
&w_j(x,0) = w_j^0(x), \quad w_{jt}(x,0) = w_j^1(x) \text{ in } \Omega, \quad j = 1,\ 2, ...,\ q,
\end{aligned}$$

where $(w_j^0, w_j^1) \in H^1(\Omega) \times L^2(\Omega)$

Assuming that c and d are positive constants, what kind of stability should we expect for the underlying semigroup?

4.3. Lamé systems with boundary damping. Given $(y_j^0, y_j^1)_j \in \left(\left[H_0^1(\Omega)\right]^N \times \left[L^2(\Omega)\right]^N\right)^q$, and a function $d \in L^\infty(\Omega)$, $d \geq 0$, consider the damped elastodynamic system

$$\begin{aligned}
&y_{jtt} - \mu_j \Delta y_j - (\mu_j + \lambda_j)\nabla \mathrm{div}(y_j) = 0 \text{ in } \Omega \times (0,\infty)\\
&\mu_j \partial_\nu y_j + (\mu_j + \lambda_j)\mathrm{div}(y_j)\nu + c y_j + d \sum_{k=1}^{q} y_{kt} = 0 \text{ on } \Gamma \times (0,\infty)\\
&y_j(x,0) = y_j^0(x), \quad y_{jt}(x,0) = y_j^1(x), \text{ in } \Omega,\\
&j = 1,\ 2, ...,\ q,
\end{aligned}$$

where, for each j, μ_j and λ_j are the Lamé constants, and c and d are positive constants.

The total energy is given, for all $t \geq 0$, by

$$2E(t) = \sum_{j=1}^{q} \int_\Omega \{|y_{jt}(x,t)|^2 + \mu_j |\nabla y_j(x,t)|^2 + (\mu_j + \lambda_j)|\mathrm{div}(y_j(x,t))|^2\}\,dx$$
$$+c \sum_{j=1}^{q} \int_\Gamma |y_j(\gamma,t)|^2\,d\gamma.$$

E is a nonincreasing function of the time variable as

$$\frac{dE}{dt} = -\int_\Gamma d\left|\sum_{k=1}^{q} y_{kt}(\gamma,t)\right|^2 d\gamma.$$

QUESTION 1. Does the energy E decay to zero as time goes to infinity?

QUESTION 2. Under which conditions is the Lamé system exponentially stable?

4.4. A Plate-wave system. Consider the damped system

$$\begin{cases} y_{tt} - \Delta y + d(x)(y_t + z_t) = 0 \text{ in } \Omega \times (0, \infty) \\ z_{tt} - \gamma \Delta z_{tt} + \Delta^2 z + d(x)(y_t + z_t) = 0 \text{ in } \Omega \times (0, \infty) \\ y = 0, \quad z = 0, \quad \partial_\nu z = 0 \text{ on } \Gamma \times (0, \infty) \\ y(0) = y^0 \in H_0^1(\Omega), \quad y_t(0) = y^1 \in L^2(\Omega), \\ z(0) = z^0 \in H_0^2(\Omega), \quad z_t(0) = z^1 \in L^2(\Omega). \end{cases}$$

The total energy is given, for all $t \geq 0$, by

$$2E(t) = \int_\Omega \{|y_t(x,t)|^2 + |\nabla y(x,t)|^2 + |z_t(x,t)|^2 + |\Delta z(x,t)|^2\} \, dx,$$

The energy E is a nonincreasing function of the time variable as

$$\frac{dE}{dt} = -\int_\Omega d(x) \, |y_t(x,t) + z_t(x,t)|^2 \, dx.$$

For this coupled system we have strong stability for all values of γ in $[0, \infty)$, and exponential stability for $\gamma = 0$. Further, we know that exponential stability fails for $\gamma > 0$. Does the energy of the Kirchhoff model decay polynomially?

4.5. Plate-wave system with structural/Kelvin-Voigt damping. Consider the new damped system

$$\begin{cases} y_{tt} - \Delta y - \operatorname{div}(d(x)\nabla(y_t + z_t)) = 0 \text{ in } \Omega \times (0, \infty) \\ z_{tt} - \gamma \Delta z_{tt} + \Delta^2 z - \operatorname{div}(d(x)\nabla(y_t + z_t)) = 0 \text{ in } \Omega \times (0, \infty) \\ y = 0, \quad z = 0, \quad \partial_\nu z = 0 \text{ on } \Gamma \times (0, \infty) \\ y(0) = y^0 \in H_0^1(\Omega), \quad y_t(0) = y^1 \in L^2(\Omega), \\ z(0) = z^0 \in H_0^2(\Omega), \quad z_t(0) = z^1 \in L^2(\Omega). \end{cases}$$

It can be checked that the energy of that system is nonincreasing. Does it decay to zero? If yes, at what rate?

4.6. Timoshenko beam with structural damping. Let $L > 0$, and set $\Omega = (0, L)$, and $\omega = (l_1, l_2)$ with $0 \leq l_1 < l_2 \leq L$.

Consider the damped Timoshenko system:

$$\begin{cases} \rho_1 y_{tt} - k(y_x + z)_x - (a(x)(y_{xt} + z_t))_x = 0 \text{ in } (0, L) \times (0, \infty) \\ \rho_2 z_{tt} - \sigma z_{xx} + k(y_x + z) + a(x)(y_{xt} + z_t) = 0 \text{ in } (0, L) \times (0, \infty), \end{cases}$$

with the boundary conditions:
(DD) $y(0,t) = 0, \quad y(L,t) = 0, \quad z(0,t) = 0, \quad z(L,t) = 0, \; t > 0$
and the initial conditions: for all $x \in \Omega$,

$$y(x,0) = y^0(x), \quad y_t(x,0) = y^1(x), \quad z(x,0) = z^0(x), \quad z_t(x,0) = z^1(x).$$

The damping coefficient a is a nonnegative bounded measurable function, which is positive in ω only.

Whether that system is exponentially stable is unknown, though strong stability is much easier to prove than in the Timoshenko system with the velocities sum damping discussed earlier.

4.7. Indirect controllability a coupled system of wave equations with different principal operators. Consider the controllability problems: Given (z^0, z^1) and (w^0, w^1), and $\varepsilon > 0$, find a control h such that if (z, w) solves the system

$$\begin{cases} z_{tt} - \partial_i(b_{ij}(x)\partial_j z) + az + cw - \operatorname{div}(k_{11}z) - (l_{11}z)_t \\ -\operatorname{div}(k_{21}w) - (l_{21}w)_t = h1_\omega \text{ in } Q \\ \\ w_{tt} - \partial_i(a_{ij}(x)\partial_j w) + bz + dw - \operatorname{div}(k_{12}z) - (l_{12}z)_t \\ -\operatorname{div}(k_{22}w) - (l_{22}w)_t = 0 \text{ in } Q \\ \\ z = 0, \quad w = 0 \text{ on } \Sigma = \partial\Omega \times (0,T) \\ \\ z(0) = z^0; \quad z_t(0) = z^1 \quad w(0) = w^0; \quad w_t(0) = w^1 \text{ in } \Omega, \end{cases}$$

then (exact controllability)

$$z(T) = 0, \quad z_t(T) = 0, \quad w(T) = 0, \quad w_t(T) = 0 \text{ in } \Omega,$$

or else (approximate controllability)

$$||z(T)||_1 + ||z_t(T)||_2 \leq \varepsilon, \quad ||w(T)||_1 + ||w_t(T)||_2 \leq \varepsilon.$$

An earlier attempt to solve that controllability problem was proposed in 2012 by Dehman-Léautaud-Lerousseau: they consider two wave equations coupled in cascade internally when the two principal operators are proportional and Ω is a compact C^∞ manifold with no boundary; in particular they show that, if $\omega \cap \mathcal{O}$ satisfies GCC, then:

$$\widehat{E}(u; 0) + E_{-2}(v; 0) \leq C \int_0^T \int_\omega |u(x,t)|^2 \, dxdt,$$

where $2E_{-2}(v; 0) = ||v^0||^2_{H^{-2}(\Omega)} + ||v^1||^2_{H^{-3}(\Omega)}$.

Consider the coupled (adjoint) system

$$\begin{cases} u_{tt} - \partial_i(b_{ij}(x)\partial_j u) + au + bv + k_{11} \cdot \nabla u + l_{11}u_t \\ +k_{12} \cdot \nabla v + l_{12}v_t = 0 \text{ in } Q \\ \\ v_{tt} - \partial_i(a_{ij}(x)\partial_j v) + cu + dv + k_{21} \cdot \nabla u + l_{21}u_t \\ +k_{22} \cdot \nabla v + l_{22}v_t = 0 \text{ in } Q \\ \\ u = 0, \quad v = 0 \text{ on } \Sigma = \partial\Omega \times (0,T) \\ \\ u(0) = u^0; \quad u_t(0) = u^1 \quad v(0) = v^0; \quad v_t(0) = v^1 \text{ in } \Omega. \end{cases}$$

The coupled system is well-posed in $H_0^1(\Omega) \times L^2(\Omega) \times L^2(\Omega) \times H^{-1}(\Omega)$.

What type of observability inequality do we have for that system? Introduce the energies:

$$E_u(t) = \frac{1}{2} \int_\Omega \{|u_t(x,t)|^2 + (b_{ij}(x)\partial_j u(x,t)\partial_i u(x,t)\} \, dx,$$

$$\widehat{E}_u(t) = \frac{1}{2} \left(||u(.,t)||^2_{L^2(\Omega)} + ||u_t(.,t)||^2_{H^{-1}(\Omega)} \right).$$

For each $t \in [0, T]$, set

$$E(t) = E_u(t) + \widehat{E}_v(t), \quad \widehat{E}(t) = \widehat{E}_u(t) + \widehat{E}_v(t).$$

Is it possible to impose conditions on the coefficients a_{ij} and b_{ij} such that we have the observability estimate

$$E(0) \leq C \int_0^T \int_\omega (r^2 |u_t|^2 + |u|^2)\, dx dt$$

4.8. Mindlin-Timoshenko plate. Consider the Mindlin-Timoshenko system with boundary damping

$$\begin{aligned}
&\rho_1 y_{tt} - k\mathrm{div}(\nabla y + z) = 0 \text{ in } \Omega \times (0, \infty)\\
&\rho_2 z_{tt} - \mu \Delta z - (\lambda + \mu)\nabla \mathrm{div} z + k(\nabla y + z) = 0 \text{ in } \Omega \times (0, \infty)\\
&y = 0, \quad \mu \partial_\nu z + (\lambda + \mu)\mathrm{div}(z)\nu + z_t = 0 \text{ on } \partial\Omega \times (0, \infty)\\
&y(.,0) = y^0 \in H_0^1(\Omega), \quad y_t(.,0) = y^1 \in L^2(\Omega),\\
&z(.,0) = z^0 \in [H_0^1(\Omega)]^N, \quad z_t(.,0) = z^1 \in [L^2(\Omega)]^N.
\end{aligned}$$

The constants ρ_1, ρ_2, k, and μ are physical constants and are all positive. In particular, the constants λ and μ are the Lamé constants with $\lambda + \mu > 0$.

A work of Nicaise et collaborators (2015) shoes that the indirectly damped Timoshenko beam, ($N = 1$), is only polynomially stable under some parameter constraints when the dissipation occurs only at the left endpoint.

Suppose that:

$$(*) \qquad \frac{k}{\rho_1} = \frac{2\mu + \lambda}{\rho_2}.$$

QUESTIONS. Is the Mindlin-Timoshenko system exponentially stable under $(*)$ when the damping acts on the whole boundary?

References

[1] F. Alabau-Boussouira, *A two-level energy method for indirect boundary observability and controllability of weakly coupled hyperbolic systems*, SIAM J. Control Optim. **42** (2003), no. 3, 871–906, DOI 10.1137/S0363012902402608. MR2002139

[2] F. Alabau-Boussouira, *Asymptotic behavior for Timoshenko beams subject to a single nonlinear feedback control*, NoDEA Nonlinear Differential Equations Appl. **14** (2007), no. 5-6, 643–669, DOI 10.1007/s00030-007-5033-0. MR2374204

[3] F. Alabau-Boussouira, *A unified approach via convexity for optimal energy decay rates of finite and infinite dimensional vibrating damped systems with applications to semi-discretized vibrating damped systems*, J. Differential Equations **248** (2010), no. 6, 1473–1517, DOI 10.1016/j.jde.2009.12.005. MR2593051

[4] F. Alabau-Boussouira and K. Ammari, *Sharp energy estimates for nonlinearly locally damped PDEs via observability for the associated undamped system*, J. Funct. Anal. **260** (2011), no. 8, 2424–2450, DOI 10.1016/j.jfa.2011.01.003. MR2772377

[5] F. Alabau and V. Komornik, *Boundary observability, controllability, and stabilization of linear elastodynamic systems*, SIAM J. Control Optim. **37** (1999), no. 2, 521–542, DOI 10.1137/S0363012996313835. MR1665070

[6] F. Alabau-Boussouira and M. Léautaud, *Indirect controllability of locally coupled systems under geometric conditions* (English, with English and French summaries), C. R. Math. Acad. Sci. Paris **349** (2011), no. 7-8, 395–400, DOI 10.1016/j.crma.2011.02.004. MR2788376

[7] F. D. Araruna, P. Braz E Silva, and E. Zuazua, *Asymptotic limits and stabilization for the 1D nonlinear Mindlin-Timoshenko system*, J. Syst. Sci. Complex. **23** (2010), no. 3, 414–430, DOI 10.1007/s11424-010-0137-8. MR2679535

[8] W. Arendt and C. J. K. Batty, *Tauberian theorems and stability of one-parameter semigroups*, Trans. Amer. Math. Soc. **306** (1988), no. 2, 837–852, DOI 10.2307/2000826. MR933321

[9] S. A. Avdonin and S. A. Ivanov, *Families of exponentials*, Cambridge University Press, Cambridge, 1995. The method of moments in controllability problems for distributed parameter systems; Translated from the Russian and revised by the authors. MR1366650

[10] S. A. Avdonin, S. A. Ivanov, and D. L. Russell, *Exponential bases in Sobolev spaces in control and observation problems for the wave equation*, Proc. Roy. Soc. Edinburgh Sect. A **130** (2000), no. 5, 947–970, DOI 10.1017/S0308210500000512. MR1800086

[11] C. Bardos, G. Lebeau, and J. Rauch, *Sharp sufficient conditions for the observation, control, and stabilization of waves from the boundary*, SIAM J. Control Optim. **30** (1992), no. 5, 1024–1065, DOI 10.1137/0330055. MR1178650

[12] M. Bellassoued, *Decay of solutions of the wave equation with arbitrary localized nonlinear damping*, J. Differential Equations **211** (2005), no. 2, 303–332, DOI 10.1016/j.jde.2004.12.010. MR2125545

[13] C. D. Benchimol, *A note on weak stabilizability of contraction semigroups*, SIAM J. Control Optim. **16** (1978), no. 3, 373–379, DOI 10.1137/0316023. MR490298

[14] A. Borichev and Y. Tomilov, *Optimal polynomial decay of functions and operator semigroups*, Math. Ann. **347** (2010), no. 2, 455–478, DOI 10.1007/s00208-009-0439-0. MR2606945

[15] N. Burq, *Contrôlabilité exacte des ondes dans des ouverts peu réguliers* (French, with English and French summaries), Asymptot. Anal. **14** (1997), no. 2, 157–191. MR1451210

[16] N. Burq and P. Gérard, *Condition nécessaire et suffisante pour la contrôlabilité exacte des ondes* (French, with English and French summaries), C. R. Acad. Sci. Paris Sér. I Math. **325** (1997), no. 7, 749–752, DOI 10.1016/S0764-4442(97)80053-5. MR1483711

[17] P. Cannarsa, V. Komornik, and P. Loreti, *One-sided and internal controllability of semilinear wave equations with infinitely iterated logarithms*, Discrete Contin. Dyn. Syst. **8** (2002), no. 3, 745–756, DOI 10.3934/dcds.2002.8.747. MR1897879

[18] M. M. Cavalcanti, V. N. Domingos Cavalcanti, F. A. Falcão Nascimento, I. Lasiecka, and J. H. Rodrigues, *Uniform decay rates for the energy of Timoshenko system with the arbitrary speeds of propagation and localized nonlinear damping*, Z. Angew. Math. Phys. **65** (2014), no. 6, 1189–1206, DOI 10.1007/s00033-013-0380-7. MR3279525

[19] G. Chen, *Control and stabilization for the wave equation in a bounded domain*, SIAM J. Control Optim. **17** (1979), no. 1, 66–81, DOI 10.1137/0317007. MR516857

[20] G. Chen, *Energy decay estimates and exact boundary value controllability for the wave equation in a bounded domain*, J. Math. Pures Appl. (9) **58** (1979), no. 3, 249–273. MR544253

[21] G. Chen, *Control and stabilization for the wave equation in a bounded domain. II*, SIAM J. Control Optim. **19** (1981), no. 1, 114–122, DOI 10.1137/0319009. MR603084

[22] W. C. Chewning, *Controllability of the nonlinear wave equation in several space variables*, SIAM J. Control Optim. **14** (1976), no. 1, 19–25, DOI 10.1137/0314002. MR638222

[23] C. M. Dafermos, *Wave equations with weak damping*, SIAM J. Appl. Math. **18** (1970), 759–767, DOI 10.1137/0118068. MR262711

[24] T. Duyckaerts, X. Zhang, and E. Zuazua, *On the optimality of the observability inequalities for parabolic and hyperbolic systems with potentials*, Ann. Inst. H. Poincaré Anal. Non Linéaire **25** (2008), no. 1, 1–41, DOI 10.1016/j.anihpc.2006.07.005. MR2383077

[25] H. O. Fattorini, *Local controllability of a nonlinear wave equation*, Math. Systems Theory **9** (1975), no. 1, 30–45, DOI 10.1007/BF01698123. MR0430923

[26] H. D. Fernández Sare, *On the stability of Mindlin-Timoshenko plates*, Quart. Appl. Math. **67** (2009), no. 2, 249–263, DOI 10.1090/S0033-569X-09-01110-2. MR2514634

[27] X. Fu, J. Yong, and X. Zhang, *Exact controllability for multidimensional semilinear hyperbolic equations*, SIAM J. Control Optim. **46** (2007), no. 5, 1578–1614, DOI 10.1137/040610222. MR2361985

[28] X. Fu, *Sharp decay rates for the weakly coupled hyperbolic system with one internal damping*, SIAM J. Control Optim. **50** (2012), no. 3, 1643–1660, DOI 10.1137/110833051. MR2968070

[29] A. V. Fursikov and O. Yu. Imanuvilov, *Controllability of evolution equations*, Lecture Notes Series, vol. 34, Seoul National University, Research Institute of Mathematics, Global Analysis Research Center, Seoul, 1996. MR1406566

[30] C. G. Gal and L. Tebou, *Carleman inequalities for wave equations with oscillatory boundary conditions and application*, SIAM J. Control Optim. **55** (2017), no. 1, 324–364, DOI 10.1137/15M1032211. MR3606412

[31] J. S. Gibson, *A note on stabilization of infinite-dimensional linear oscillators by compact linear feedback*, SIAM J. Control Optim. **18** (1980), no. 3, 311–316, DOI 10.1137/0318022. MR569020

[32] K. D. Graham and D. L. Russell, *Boundary value control of the wave equation in a spherical region*, SIAM J. Control **13** (1975), 174–196. MR0355756

[33] A. Hajej, Z. Hajjej, and L. Tebou, *Indirect stabilization of weakly coupled Kirchhoff plate and wave equations with frictional damping*, J. Math. Anal. Appl. **474** (2019), no. 1, 290–308, DOI 10.1016/j.jmaa.2019.01.046. MR3912904

[34] A. Haraux, *Contrôlabilité exacte d'une membrane rectangulaire au moyen d'une fonctionnelle analytique localisée* (French, with English summary), C. R. Acad. Sci. Paris Sér. I Math. **306** (1988), no. 3, 125–128. MR929104

[35] A. Haraux, *On a completion problem in the theory of distributed control of wave equations*, Nonlinear partial differential equations and their applications. Collège de France Seminar, Vol. X (Paris, 1987), Pitman Res. Notes Math. Ser., vol. 220, Longman Sci. Tech., Harlow, 1991, pp. 241–271. MR1131826

[36] A. Haraux, *Une remarque sur la stabilisation de certains systèmes du deuxième ordre en temps* (French, with English summary), Portugal. Math. **46** (1989), no. 3, 245–258. MR1021188

[37] A. Haraux, *A generalized internal control for the wave equation in a rectangle*, J. Math. Anal. Appl. **153** (1990), no. 1, 190–216, DOI 10.1016/0022-247X(90)90273-I. MR1080126

[38] A. Haraux, *An alternative functional approach to exact controllability of reversible systems*, Port. Math. (N.S.) **61** (2004), no. 4, 399–437. MR2113557

[39] L. F. Ho, *Observabilité frontière de l'équation des ondes* (English, with French summary), C. R. Acad. Sci. Paris Sér. I Math. **302** (1986), no. 12, 443–446. MR838598

[40] L. F. Ho, *Exact controllability of second order hyperbolic systems with control in the Dirichlet boundary conditions*, J. Math. Pures Appl. (9) **66** (1987), no. 4, 363–368. MR928269

[41] F. L. Huang, *Characteristic conditions for exponential stability of linear dynamical systems in Hilbert spaces*, Ann. Differential Equations **1** (1985), no. 1, 43–56. MR834231

[42] O. Yu. Imanuvilov, *On Carleman estimates for hyperbolic equations*, Asymptot. Anal. **32** (2002), no. 3-4, 185–220. MR1993649

[43] O. Yu. Imanuvilov and J.-P. Puel, *Global Carleman estimates for weak solutions of elliptic nonhomogeneous Dirichlet problems*, Int. Math. Res. Not. **16** (2003), 883–913, DOI 10.1155/S107379280321117X. MR1959940

[44] O. Yu. Imanuvilov and M. Yamamoto, *Carleman estimate for a stationary isotropic Lamé system and the applications*, Appl. Anal. **83** (2004), no. 3, 243–270, DOI 10.1080/00036810310001632772. MR2033238

[45] V. Komornik, *Exact controllability in short time for the wave equation*, Ann. Inst. H. Poincaré Anal. Non Linéaire **6** (1989), no. 2, 153–164. MR991876

[46] V. Komornik, *Exact controllability and stabilization*, RAM: Research in Applied Mathematics, Masson, Paris; John Wiley & Sons, Ltd., Chichester, 1994. The multiplier method. MR1359765

[47] V. Komornik and P. Loreti, *Fourier series in control theory*, Springer Monographs in Mathematics, Springer-Verlag, New York, 2005. MR2114325

[48] J. Lagnese, *Control of wave processes with distributed controls supported on a subregion*, SIAM J. Control Optim. **21** (1983), no. 1, 68–85, DOI 10.1137/0321004. MR688440

[49] I. Lasiecka and R. Triggiani, *Exact controllability of the wave equation with Neumann boundary control*, Appl. Math. Optim. **19** (1989), no. 3, 243–290, DOI 10.1007/BF01448201. MR974187

[50] I. Lasiecka and R. Triggiani, *Exact controllability of semilinear abstract systems with application to waves and plates boundary control problems*, Appl. Math. Optim. **23** (1991), no. 2, 109–154, DOI 10.1007/BF01442394. MR1086465

[51] I. Lasiecka and R. Triggiani, *Carleman estimates and exact boundary controllability for a system of coupled, nonconservative second-order hyperbolic equations*, Partial differential equation methods in control and shape analysis (Pisa), Lecture Notes in Pure and Appl. Math., vol. 188, Dekker, New York, 1997, pp. 215–243. MR1452894

[52] I. Lasiecka and R. Triggiani, *Global exact controllability of semilinear wave equations by a double compactness/uniqueness argument*, Discrete Contin. Dyn. Syst. **suppl.** (2005), 556–565. MR2192714

[53] I. Lasiecka, R. Triggiani, and P.-F. Yao, *An observability estimate in $L_2(\Omega) \times H^{-1}(\Omega)$ for second-order hyperbolic equations with variable coefficients*, Control of distributed parameter and stochastic systems (Hangzhou, 1998), Kluwer Acad. Publ., Boston, MA, 1999, pp. 71–78. MR1777397

[54] I. Lasiecka, R. Triggiani, and P.-F. Yao, *Inverse/observability estimates for second-order hyperbolic equations with variable coefficients*, J. Math. Anal. Appl. **235** (1999), no. 1, 13–57, DOI 10.1006/jmaa.1999.6348. MR1758667

[55] I. Lasiecka, R. Triggiani, and X. Zhang, *Nonconservative wave equations with unobserved Neumann B.C.: global uniqueness and observability in one shot*, Differential geometric methods in the control of partial differential equations (Boulder, CO, 1999), Contemp. Math., vol. 268, Amer. Math. Soc., Providence, RI, 2000, pp. 227–325, DOI 10.1090/conm/268/04315. MR1804797

[56] G. Lebeau and L. Robbiano, *Contrôle exact de l'équation de la chaleur* (French), Comm. Partial Differential Equations **20** (1995), no. 1-2, 335–356, DOI 10.1080/03605309508821097. MR1312710

[57] L. Li and X. Zhang, *Exact controllability for semilinear wave equations*, J. Math. Anal. Appl. **250** (2000), no. 2, 589–597, DOI 10.1006/jmaa.2000.6998. MR1786083

[58] J.-L. Lions, *Exact controllability, stabilization and perturbations for distributed systems*, SIAM Rev. **30** (1988), no. 1, 1–68, DOI 10.1137/1030001. MR931277

[59] J.-L. Lions, *Contrôlabilité exacte, perturbations et stabilisation de systèmes distribués. Tome 1* (French), Recherches en Mathématiques Appliquées [Research in Applied Mathematics], vol. 8, Masson, Paris, 1988. Contrôlabilité exacte. [Exact controllability]; With appendices by E. Zuazua, C. Bardos, G. Lebeau and J. Rauch. MR953547

[60] J.-L. Lions, *Contrôlabilité exacte, perturbations et stabilisation de systèmes distribués. Tome 2* (French), Recherches en Mathématiques Appliquées [Research in Applied Mathematics], vol. 9, Masson, Paris, 1988. Perturbations. [Perturbations]. MR963060

[61] J.-L. Lions, *An introduction to the methods based on uniqueness for exact controllability of distributed systems*, Control of partial differential equations (Santiago de Compostela, 1987), Lect. Notes Control Inf. Sci., vol. 114, Springer, Berlin, 1989, pp. 35–46, DOI 10.1007/BFb0002578. MR987964

[62] K. Liu, *Locally distributed control and damping for the conservative systems*, SIAM J. Control Optim. **35** (1997), no. 5, 1574–1590, DOI 10.1137/S0363012995284928. MR1466917

[63] W.-J. Liu, *Exact distributed controllability for the semilinear wave equation*, Portugal. Math. **57** (2000), no. 4, 493–508. MR1807358

[64] K. Liu, M. Yamamoto, and X. Zhang, *Observability inequalities by internal observation and their applications*, J. Optim. Theory Appl. **116** (2003), no. 3, 621–645, DOI 10.1023/A:1023069420681. MR1977754

[65] L. Markus, *Controllability of nonlinear processes*, J. SIAM Control Ser. A **3** (1965), 78–90. MR186487

[66] P. Martinez, *Stabilisation de systèmes distribués semilinéaires: domaines presque étoilés et inégalités intégrales généralisées* (French, with French summary), Prépublication de l'Institut de Recherche Mathématique Avancée [Prepublication of the Institute of Advanced Mathematical Research], vol. 1998/44, Université Louis Pasteur, Département de Mathématique, Institut de Recherche Mathématique Avancée, Strasbourg, 1998. Thèse, Université Louis Pasteur (Strasbourg I), Strasbourg, 1998. MR1688957

[67] P. Martinez and J. Vancostenoble, *Exact controllability in "arbitrarily short time" of the semilinear wave equation*, Discrete Contin. Dyn. Syst. **9** (2003), no. 4, 901–924, DOI 10.3934/dcds.2003.9.901. MR1975359

[68] L. Miller, *Geometric bounds on the growth rate of null-controllability cost for the heat equation in small time*, J. Differential Equations **204** (2004), no. 1, 202–226, DOI 10.1016/j.jde.2004.05.007. MR2076164

[69] S. Nicaise, *Internal stabilization of a Mindlin-Timoshenko model by interior feedbacks*, Math. Control Relat. Fields **1** (2011), no. 3, 331–352, DOI 10.3934/mcrf.2011.1.331. MR2846089

[70] A. Osses, *A rotated multiplier applied to the controllability of waves, elasticity, and tangential Stokes control*, SIAM J. Control Optim. **40** (2001), no. 3, 777–800, DOI 10.1137/S0363012998345615. MR1871454

[71] A. Pazy, *Semigroups of linear operators and applications to partial differential equations*, Applied Mathematical Sciences, vol. 44, Springer-Verlag, New York, 1983. MR710486

[72] J. Prüss, *On the spectrum of C_0-semigroups*, Trans. Amer. Math. Soc. **284** (1984), no. 2, 847–857, DOI 10.2307/1999112. MR743749

[73] L. Rosier and L. de Teresa, *Exact controllability of a cascade system of conservative equations* (English, with English and French summaries), C. R. Math. Acad. Sci. Paris **349** (2011), no. 5-6, 291–296, DOI 10.1016/j.crma.2011.01.014. MR2783322

[74] A. Ruiz, *Unique continuation for weak solutions of the wave equation plus a potential*, J. Math. Pures Appl. (9) **71** (1992), no. 5, 455–467. MR1191585

[75] D. L. Russell, *Nonharmonic Fourier series in the control theory of distributed parameter systems*, J. Math. Anal. Appl. **18** (1967), 542–560, DOI 10.1016/0022-247X(67)90045-5. MR211044

[76] D. L. Russell, *Boundary value control of the higher-dimensional wave equation*, SIAM J. Control **9** (1971), 29–42. MR0274917

[77] D. L. Russell, *Boundary value control theory of the higher-dimensional wave equation. II*, SIAM J. Control **9** (1971), 401–419. MR0294896

[78] D. L. Russell, *Control theory of hyperbolic equations related to certain questions in harmonic analysis and spectral theory (an outline)*, Ordinary differential equations (Proc. Conf., Math. Res. Center, Naval Res. Lab., Washington, D. C., 1971), Academic Press, New York, 1972, pp. 241–263. MR0428179

[79] D. L. Russell, *A unified boundary controllability theory for hyperbolic and parabolic partial differential equations*, Studies in Appl. Math. **52** (1973), 189–211, DOI 10.1002/sapm1973523189. MR341256

[80] D. L. Russell, *Exact boundary value controllability theorems for wave and heat processes in star-complemented regions*, Differential games and control theory (Proc. NSF—CBMS Regional Res. Conf., Univ. Rhode Island, Kingston, R.I., 1973), Dekker, New York, 1974, pp. 291–319. Lecture Notes in Pure Appl. Math., Vol. 10. MR0467472

[81] D. L. Russell, *Controllability and stabilizability theory for linear partial differential equations: recent progress and open questions*, SIAM Rev. **20** (1978), no. 4, 639–739, DOI 10.1137/1020095. MR508380

[82] D. L. Russell, *The Dirichlet-Neumann boundary control problem associated with Maxwell's equations in a cylindrical region*, SIAM J. Control Optim. **24** (1986), no. 2, 199–229, DOI 10.1137/0324012. MR826513

[83] L. Tebou, *Locally distributed desensitizing controls for the wave equation* (English, with English and French summaries), C. R. Math. Acad. Sci. Paris **346** (2008), no. 7-8, 407–412, DOI 10.1016/j.crma.2008.02.019. MR2417559

[84] L. Tebou, *Some results on the controllability of coupled semilinear wave equations: the desensitizing control case*, SIAM J. Control Optim. **49** (2011), no. 3, 1221–1238, DOI 10.1137/100803080. MR2806582

[85] L. Tebou, *Simultaneous observability and stabilization of some uncoupled wave equations* (English, with English and French summaries), C. R. Math. Acad. Sci. Paris **350** (2012), no. 1-2, 57–62, DOI 10.1016/j.crma.2011.12.001. MR2887836

[86] L. Tebou, *Sharp observability estimates for a system of two coupled nonconservative hyperbolic equations*, Appl. Math. Optim. **66** (2012), no. 2, 175–207, DOI 10.1007/s00245-012-9168-y. MR2965263

[87] L. Tebou, Simultaneous stabilization of some interacting plate and membrane. EECT, 2 (2013), 153-172.

[88] L. Tebou, *Simultaneous controllability of some uncoupled semilinear wave equations*, Discrete Contin. Dyn. Syst. **35** (2015), no. 8, 3721–3743, DOI 10.3934/dcds.2015.35.3721. MR3320143

[89] L. Tebou, *A localized nonstandard stabilizer for the Timoshenko beam* (English, with English and French summaries), C. R. Math. Acad. Sci. Paris **353** (2015), no. 3, 247–253, DOI 10.1016/j.crma.2015.01.004. MR3306493

[90] L. Tebou, *Stabilization of some elastodynamic systems with localized Kelvin-Voigt damping*, Discrete Contin. Dyn. Syst. **36** (2016), no. 12, 7117–7136, DOI 10.3934/dcds.2016110. MR3567835

[91] L. Tebou, *Indirect stabilization of a Mindlin-Timoshenko plate*, J. Math. Anal. Appl. **449** (2017), no. 2, 1880–1891, DOI 10.1016/j.jmaa.2017.01.015. MR3601621

[92] R. Triggiani, *Lack of uniform stabilization for noncontractive semigroups under compact perturbation*, Proc. Amer. Math. Soc. **105** (1989), no. 2, 375–383, DOI 10.2307/2046953. MR953013

[93] R. Triggiani, *Carleman estimates and exact boundary controllability for a system of coupled non-conservative Schrödinger equations*, Rend. Istit. Mat. Univ. Trieste **28** (1996), no. suppl., 453–504 (1997). Dedicated to the memory of Pierre Grisvard. MR1602299

[94] R. Triggiani and P. F. Yao, *Carleman estimates with no lower-order terms for general Riemann wave equations. Global uniqueness and observability in one shot*, Appl. Math. Optim. **46** (2002), no. 2-3, 331–375, DOI 10.1007/s00245-002-0751-5. Special issue dedicated to the memory of Jacques-Louis Lions. MR1944764

[95] J. Vancostenoble and E. Zuazua, *Hardy inequalities, observability, and control for the wave and Schrödinger equations with singular potentials*, SIAM J. Math. Anal. **41** (2009), no. 4, 1508–1532, DOI 10.1137/080731396. MR2556573

[96] H. E. Wilhelm, T. J. van der Werff, Nonlinear wave equations for chemical reactions with diffusion in multicomponent systems. J. Chem. Phys. 67 (1977), 3382-3387.

[97] A. Wehbe and W. Youssef, *Stabilization of the uniform Timoshenko beam by one locally distributed feedback*, Appl. Anal. **88** (2009), no. 7, 1067–1078, DOI 10.1080/00036810903156149. MR2561475

[98] P.-F. Yao, *On the observability inequalities for exact controllability of wave equations with variable coefficients*, SIAM J. Control Optim. **37** (1999), no. 5, 1568–1599, DOI 10.1137/S0363012997331482. MR1710233

[99] X. Zhang, *Explicit observability estimate for the wave equation with potential and its application*, R. Soc. Lond. Proc. Ser. A Math. Phys. Eng. Sci. **456** (2000), no. 1997, 1101–1115, DOI 10.1098/rspa.2000.0553. MR1809954

[100] X. Zhang, *Explicit observability inequalities for the wave equation with lower order terms by means of Carleman inequalities*, SIAM J. Control Optim. **39** (2000), no. 3, 812–834, DOI 10.1137/S0363012999350298. MR1786331

[101] E. Zuazua, *Exact controllability for the semilinear wave equation* (English, with French summary), J. Math. Pures Appl. (9) **69** (1990), no. 1, 1–31. MR1054122

[102] E. Zuazua, *Exact boundary controllability for the semilinear wave equation* (English, with French summary), Nonlinear partial differential equations and their applications. Collège de France Seminar, Vol. X (Paris, 1987), Pitman Res. Notes Math. Ser., vol. 220, Longman Sci. Tech., Harlow, 1991, pp. 357–391. MR1131832

[103] E. Zuazua, *Exact controllability for semilinear wave equations in one space dimension* (English, with English and French summaries), Ann. Inst. H. Poincaré Anal. Non Linéaire **10** (1993), no. 1, 109–129, DOI 10.1016/S0294-1449(16)30221-9. MR1212631

Department of Mathematics and Statistics, Florida International University, Modesto Maidique Campus, Miami, Florida 33199

Email address: `teboul@fiu.edu`

Contemporary Mathematics
Volume **757**, 2020
https://doi.org/10.1090/conm/757/15242

Observability and stabilization of the wave equation with moving boundary or pointwise feedback

Kaïs Ammari, Ahmed Bchatnia, and Karim El Mufti

ABSTRACT. We deal with the one dimensional wave equation in a bounded interval with a moving interior or a boundary point $a(t)$. Here $a(t)$ is assumed to move slower than the light and periodically. We survey the results of observation and stabilization problems.

CONTENTS

1. Introduction and preliminaries

We survey the results of observability and stabilization for the one dimensional wave equation with moving boundary and with a moving interior point established in [**1, 2**], except subection 2.2 which is a novelty. Indeed, we consider the following problems:

$$\left\{ \begin{array}{l} u_{tt} - u_{xx} = 0 \quad \text{for} \quad 0 < x < a(t),\, t > 0, \\ u(x,0) = \phi(x), u_t(x,0) = \psi(x), \;\; 0 < x < a(0), \end{array} \right. \tag{1.1}$$

$(\phi, \psi) \in H^1((0, a(0))) \times L^2((0, a(0)))$, with Dirichlet boundary conditions

$$u(0,t) = 0 \quad \text{and} \quad u(a(t),t) = 0,\, t > 0, \tag{1.2}$$

or with mixed boundary conditions for which (1.2) is replaced by

$$u(0,t) = 0 \quad \text{and} \quad u_x(a(t),t) = 0,\, t > 0, \tag{1.3}$$

the subscripts denote partial differentiations, here a is a strictly positive real function which is continuous and 1-periodic.

2020 *Mathematics Subject Classification.* Primary 37E45, 93B07, 93D15.
Key words and phrases. Observability, stabilizability, rotation number.

We prove the following observability inequalities:

$$\int_0^T |u_x(a(t),t)|^2 \, dt \geq C^* \left(\|\phi\|^2_{H_0^1(0,a(0))} + \|\psi\|^2_{L^2(0,a(0))} \right), \tag{1.4}$$

for (1.1)-(1.2) and

$$\int_0^T |u_t(a(t),t)|^2 \, dt \geq C^* \left(\|\phi\|^2_{H_l^1(0,a(0))} + \|\psi\|^2_{L^2(0,a(0))} \right), \tag{1.5}$$

for (1.1)-(1.3), where $H_l^1(0,a(0)) = \left\{ f \in H^1(0,a(0)) \text{ such that } f(0) = 0 \right\}$.

After we consider :

$$\begin{cases} u_{tt} - u_{xx} = 0 \quad \text{for} \quad 0 < x < 1, \, t > 0, \\ u(x,0) = \phi(x), u_t(x,0) = \psi(x), \quad 0 < x < 1, \end{cases} \tag{1.6}$$

$(\phi, \psi) \in H^1(0,1) \times L^2(0,1)$, with Dirichlet boundary conditions

$$u(0,t) = 0 \quad \text{and} \quad u(1,t) = 0, \, t > 0. \tag{1.7}$$

We prove that the inequality (1.4) is satisfied for (1.6) and an associated curve $0 < a(t) < 1$.

In [**9**] the author gives several examples of curves $a(t)$ for exact controllability related to the system:

$$\begin{cases} u_{tt} - u_{xx} = f(t)\delta_{a(t)} \quad \text{for} \quad 0 < x < 1, \, t > 0, \\ u(0,t) = u(1,t) = 0, \, t > 0, \\ u(x,0) = \phi(x), u_t(x,0) = \psi(x), \quad 0 < x < 1, \end{cases}$$

via the observability estimate:

$$\int_0^T \left| \frac{d}{dt}[\varphi(a(t),t)] \right|^2 dt \geq C^* \left(\|\phi\|^2_{H_0^1(0,1)} + \|\psi\|^2_{L^2(0,1)} \right), \tag{1.8}$$

where φ is the solution of the associated conservative system:

$$\begin{cases} \varphi_{tt} - \varphi_{xx} = 0 \quad \text{for} \quad 0 < x < 1, \, t > 0, \\ \varphi(0,t) = \varphi(1,t) = 0, \, t > 0, \\ \varphi(x,0) = \phi(x), \varphi_t(x,0) = \psi(x), \quad 0 < x < 1, \end{cases}$$

and T is given by an optical geometric condition (for more details see [**7**] and [**15**] for the generalization to time-dependent observation domain) requiring that any ray, starting anywhere in the domain with any initial direction, must meet the control zone before the time T.

Here we introduce a new approach that provides (1.4) for another class of curves $a(t)$.

Finaly, we consider the system:

$$\begin{cases} u_{tt} - u_{xx} = 0 \quad \text{for} \quad 0 < x < a(t), \, t > 0, \\ u(0,t) = 0 \quad \text{and} \quad u_t(a(t),t) + f(t)u_x(a(t),t) = 0, \, t > 0, \\ u(x,0) = \phi(x), u_t(x,0) = \psi(x), \quad 0 < x < a(0), \end{cases} \tag{1.9}$$

$(\phi, \psi) \in H_l^1((0,a(0))) \times L^2((0,a(0)))$ and where a is a strictly positive real function which is continuous, 1-periodic and $f \in L^\infty(\mathbb{R}_+^*)$.

Denote by

$$E_u(t) = \frac{1}{2} \int_0^{a(t)} \left[|u_t(x,t)|^2 + |u_x(x,t)|^2 \right] dx$$

the energy of the field u. Our major concern will be to detect the feedback $f(t)$ necessary to obtain the exponential decay of $E_u(t)$.

The mechanism of the problem was discovered by Dittrich, Duclos and Gonzalez [**12**] where they used a direct estimates on the energy. A detailed survey is given in Gonzalez [**17**].

First, we denote by D_p the set of continuous functions of the form $x + g(x)$, where $g(x)$ is a 1-periodic continuous function.

PROPOSITION 1.1. ([**18**, Herman],[**22**, Yamaguchi]) *Let a be a 1-periodic function. Then*

$$F := (I + a) \circ (I - a)^{-1} \tag{1.10}$$

belongs to D_p. Moreover, the rotation number $\rho(F)$ defined by

$$\rho(F) = \lim_{n\to\infty} \frac{F^n(x) - x}{n}$$

exists, and the limit is equal for all $x \in \mathbb{R}$.

After, we construct a transformation of the time-dependent domain $[0, a(t)] \times \mathbb{R}$ onto $[0, \rho(F)/2] \times \mathbb{R}$ that preserves the D'Alembertian form of the wave equation. For this purpose, we use:

PROPOSITION 1.2. ([**16**, Lemma 3.3]) *Assume that $a(t)$ is a 1-periodic function, $a(t) > 0$, $a \in Lip(\mathbb{R})$ such that the Lipschitz constant of a, $L(a) \in [0, 1)$. Assume also that a is a piecewise differentiable function and $|a'(t)| < 1$ for all $t \in \mathbb{R}$ and $\rho(F) \in \mathbb{R} \setminus \mathbb{Q}$. There exists a function $H \in D_p$ such that*

$$H \circ F(\xi) = H(\xi) + \rho(F). \tag{1.11}$$

For stating our main results, we need to specify the following hypothesis on H.

ASSUMPTION 1.3. Assume that H is a piecewise differentiable function and there exist $\lambda_1 > 0$ and $\lambda_2 > 0$ such that

$$\lambda_1 \leq H'(t) \leq \lambda_2, \; t \in \mathbb{R}. \tag{1.12}$$

ASSUMPTION 1.4. The function $b(t) := \dfrac{H'(-a(t) + t) - H'(a(t) + t)}{H'(-a(t) + t) + H'(a(t) + t)}$ satisfies

$$c_1 \leq b(t) \leq c_2, \quad c_1, c_2 > 0, \text{ for all } t \in \mathbb{R}. \tag{1.13}$$

The next example is given in [**16**], where Assumption 1.3 is guaranteed.

Let a be continuous and 1-periodic on $\mathbb{R}$, $a > 0$, be such that

$$a(t) := \begin{cases} \alpha t + \frac{\alpha(1-\alpha)(1+\beta)}{2(\alpha-\beta)} & \text{if } \frac{\alpha(1+\beta)}{2(\alpha-\beta)} \leq t \leq \frac{\alpha(1+\beta)-2\beta}{2(\alpha-\beta)}, \\ \beta t - \beta + \frac{\alpha(1-\beta^2)}{2(\alpha-\beta)} & \text{if } \frac{\alpha(1+\beta)-2\beta}{2(\alpha-\beta)} \leq t \leq \frac{\alpha(3+\beta)-2\beta}{2(\alpha-\beta)}, \end{cases} \tag{1.14}$$

with $-1 < \beta < 0 < \alpha < 1$.

The function F is defined by:

$$F(x) := (I + a) \circ (I - a)^{-1}(x) = \begin{cases} l_1 x + F_0 & \text{if } 0 \leq x \leq x_0, \\ l_2 x + F_0 + 1 - l_2 & \text{if } x_0 < x < 1, \end{cases}$$

with $l_1 := \frac{1+\alpha}{1-\alpha}$, $l_2 := \frac{1+\beta}{1-\beta}$, $F_0 := \frac{l_2(l_1-1)}{l_1-l_2}$ and $x_0 := \frac{1-l_2}{l_1-l_2}$.

We extend F through the formula: $F(x + 1) = F(x) + 1$ for any $x \in \mathbb{R}$.

The function H given by (1.11) and which is 1-periodic is done by

$$H(x) = h_0 \ln(|x + h_1|) + h_2, \; 0 \leq x < 1,$$

where $h_0 = \frac{1}{\ln\left(\frac{l_1}{l_2}\right)}$, $h_1 = \frac{l_2}{l_1 - l_2}$ and $h_2 = -\ln(h_1)$, also the rotation number is given by the expression:

$$\rho(F) = \frac{\ln l_1}{\ln\left(\frac{l_1}{l_2}\right)}. \tag{1.15}$$

H satisfies the inequalities:

$$\frac{1}{\ln(\frac{l_1}{l_2})}\frac{l_1 - l_2}{l_1} \le H'(x) \le \frac{1}{\ln(\frac{l_1}{l_2})}\frac{l_1 - l_2}{l_2}. \tag{1.16}$$

Note that for $a(t)$ giving by (1.14) which also 1-periodic,

$$b(t) = \frac{a(t)}{t + \frac{l_2}{l_1 - l_2}}.$$

We use the expression of a in $[\frac{\alpha(1+\beta)}{2(\alpha-\beta)}, \frac{\alpha(3+\beta)-2\beta}{2(\alpha-\beta)}]$, and we deduce the a satisfies for all $t \in \mathbb{R}$

$$2\,a_{min}\,\frac{\alpha-\beta}{1-\beta} \le 2\,a(t)\,\frac{\alpha-\beta}{1-\beta} \le b(t) \le 2\,a(t)\,\frac{\alpha-\beta}{1+\beta} \le 2\,a_{max}\,\frac{\alpha-\beta}{1+\beta}, \tag{1.17}$$

and Assumption 1.4 is satisfied.

The paper is organized as follows: In section 2 we give the results on the observability. In section 3 we prove the result on stabilization of the wave equation with moving boundary.

2. Observability of the wave equation

2.1. Observability of the wave equation with moving boundary. We consider problems (1.1) with Dirichlet boundary conditions or mixed boundary conditions, here a is a strictly positive real function which is continuous, 1-periodic.

Our major concern will be to prove the observability inequalities (1.4) for (1.1)-(1.2) and (1.5) for (1.1)-(1.3).

Note that if a is a constant, the observability inequality (1.4) holds for some positive constant C^*, if $T \ge 2a$ for the Dirichlet problem. Also, (1.5) holds for the mixed problem.

We prove the Neumann observability for (1.6) with the Dirichlet boundary condition (1.7).

THEOREM 2.1 (Neumann observability). *Under the Assumption* 1.3 *and for all* $T \ge \rho(F)$ *there exists a constant* $C^* > 0$ *such that for all* u *solution of the system* (1.1) *with the Dirichlet boundary condition* (1.2) *and initial data* $(\phi, \psi) \in H_0^1(0, a(0)) \times L^2(0, a(0))$, *we have*

$$\int_0^T |\, u_x(a(t), t)|^2 \, dt \ge C^* \left(\|\phi\|^2_{H_0^1(0,a(0))} + \|\psi\|^2_{L^2(0,a(0))} \right). \tag{2.1}$$

REMARK 2.2. Here the observability time $\rho(F)$ is optimal, in the sense: if a is constant equal to a, it's given by $2a = \rho(F)$.

We treat also the Dirichlet observability for (1.1) with the Neumann boundary condition (1.3).

THEOREM 2.3 (Dirichlet observability). *Under the assumptions* 1.3 *and* 1.4 *we have that for all* $T \geq \rho(F)$, *there exists a constant* $C^* > 0$ *such that for all solution* u *of the system* (1.1) *with the mixed boundary condition* (1.3) *and initial data* $(\phi, \psi) \in H_l^1(0, a(0)) \times L^2(0, a(0))$,

$$\int_0^T |u_t(a(t), t)|^2 \, dt \geq C^* \left(\|\phi\|^2_{H_l^1(0,a(0))} + \|\psi\|^2_{L^2(0,a(0))} \right). \tag{2.2}$$

REMARK 2.4. The observability time is optimal here for the same reason as in Remark 2.2 and Theorem 2.1.

The exact controllability problem for the system

$$\begin{cases} u_{tt} - u_{xx} = 0 \quad \text{for} \quad 0 < x < a(t), \, t > 0, \\ u(0, t) = 0 \quad \text{and} \quad u(a(t), t) = r(t), \, t > 0, \\ u(x, 0) = \phi(x), u_t(x, 0) = \psi(x) \;\; 0 < x < a(0) \end{cases} \tag{2.3}$$

at time T is the following:

for each $(\phi, \psi) \in L^2(0, a(0)) \times H^{-1}(0, a(0))$, find $r \in L^2(0, T)$ such that the corresponding solution to (2.3) satisfies $u(., T) = 0, u_t(., T) = 0$ in $(0, a(T))$.

Based on the observability estimate mentioned above and for a given by (1.14), we get:

COROLLARY 2.5. *There exists* $r \in L^2(0, T)$ *such that the system* (2.3) *is exactly controllable at time* $T := |e^{\frac{\rho(F) - h_2}{h_0}} - h_1|$. *Note that* T *is optimal in the sense that this time is derived from an optimal time of observability.*

PROOF OF THEOREM 2.1. We begin, by defining a domain transformation $\Phi : \mathbb{R}^2 \to \mathbb{R}^2$, using H given by (1.11), as follows:

$$\begin{cases} \xi = (H(x + t) - H(-x + t))/2, \\ \tau = (H(x + t) + H(-x + t))/2, \end{cases} \tag{2.4}$$

for $(x, t) \in \mathbb{R}^2$.

The propositions below can essentially be found in [**23**] (see also the references therein), we reproduce them for the reader's convenience and because our presentation is synthetic.

PROPOSITION 2.6. ([**23**, Yamaguchi]) *The transformation* Φ *is a bijection of* $[0, a(t)] \times \mathbb{R}$ *to* $[0, \rho(F)/2] \times \mathbb{R}$ *and* Φ *maps the boundaries* $x = 0$ *and* $x = a(t)$ *onto the boundaries* $\xi = 0$ *and* $\xi = \rho(F)/2$.

PROPOSITION 2.7. ([**23**, Yamaguchi]) *Let* $u(x, t)$ *satisfying* $(\partial_t^2 - \partial_x^2)u(x, t) = 0$ *and* $V(\xi, \tau)$ *defined by* $u(\Phi^{-1}(\xi, \tau))$. *Then the following identity holds*

$$(\partial_t^2 - \partial_x^2)u(x, t) = K(\xi, \tau)(\partial_\tau^2 - \partial_\xi^2)V(\xi, \tau)$$

where $K(\xi, \tau)$ *is defined by*

$$4H' \circ H^{-1}(\xi + \tau) H' \circ H^{-1}(-\xi + \tau) \circ H^{-1}(\xi + \tau).$$

The next lemma will be very useful for the proof of Theorem 2.1 and Theorem 2.3.

LEMMA 2.8. *Denote by*

$$E_u(t) = \frac{1}{2} \int_0^{a(t)} \left[|u_t(x, t)|^2 + |u_x(x, t)|^2 \right] dx$$

the energy of the field u, and

$$E_V(\tau) = \int_0^{\rho(F)/2} \left(|V_\xi(\xi,\tau)|^2 + |V_\tau(\xi,\tau)|^2 \right) d\xi,$$

the energy of the field V. Under Assumption 1.3, there are two positive constants C_1 and C_2 such that

$$C_1 E_V\left(H(a(t)+t) - \frac{\rho(F)}{2}\right) \leq E_u(t) \leq C_2 E_V\left(H(a(t)+t) - \frac{\rho(F)}{2}\right). \tag{2.5}$$

Note that we can write $H(a(t)+t) = a(t)+t+g\big(a(t)+t\big)$, where $g(x)$ is a 1-periodic continuous function.

PROOF OF LEMMA 2.8. We calculate,

$$\partial_t u = \partial_\xi V \partial_t \xi + \partial_\tau V \partial_t \tau \quad \text{and} \quad \partial_x u = \partial_\xi V \partial_x \xi + \partial_\tau V \partial_x \tau$$

and so,

$$\begin{aligned} E_u(t) &= \frac{1}{2}\int_0^{a(t)} \left[|u_t(x,t)|^2 + |u_x(x,t)|^2 \right] dx \\ &= \frac{1}{2}\int_0^{a(t)} \left\{ [V_\xi \xi_t + V_\tau \tau_t]^2 + |V_\xi \xi_x + V_\tau \tau_x|^2 \right\} dx. \end{aligned}$$

Make use of:

$$\xi_x = (\partial_x \xi) = (\partial_t \tau) = \tau_t = [H'(x+t) + H'(-x+t)]/2,$$

$$\xi_t = (\partial_t \xi) = (\partial_x \tau) = \tau_x = [H'(x+t) - H'(-x+t)]/2.$$

Hence,

$$\xi_t^2 + \xi_x^2 = \tau_t^2 + \tau_x^2 = \frac{1}{2}\left[|H'(x+t)|^2 + |H'(-x+t)|^2 \right],$$

$$\xi_t \tau_t = \xi_x \tau_x = \frac{1}{4}[(H'(x+t))^2 - (H'(-x+t))^2].$$

Back to the energy,

$$\begin{aligned} E_u(t) &= \int_0^{a(t)} \frac{1}{4}\left[|V_\xi|^2 + |V_\tau|^2 \right] \left[|H'(x+t)|^2 + |H'(-x+t)|^2 \right] dx \\ &\quad + \int_0^{a(t)} \frac{1}{2}[V_\xi V_\tau] \left[|H'(x+t)|^2 - |H'(-x+t)|^2 \right] dx \\ &= \int_0^{a(t)} \frac{1}{4}[|V_\xi|^2 + |V_\tau|^2 + 2V_\xi V_\tau]|H'(x+t)|^2 dx \\ &\quad + \int_0^{a(t)} \frac{1}{4}[|V_\xi|^2 + |V_\tau|^2 - 2V_\xi V_\tau]|H'(-x+t)|^2 dx. \end{aligned}$$

Also, differentiating $\xi = (H(x+t) - H(-x+t))/2$ on a slice $t =$constant, we obtain

$$d\xi = \frac{H'(x+t) + H'(-x+t)}{2} dx.$$

Thanks to Assumption 1.3 and the inequality $|V_\xi V_\tau| \leq 1/2\left(V_\xi^2 + V_\tau^2\right)$ we get:

$$C_1 E_V(\tau) \leq E_u(t) \leq C_2 E_V(\tau), \tag{2.6}$$

for positive constants C_1, C_2. □

Applying the transformation Φ, the system (1.9) becomes:

$$\begin{cases} \partial_\tau^2 V - \partial_\xi^2 V = 0, \quad \text{for} \quad 0 < \xi < \rho(F)/2,\ \tau \in \mathbb{R}, \\ V(0,\tau) = 0, \ \ V(\rho(F)/2,\tau) = 0, \ \ \tau \in \mathbb{R}, \\ V(\xi,0) = \phi_1(\xi), \ \ V_\tau(\xi,0) = \psi_1(\xi), \ \ \xi \in (0,\rho(F)/2). \end{cases} \tag{2.7}$$

We need the following lemma which is the classical result for the observability of the 1-d wave equation on the interval $[0, \rho(F)/2]$.

LEMMA 2.9. (**[11, 14]**) *If $T \geq \rho(F)$, there exists $C(T) > 0$ such that for all $(\phi_1, \psi_1) \in H_0^1(0, \rho(F)/2) \times L^2(0, \rho(F)/2)$ we have*

$$C(T) \int_0^T |V_\xi(\rho(F)/2,\tau)|^2 d\tau \geq \|\phi_1\|^2_{H_0^1(0,\rho(F)/2)} + \|\psi_1\|^2_{L^2(0,\rho(F)/2)}.$$

We consider (1.1)-(1.2) and state:

$$\partial_x u = \partial_\xi V \partial_x \xi + \partial_\tau V \partial_x \tau.$$

Thanks to $\partial_\tau V(\rho(F)/2,\tau) = 0$, we get

$$\partial_x u(a(t),t) = \partial_\xi V(\rho(F)/2,\tau)\partial_x\xi(a(t),t).$$

Since $\partial_x \xi = [H'(x+t) + H'(-x+t)]/2$, it follows that:

$$\begin{aligned} |\partial_x u(a(t),t)|^2 &= |\partial_\xi V(\rho(F)/2,\tau)\partial_x\xi(a(t),t)|^2 \\ &= \tfrac{1}{4}\Big(\partial_\xi V(\rho(F)/2,\tau)[H'(x+t) + H'(-x+t)]\Big)^2. \end{aligned}$$

Make use of Young inequalities, (2.1) is a consequence of Lemma 2.8 and Lemma 2.9. □

PROOF OF COROLLARY 2.5. Let us consider

$$\begin{cases} \partial_\tau^2 V - \partial_\xi^2 V = 0, \quad \text{for} \quad 0 < \xi < \rho(F)/2,\ \tau \in \mathbb{R}, \\ V(0,\tau) = 0, \ \ V(\rho(F)/2,\tau) = 0, \ \ \tau \in \mathbb{R}, \\ V(\xi,0) = V_0(\xi), \ \ V_\tau(\xi,0) = V_1(\xi), \ \ \xi \in (0,\rho(F)/2). \end{cases} \tag{2.8}$$

The system (2.3) is exactly observable at time $\rho(F)$, i.e., there exists $C > 0$ such that for all $T \geq \rho(F)$, we have

$$C(T) \int_0^T |V_\xi(\rho(F)/2,\tau)|^2 d\tau \geq \|\phi_1\|^2_{H_0^1(0,\rho(F)/2)} + \|\psi_1\|^2_{L^2(0,\rho(F)/2)},$$

and so the problem

$$\begin{cases} \partial_\tau^2 \tilde{V} - \partial_\xi^2 \tilde{V} = 0, \qquad 0 < \xi < \rho(F)/2,\ \tau \in \mathbb{R}, \\ \tilde{V}(0,\tau) = 0, \ \ \tilde{V}(\rho(F)/2,\tau) = g(\tau), \ \ \tau \in \mathbb{R}, \\ \tilde{V}(\xi,0) = \tilde{V}_0(\xi), \ \ \tilde{V}_\tau(\xi,0) = \tilde{V}_1(\xi), \ \ \xi \in (0,\rho(F)/2) \end{cases} \tag{2.9}$$

is exactly controllable at $\rho(F)$, i.e., for all $(\tilde{V}_0, \tilde{V}_1) \in L^2(0,\rho(F)/2) \times H^{-1}(0,\rho(F)/2)$, there exists $g \in L^2(0,\rho(F))$ such that $\tilde{V}(\xi,\tau) = 0$ for all $\tau \geq \rho(F)$.

Where, $g := V_\xi(\rho(F)/2,\tau)\chi_{(0,\rho(F))}(\tau)$.

So the transformed system

$$\begin{cases} \partial_t^2 u - \partial_x^2 u = 0, \qquad 0 < x < a(t),\ t \in \mathbb{R}, \\ u(0,t) = 0, \ \ u(a(t),t) = r(t), \ \ t \in \mathbb{R}, \\ u(x,0) = u_0(x), \ \ u_t(x,0) = u_1(x), \ \ x \in (0,a(0)) \end{cases}$$

is exactly controllable with a time of control $T := |e^{\frac{\rho(F)-h_2}{h_0}} - h_1|$ and a control $r(t)$ is given by $r(t) = g\left(\frac{H(a(t)+t)+H(-a(t)+t)}{2}\right)$. □

PROOF OF THEOREM 2.3. Using Φ given by (2.4), we transform the system (1.1)-(1.3) into:

$$\begin{cases} \partial_\tau^2 V - \partial_\xi^2 V = 0, \quad \text{for} \quad 0 < \xi < \rho(F)/2,\ \tau \in \mathbb{R}, \\ V(0,\tau) = 0,\ \ V_\xi(\rho(F)/2,\tau) + b(t(\tau))V_\tau(\rho(F)/2,\tau) = 0,\ \ \tau \in \mathbb{R}, \\ V(\xi,0) = \phi_2(\xi),\ \ V_\tau(\xi,0) = \psi_2(\xi),\ \ \xi \in (0,\omega/2). \end{cases} \tag{2.10}$$

For the proof of Theorem 2.3, we need the following lemmas.

LEMMA 2.10. *There exist positive constants C and ω such that*

$$E_V(\tau) \leq Ce^{-\omega\tau}E_V(0). \tag{2.11}$$

PROOF. Define the Lyapunov function:

$$E_1(\tau) = \frac{1}{2}\int_0^{\rho(F)} [V_\xi^2(\xi,\tau) + V_\tau^2(\xi,\tau)]d\xi + \delta\int_0^{\rho(F)} \xi V_\xi(\xi,\tau)V_\tau(\xi,\tau)d\xi.$$

We obtain for $\delta < \frac{1}{\rho(F)}$,

$$0 < (1-\delta\rho(F))E_V(\tau) \leq E_1(\tau) \leq (1+\delta\rho(F))E_V(\tau). \tag{2.12}$$

We derive E_1 with respect to τ, we get

$$\begin{aligned} &E_1'(\tau) \\ = \ &[V_\xi V_\tau]_{\xi=0}^{\xi=\rho(F)} - \frac{\delta}{2}\int_0^{\rho(F)} [V_\xi^2(\xi,\tau) + V_\tau^2(\xi,\tau)]d\xi + \frac{\delta}{2}[\xi(V_\xi^2 + V_\tau^2)]_{\xi=0}^{\xi=\rho(F)} \\ = \ &[\frac{\delta}{2}(1+b(t(\tau))^2) - b(t(\tau))]V_\tau^2(\rho(F),\tau) - \frac{\delta}{2}\int_0^{\rho(F)} [V_\xi^2(\xi,\tau) + V_\tau^2(\xi,\tau)]d\xi. \end{aligned}$$

We choose δ small enough, taking into account (1.17) and (2.12) we get

$$E_1'(\tau) \leq -\omega E_1(\tau).$$

The proof is complete. □

LEMMA 2.11. *If $T \geq \rho(F)$, then there exists $C(T) > 0$ such that for all $(\phi_2,\psi_2) \in H_l^1(0,\rho(F)/2) \times L^2(0,\rho(F)/2)$ we have*

$$C(T)\int_0^T |V_\xi(\rho(F)/2,\tau)|^2\, d\tau \geq \|\phi_2\|^2_{H_l^1(0,\rho(F)/2)} + \|\psi_2\|^2_{L^2(0,\rho(F)/2)}, \tag{2.13}$$

and

$$C(T)\int_0^T |V_\tau(\rho(F)/2,\tau)|^2\, d\tau \geq \|\phi_2\|^2_{H_l^1(0,\rho(F)/2)} + \|\psi_2\|^2_{L^2(0,\rho(F)/2)}. \tag{2.14}$$

PROOF. The energy identity for the system (2.10) gives :

$$E_V(T) - E_V(0) = -\int_o^T b(t(\tau))|V_\tau(\rho(F)/2,\tau)|^2 d\tau.$$

Using (1.17) and (2.11), we obtain

$$\begin{aligned}\int_0^T |V_\tau(\rho(F)/2,\tau)|^2\, d\tau &\geq C\int_0^T b(t(\tau))\,|V_\tau(\rho(F)/2,\tau)|^2\, d\tau\\ &\geq C(E_V(0)-E_V(T))\\ &\geq C\,E_V(0)\,(1-e^{-\omega T}).\end{aligned}$$

This permit to conclude the second inequality in Lemma 2.11.

For the first inequality, it suffices to use (2.10) and (1.17). □

For the proof of (2.2), we state as above:

$$\partial_t u = \partial_\xi V \partial_t \xi + \partial_\tau V \partial_t \tau.$$

Next we have:

$$\begin{aligned}|\partial_t u(a(t),t)|^2 &= \frac{1}{4}\{|\partial_\xi V(\rho(F)/2,\tau)[H'(x+t)-H'(-x+t)]\\ &+\ \partial_\tau V(\rho(F)/2,\tau)[H'(x+t)+H'(-x+t)]\}^2.\end{aligned}$$

Make use of Young inequalities, Lemma 2.8, (2.13), (2.14) and (1.16), we obtain the desired result. □

2.2. Observability of the wave equation with a moving interior point. Here, we consider the system :

$$(2.15)\qquad \begin{cases} u_{tt}-u_{xx}=0 \quad \text{for} \quad 0<x<1,\, t>0,\\ u(0,t)=0 \quad \text{and} \quad u(1,t)=0,\, t>0,\\ u(x,0)=\phi(x), u_t(x,0)=\psi(x),\ \ 0<x<1.\end{cases}$$

We prove the following observability estimate.

THEOREM 2.12. *Under the Assumption 1.3, there exist $T>0$ and a constant $C^*>0$ such that for all u solution of the system (2.15) with the Dirichlet boundary condition and initial data $(\phi,\psi)\in H_0^1(0,1)\times L^2(0,1)$, we have*

$$(2.16)\qquad \int_0^T |u_t(a(t),t)|^2\, dt \geq C^*\left(\|\phi\|^2_{H_0^1(0,1)}+\|\psi\|^2_{L^2(0,1)}\right).$$

REMARK 2.13. We mention that interior point observation in the case a is constant is not possible.

PROOF OF THEOREM 2.12. The next lemma will be very useful for the proof of Theorem 2.12.

LEMMA 2.14. *Denote by*

$$E_u(t)=\frac{1}{2}\int_0^1 \left[|u_t(x,t)|^2+|u_x(x,t)|^2\right] dx$$

the energy of the field u, and

$$E_V(\tau)=\int_0^{s(\tau)} \left(|V_\xi(\xi,\tau)|^2+|V_\tau(\xi,\tau)|^2\right) d\xi,$$

the energy of the field V. Under the Assumption 1.3, there are two positive constants C_1 and C_2 such that

(2.17)

$$C_1E_V\left(t+\frac{g(t+a(t))+g(t-a(t))}{2}\right)\leq E_u(t)\leq C_2E_V\left(t+\frac{g(t+a(t))+g(t-a(t))}{2}\right),$$

where g is a continuous and 1-periodic function.

PROOF OF LEMMA 2.14. We calculate,

$$\partial_t u = \partial_\xi V \partial_t \xi + \partial_\tau V \partial_t \tau \quad \text{and} \quad \partial_x u = \partial_\xi V \partial_x \xi + \partial_\tau V \partial_x \tau$$

and so,

$$\begin{aligned} E_u(t) &= \frac{1}{2} \int_0^1 \left[|u_t(x,t)|^2 + |u_x(x,t)|^2 \right] dx \\ &= \frac{1}{2} \int_0^1 \left\{ |V_\xi \xi_t + V_\tau \tau_t|^2 + |V_\xi \xi_x + V_\tau \tau_x|^2 \right\} dx. \end{aligned}$$

Make use of:

$$\xi_x = (\partial_x \xi) = (\partial_t \tau) = \tau_t = [H'(x+t) + H'(-x+t)]/2,$$

$$\xi_t = (\partial_t \xi) = (\partial_x \tau) = \tau_x = [H'(x+t) - H'(-x+t)]/2.$$

Hence,

$$\xi_t^2 + \xi_x^2 = \tau_t^2 + \tau_x^2 = \frac{1}{2} \left[|H'(x+t)|^2 + |H'(-x+t)|^2 \right],$$

$$\xi_t \tau_t = \xi_x \tau_x = \frac{1}{4} [(H'(x+t))^2 - (H'(-x+t))^2].$$

Back to the energy,

$$\begin{aligned} E_u(t) &= \frac{1}{4} \int_0^1 \left[|V_\xi|^2 + |V_\tau|^2 \right] \left[|H'(x+t)|^2 + |H'(-x+t)|^2 \right] dx \\ &\quad + \frac{1}{2} \int_0^1 [V_\xi V_\tau] \left[|H'(x+t)|^2 - |H'(-x+t)|^2 \right] dx \\ &= \frac{1}{4} \int_0^1 [|V_\xi|^2 + |V_\tau|^2 + 2 V_\xi V_\tau] |H'(x+t)|^2 dx \\ &\quad + \frac{1}{4} \int_0^1 [|V_\xi|^2 + |V_\tau|^2 - 2 V_\xi V_\tau] |H'(-x+t)|^2 dx. \end{aligned}$$

Also, differentiating $\xi = (H(x+t) - H(-x+t))/2$ on a slice $t =$constant, we obtain

$$d\xi = \frac{H'(x+t) + H'(-x+t)}{2} dx.$$

Thanks to Assumption 1.3 and the inequality $|V_\xi V_\tau| \leq 1/2 \left(V_\xi^2 + V_\tau^2 \right)$ we get:

$$C_1 E_V(\tau) \leq E_u(t) \leq C_2 E_V(\tau), \tag{2.18}$$

for positive constants C_1, C_2. □

Applying the transformation Φ, the system (1.6) becomes:

$$\begin{cases} \partial_\tau^2 V - \partial_\xi^2 V = 0, & \text{for} \quad 0 < \xi < s(t),\ \tau \in \mathbb{R}, \\ V(0,\tau) = 0, \ V(s(t),\tau) = 0, & \tau \in \mathbb{R}, \\ V(\xi,0) = \phi_1(\xi), \ V_\tau(\xi,0) = \psi_1(\xi), & \xi \in (0, s(t)). \end{cases} \tag{2.19}$$

We denote $\alpha(t) := t + s(t)$ and $\beta(t) := t - s(t)$. These functions are both strictly increasing bijections from $\mathbb{R}_+$ to $[\pm s(0), \infty)$ respectively. We will also consider $\gamma = \alpha \circ \beta^{-1} : [-s(0), \infty) \to [s(0), \infty)$.

We develop the solution V of the system (2.19) into a series of the form

$$V(x,t) := \sum_{n \in \mathbb{Z}} A_n \left(e^{2i\pi n \varphi(t+x)} - e^{2i\pi n \varphi(t-x)} \right), \tag{2.20}$$

where the coefficients A_n are given by the initial data (ϕ_1, ψ_1).

In order to satisfy the Dirichlet boundary condition, we need a solution φ to the functional equation

$$\varphi(t+s(t)) - \varphi(t-s(t)) = 1. \tag{2.21}$$

We need the following theorem which is the exact observability estimates of the 1-d wave equation on a non-cylindrical domain (see also [**20**] for a particular case).

THEOREM 2.15. ([**13**]) *Let s be an monotonic admissible boundary curve and φ be a C^2-solution to (2.21). Assume additionally that φ' is strictly decreasing if $s(.)$ is increasing or that φ' is increasing if $s(.)$ is decreasing, respectively.*

The solution V to the wave equation (2.19) satisfies the following double inequalities:

$$C_1(a)E_V(0) \leq \int_0^{\rho(F)/2+\gamma(-\rho(F)/2)} |V_\xi(\rho(F)/2,\tau)|^2 d\tau \leq C_2(a)E_V(0), \tag{2.22}$$

$$C_1(a)E_V(0) \leq \int_0^{\rho(F)/2+\gamma(-\rho(F)/2)} |V_\tau(\rho(F)/2,\tau)|^2 d\tau \leq C_2(a)E_V(0), \tag{2.23}$$

where the constants C_1 and C_2 depend only on a.

REMARK 2.16. Note that, if we consider a is the constant function, then $s(t)$ will not be monotonic.

We consider (2.15) and state:

$$\partial_t u = \partial_\xi V \partial_t \xi + \partial_\tau V \partial_t \tau.$$

Thanks to (2.4) we get

$$\begin{aligned}
|\partial_t u(a(t),t)|^2 &= |\partial_\xi V(\rho(F)/2,\tau)|^2 |\partial_t \xi|^2 + |\partial_\tau V(\rho(F)/2,\tau)|^2 |\partial_t \tau|^2 \\
(2.24) \qquad &+ 2\partial_\xi V(\rho(F)/2,\tau)\partial_\tau V(\rho(F)/2,\tau)\partial_t\xi\partial_t\tau \\
&= \frac{|H'(a(t)+t)|^2}{4}[|\partial_\xi V(\rho(F)/2,\tau)|^2 + |\partial_\tau V(\rho(F)/2,\tau)|^2 \\
&+ 2\partial_\xi V(\rho(F)/2,\tau)\partial_\tau V(\rho(F)/2,\tau)] \\
&+ \frac{|H'(-a(t)+t)|^2}{4}[|\partial_\xi V(\rho(F)/2,\tau)|^2 + |\partial_\tau V(\rho(F)/2,\tau)|^2 \\
&- 2\partial_\xi V(\rho(F)/2,\tau)\partial_\tau V(\rho(F)/2,\tau)] \\
&+ \frac{H'(a(t)+t)H'(-a(t)+t)}{2}[-|\partial_\xi V(\rho(F)/2,\tau)|^2 + |\partial_\tau V(\rho(F)/2,\tau)|^2].
\end{aligned}$$

Differentiating $\tau = (H(x+t)+H(-x+t))/2$ on a slice x =constant, we obtain

$$d\tau = \frac{H'(x+t)+H'(-x+t)}{2}\, dt. \tag{2.25}$$

Make use of the inequality $2(a+b)^2 \geq a^2+b^2$, (2.17), (2.22), (2.23) and (2.25) we get:

$$\int_0^T |u_t(a(t),t)|^2 \, dt \geq C^* \left(\|\phi\|^2_{H_0^1(0,1)} + \|\psi\|^2_{L^2(0,1)} \right), \tag{2.26}$$

with $T = \Phi^{-1}\left(\rho(F)/2 + \gamma(-\rho(F)/2)\right)$. □

3. Stabilization of the wave equation with moving boundary

In this section, we analyze the stabilization property of solutions for the wave equation with a moving boundary. The mechanism of the problem was discovered by Dittrich, Duclos and Gonzalez [**12**] where they used a direct estimates on the energy. A detailed survey is given in Gonzalez [**17**]. Here, we consider the following system:

$$\left\{\begin{array}{l} u_{tt}-u_{xx}=0 \quad \text{for} \quad 0<x<a(t),\, t>0, \\ u(0,t)=0 \quad \text{and} \quad u_t(a(t),t)+f(t)u_x(a(t),t)=0,\, t>0, \\ u(x,0)=\phi(x), u_t(x,0)=\psi(x), \;\; 0<x<a(0), \end{array}\right. \tag{3.1}$$

$(\phi,\psi)\in H^1_l((0,a(0)))\times L^2((0,a(0)))$, where

$$H^1_l((0,a(0)))=\left\{v\in H^1((0,a(0))),\, v(0)=0\right\}.$$

Here a is a strictly positive real function which is 1-periodic (i.e., $a(t+1)=a(t), \forall\, t>0$) and $f\in L^\infty\left(\mathbb{R}^*_+\right)$.

Denote by

$$E_u(t)=\frac{1}{2}\int_0^{a(t)}\left[\left|u_t(x,t)\right|^2+\left|u_x(x,t)\right|^2\right]dx$$

the energy of the field u. Our major concern will be to detect the feedback $f(t)$ necessary to obtain the exponential decay of $E_u(t)$.

On the existence of solutions to system (3.1), we refer the reader to [**17**].

Here we prove the exponential stability for the solutions of (3.1).

THEOREM 3.1 (Exponential stability). *Let*

$$f(t)=\frac{(\mu-1)H'(a(t)+t)+(\mu+1)H'(-a(t)+t)}{(1-\mu)H'(a(t)+t)+(\mu+1)H'(-a(t)+t)} \tag{3.2}$$

where μ is a nonnegative constant and assume that there exist $\lambda_1>0$ and $\lambda_2>0$ such that

$$\lambda_1\le H'(t)\le\lambda_2,\; t\in\mathbb{R}. \tag{3.3}$$

Then, in the case where $\mu\neq 1$, there exists a positive constant C such that

$$E_u(t)\le C\,e^{-\omega t}E_u(0), \tag{3.4}$$

for every solution u of (3.1) with initial data $(\phi,\psi)\in H^1_l((0,a(0)))\times L^2((0,a(0)))$ and where $\omega=\ln\left(\left|\frac{1+\mu}{1-\mu}\right|\right)$.

In the case $\mu=1$ which corresponds to $f(t)=1$, we obtain

$$E_u(t)=0,\; \text{for all } t\ge T_0=:\left(I+a\right)^{-1}\circ H^{-1}\left(\frac{3\rho(F)}{2}\right). \tag{3.5}$$

PROOF OF THEOREM 3.1. Now we consider the system:

$$\left\{\begin{array}{l} u_{tt}-u_{xx}=0 \quad \text{for} \quad 0<x<a(t),\, t>0, \\ u(0,t)=0 \quad \text{and} \quad u_t(a(t),t)+f(t)\,u_x(a(t),t)=0,\, t>0, \\ u(x,0)=\phi_1(x), u_t(x,0)=\psi_1(x), \;\; 0<x<a(0) \end{array}\right. \tag{3.6}$$

where $f(t)=\dfrac{(\mu-1)H'(a(t)+t)+(\mu+1)H'(-a(t)+t)}{(1-\mu)H'(a(t)+t)+(\mu+1)H'(-a(t)+t)}$ and μ is a nonnegative constant.

PROPOSITION 3.2. *The transformation of system* (3.6) *is*

$$\begin{cases} v_{\tau\tau} - v_{\xi\xi} = 0 \quad \textit{for} \quad 0 < \xi < \rho(F)/2, \, \tau > 0, \\ v(0, \tau) = 0 \quad \textit{and} \quad v_\tau(\rho(F)/2, \tau) + \mu v_\xi(\rho(F)/2, \tau) = 0, \, \tau > 0, \\ v(\xi, 0) = \phi_1(\xi), v_\tau(\xi, 0) = \psi_1(\xi), \ \ 0 < \xi < \rho(F)/2). \end{cases} \tag{3.7}$$

PROOF. We have:

$$\begin{cases} u_x(a(t), t) = V_\xi(\rho(F)/2, \tau)\xi_x(a(t), t) + V_\tau(\rho(F)/2, \tau)\tau_x(a(t), t), \\ u_t(a(t), t) = V_\xi(\rho(F)/2, \tau)\xi_t(a(t), t) + V_\tau(\rho(F)/2, \tau)\tau_t(a(t), t). \end{cases} \tag{3.8}$$

Starting from (3.6) and make use of:

$$\xi_x = (\partial_x \xi) = (\partial_t \tau) = \tau_t = [H'(x + t) + H'(-x + t)]/2,$$

$$\xi_t = (\partial_t \xi) = (\partial_x \tau) = \tau_x = [H'(x + t) - H'(-x + t)]/2,$$

we conclude:

$$u_t(a(t), t) + \frac{\xi_t - \mu\tau_t}{\mu\tau_x - \xi_x} u_x(a(t), t) = 0.$$

So we have

$$\xi_t u_x(a(t), t) - \xi_x u_t(a(t), t) + \mu(\tau_x u_t(a(t), t) - \tau_t u_x(a(t), t)) = 0.$$

Finally we get: $v_\tau(\rho(F)/2, \tau) + \mu v_\xi(\rho(F)/2, \tau) = 0$.

Note that $f(t) = 1$ in the special case $\mu = 1$. □

The stabilization of system (3.1) is a direct combination of Proposition 3.2, Lemma 2.8 and the following Lemma 3.3.

LEMMA 3.3. *For* $\mu \neq 1$, *there exists a positive constant* C *such that*

$$E_V(\tau) \leq C e^{-\ln(|\frac{1+\mu}{1-\mu}|)\tau} E_V(0), \, \forall \tau > 0, \tag{3.9}$$

where V *is the solution of the following system:*

$$\begin{cases} V_{\tau\tau} - V_{\xi\xi} = 0 \quad \textit{for} \quad 0 < \tau < \rho(F)/2, \, \tau > 0, \\ V(0, \tau) = 0 \quad \textit{and} \quad V_\tau(\rho(F)/2, \tau) + \mu V_\xi(\rho(F)/2, \tau) = 0, \, \tau > 0, \\ V(\xi, 0) = \phi_1(\xi), V_\tau(\xi, 0) = \psi_1(\xi), \ \ 0 < \xi < \rho(F)/2. \end{cases} \tag{3.10}$$

It is well known that for $\mu = 1$, $E_V(\tau) = 0$ *for all* $\tau \geq \rho(F)$, *see* **[10]** *for more details.* □

4. Moving pointwise stabilization

We consider the following problem:

$$\begin{cases} u_{tt} - u_{xx} + [f_1(t)u_t + f_2(t)u_x]\delta_{a(t)} = 0 \quad \text{for} \quad 0 < x < b(t), \, t > 0, \\ u(0, t) = 0 \quad \text{and} \quad u(b(t), t) = 0, \, t > 0, \\ u(x, 0) = \phi_1(x), u_t(x, 0) = \psi_1(x) \ \ 0 < x < b(0), \end{cases} \tag{4.1}$$

where $\delta_{a(t)}$ denotes the Dirac mass concentrated in the point $a(t)$.

The aim of this section is to determine the functions f_1, f_2 and b to get after transformation the vibrations of a string with the static pointwise damping and conclude the asymptotic behavior of the energy.

PROPOSITION 4.1. *The transformation of the system:*

$$
\left\{
\begin{array}{l}
u_{tt} - u_{xx} + K\left(\frac{H(a(t)+t)-H(-a(t)+t)}{2}, \frac{H(a(t)+t)+H(-a(t)+t)}{2}\right) \\
\cdot\left[(\frac{1}{H'(a(t)+t)} - \frac{1}{H'(-a(t)+t)})u_t + (\frac{1}{H'(a(t)+t)} + \frac{1}{H'(-a(t)+t)})u_x\right]\delta_{a(t)} = 0 \\
\quad for \quad 0 < x < b(t) = \Lambda_t^{-1}(1) - t, \, t > 0, \\
u(0,t) = 0 \quad and \quad u(b(t),t) = 0, \, t > 0, \\
u(x,0) = \phi_1(x), u_t(x,0) = \psi_1(x), \;\; 0 < x < b(0)
\end{array}
\right. \tag{4.2}
$$

is

$$
\left\{
\begin{array}{l}
v_{\tau\tau} - v_{\xi\xi} + v_\tau \delta_{\frac{\rho(F)}{2}} = 0 \quad for \quad 0 < \xi < 1, \, \tau > 0, \\
v(0,\tau) = 0 \quad and \quad v(1,\tau) = 0, \, \tau > 0, \\
v(\xi,0) = \phi(\xi), v_\tau(\xi,0) = \psi(\xi), \;\; 0 < \xi < 1,
\end{array}
\right. \tag{4.3}
$$

where Λ_t *is defined by* $\Lambda_t(y) = \frac{H(y+t)-H(-y+t)}{2}$.

PROOF. We recall that if $u(x,t)$ satisfies $(\partial_t^2 - \partial_x^2)u(x,t) = 0$ and $v(\xi,\tau)$ is defined by $u(\Phi^{-1}(\xi,\tau))$. Then the following identity holds

$$
(\partial_t^2 - \partial_x^2)u(x,t) = K(\xi,\tau)(\partial_\tau^2 - \partial_\xi^2)v(\xi,\tau) \tag{4.4}
$$

where $K(\xi,\tau)$ is defined by

$$
4H' \circ H^{-1}(\xi+\tau)H' \circ H^{-1}(-\xi+\tau) \circ H^{-1}(\xi+\tau).
$$

On the other hand, make use of (3.8) we obtain:

$$
\begin{array}{rcl}
v_\tau & = & \frac{\xi_t u_x - \xi_x u_t}{\xi_t^2 - \xi_x^2} \\
& = & (\frac{1}{H'(a(t)+t)} - \frac{1}{H'(-a(t)+t)})u_t + (\frac{1}{H'(a(t)+t)} + \frac{1}{H'(-a(t)+t)})u_x.
\end{array} \tag{4.5}
$$

We combine (4.4) and (4.5) to obtain the result of the Proposition 4.1. □

REMARK 4.2. If we return to the Example 1.14 and after some computation we get that the transformation of the system:

$$
\left\{
\begin{array}{l}
u_{tt} - u_{xx} + \left(\frac{8}{a(t)+t+h_1} \cdot \frac{a(t)+t}{-a(t)+t+h_1}\right) \cdot [a(t)u_t + (t+h_1)u_x]\,\delta_{a(t)} = 0 \\
\quad \text{for} \quad 0 < x < b(t) = (t+h_1)e^{\frac{h_2}{h_0}}\tanh(\frac{1}{h_0}), \, t > 0, \\
u(0,t) = 0 \quad \text{and} \quad u(b(t),t) = 0, \, t > 0, \\
u(x,0) = \phi_1(x), u_t(x,0) = \psi_1(x), \;\; 0 < x < b(0),
\end{array}
\right. \tag{4.6}
$$

is system (4.3).

As above and according to [**3**, **4**] we have the following:

- $\lim\limits_{t\to+\infty} E_u(t) = 0, \, \forall\, (\phi_1,\psi_1) \in H_0^1(0,b(0)) \times L^2(0,b(0)) \Leftrightarrow \rho(F) \notin \mathbb{Q}$.
- For any $\rho(F)/2 \in (0,1)$, system (4.2) is not exponentially stable in $H_0^1(0,b(0)) \times L^2(0,b(0))$.

  Denote by $\mathcal{S}$ the set of all numbers $\xi \in (0,1)$ such that $\xi \notin \mathbb{Q}$ and if $[0, a_1, ..., a_n, ...]$ is the expansion of ξ as a continued fraction, then (a_n) is bounded. Let us notice that $\mathcal{S}$ is obviously uncountable and, by classical results on diophantine approximation (cf. [**8**], p. 120), its Lebesgue measure is equal to zero. In particular, by Euler-Lagrange theorem (see [**19**], p. 57) $\mathcal{S}$ contains all $\xi \in (0,1)$ such that ξ is an irrational quadratic number (i.e. satisfying a second degree equation with rational coefficients).

- For all $\rho(F)/2 \in \mathcal{S}$ and for all $(\phi_1, \psi_1) \in \mathcal{D}$ we have according to [**5,6**] that there exists $C > 0$ such that:

$$E_u(t) \leq \frac{C}{t}, \forall t > 0.$$

- If $\varepsilon > 0$ then, for almost all $\rho(F)/2 \in (0,1)$ and for all $(\phi_1, \psi_1) \in \mathcal{D}$ we have according to [**5,6**] that there exists $C > 0$ such that:

$$E_u(t) \leq \frac{C}{t^{\frac{1}{1+\varepsilon}}}, \forall t > 0,$$

where

$$\mathcal{D} := \Big\{(\varphi, \psi) \in [H^2(0, b(0)) \cap H_0^1(0, b(0))] \times H_0^1(0, b(0)),$$

$$\left(\frac{1}{H'(a(0))} - \frac{1}{H'(-a(0))}\right) \psi(a(0)) + \left(\frac{1}{H'(a(0))} + \frac{1}{H'(-a(0))}\right) \frac{d\varphi}{dx}(a(0)) = 0\Big\}.$$

5. Further comments: The quasi-periodic case

One can try to generalize the previous results to the case when a is no longer periodic but has some sort of quasiperiodicity [1].

The problem is much more complicated, since there is no rotation number. However, in [**21**] the author uses a weaker notion of upper (resp. lower) rotation number of F at every point x as follows:

$$\overline{\rho}(F) = \limsup_{n \to +\infty} \frac{F^n(x) - x}{n}$$

$$(\text{resp. } \underline{\rho}(F) = \liminf_{n \to +\infty} \frac{F^n(x) - x}{n}).$$

As a consequence, it is shown that under the same Diophantine condition (see [**8,19**]) satisfied by $\overline{\rho}(F)$ (resp. $\underline{\rho}(F)$), the rotation number of F exists and coincides with the lower (resp. upper) rotation number.

LEMMA 5.1. *Assume that $a(t)$ is an $\eta-$q.p function, $\hat{a}(\theta)$ is real analytic and satisfy $|\hat{a'}(\theta)| < 1$ for $\eta, \theta \in \mathbb{R}^m$ and set $\beta = (\frac{2\pi}{\eta_1}, ..., \frac{2\pi}{\eta_m})$. Assume also that there exists $C_0 > 0$ depending on β such that $|(k, \beta) + \pi l/\overline{\rho}(F)| > \frac{C_0}{|k|^{m+1}}$. Then, there exists a real analytic function $H(\xi) = \xi + h(\xi)$, where $h(\xi)$ is an η-q.p. function, such that*

$$H^{-1} \circ F \circ H(\xi) = \xi + \overline{\rho}(F). \tag{5.1}$$

REMARK 5.2. Thanks to Lemma 5.1, all the privious results can be extended by similar arguments.

---

[1] A function $a(t), t \in \mathbb{R}$ is called quasi-periodic with basic frequencies $\omega = (\omega_1, ..., \omega_m) \in \mathbb{R}^m$ (briefly $2\pi/\omega-$q.p.) if there exists a continuous function $\hat{a}(\theta), \theta = (\theta_1, ..., \theta_m) \in \mathbb{R}^m$ that is 2π-periodic in each $\theta_i, i = 1, ..., m$ such that $a(t) = \hat{a}(\omega t)$ holds. $\hat{a}(\theta)$, is called the corresponding function and $\frac{2\pi}{\omega} = \left(\frac{2\pi}{\omega_1}, ..., \frac{2\pi}{\omega_m}\right)$ is called the basic periods of a.

References

[1] K. Ammari, A. Bchatnia, and K. El Mufti, *A remark on observability of the wave equation with moving boundary*, J. Appl. Anal. **23** (2017), no. 1, 43–51, DOI 10.1515/jaa-2017-0007. MR3657928

[2] K. Ammari, A. Bchatnia, and K. El Mufti, *Stabilization of the wave equation with moving boundary*, Eur. J. Control **39** (2018), 35–38, DOI 10.1016/j.ejcon.2017.10.004. MR3747447

[3] K. Ammari, A. Henrot, and M. Tucsnak, *Optimal location of the actuator for the pointwise stabilization of a string* (English, with English and French summaries), C. R. Acad. Sci. Paris Sér. I Math. **330** (2000), no. 4, 275–280, DOI 10.1016/S0764-4442(00)00113-0. MR1753293

[4] K. Ammari, A. Henrot, and M. Tucsnak, *Asymptotic behaviour of the solutions and optimal location of the actuator for the pointwise stabilization of a string*, Asymptot. Anal. **28** (2001), no. 3-4, 215–240. MR1878795

[5] K. Ammari and S. Nicaise, *Stabilization of elastic systems by collocated feedback*, Lecture Notes in Mathematics, vol. 2124, Springer, Cham, 2015. MR3308343

[6] K. Ammari and M. Tucsnak, *Stabilization of second order evolution equations by a class of unbounded feedbacks*, ESAIM Control Optim. Calc. Var. **6** (2001), 361–386, DOI 10.1051/cocv:2001114. MR1836048

[7] C. Bardos, G. Lebeau, and J. Rauch, *Sharp sufficient conditions for the observation, control, and stabilization of waves from the boundary*, SIAM J. Control Optim. **30** (1992), no. 5, 1024–1065, DOI 10.1137/0330055. MR1178650

[8] J. W. S. Cassels, *An introduction to Diophantine approximation*, Cambridge Tracts in Mathematics and Mathematical Physics, No. 45, Cambridge University Press, New York, 1957. MR0087708

[9] C. Castro, *Exact controllability of the 1-D wave equation from a moving interior point*, ESAIM Control Optim. Calc. Var. **19** (2013), no. 1, 301–316, DOI 10.1051/cocv/2012009. MR3023072

[10] S. Cox and E. Zuazua, *The rate at which energy decays in a damped string*, Comm. Partial Differential Equations **19** (1994), no. 1-2, 213–243, DOI 10.1080/03605309408821015. MR1257004

[11] R. Dáger and E. Zuazua, *Wave propagation, observation and control in 1-d flexible multi-structures*, Mathématiques & Applications (Berlin) [Mathematics & Applications], vol. 50, Springer-Verlag, Berlin, 2006. MR2169126

[12] J. Dittrich, P. Duclos, and N. Gonzalez, *Stability and instability of the wave equation solutions in a pulsating domain*, Rev. Math. Phys. **10** (1998), no. 7, 925–962, DOI 10.1142/S0129055X98000306. MR1653737

[13] B. H. Haak and D.-T. Hoang, *Exact observability of a 1-dimensional wave equation on a noncylindrical domain*, SIAM J. Control Optim. **57** (2019), no. 1, 570–589, DOI 10.1137/17M112960X. MR3911710

[14] V. Komornik and P. Loreti, *Fourier series in control theory*, Springer Monographs in Mathematics, Springer-Verlag, New York, 2005. MR2114325

[15] J. Le Rousseau, G. Lebeau, P. Terpolilli, and E. Trélat, *Geometric control condition for the wave equation with a time-dependent observation domain*, Anal. PDE **10** (2017), no. 4, 983–1015, DOI 10.2140/apde.2017.10.983. MR3649373

[16] N. Gonzalez, *An example of pure stability for the wave equation with moving boundary*, J. Math. Anal. Appl. **228** (1998), no. 1, 51–59, DOI 10.1006/jmaa.1998.6113. MR1659940

[17] Nicolas Gonzalez, *L'équation des ondes dans un domaine dépendant du temps*, Ph.D. Thesis, University of Toulon and Czech Technical University, 1997.

[18] M.-R. Herman, *Sur la conjugaison différentiable des difféomorphismes du cercle à des rotations* (French), Inst. Hautes Études Sci. Publ. Math. **49** (1979), 5–233. MR538680

[19] S. Lang, *Introduction to Diophantine approximations*, 2nd ed., Springer-Verlag, New York, 1995. MR1348400

[20] A. Sengouga, *Observability and controllability of the 1-D wave equation in domains with moving boundary*, Acta Appl. Math. **157** (2018), 117–128, DOI 10.1007/s10440-018-0166-1. MR3850022

[21] M. Yamaguchi, *One dimensional wave equations in domain with quasiperiodically moving boundaries and quasiperiodic dynamical systems*, J. Math. Kyoto Univ. **45** (2005), no. 1, 57–97, DOI 10.1215/kjm/1250282968. MR2138801

[22] M. Yamaguchi, *Periodic solutions of nonlinear equations of string with periodically oscillating boundaries*, Funkcial. Ekvac. **45** (2002), no. 3, 397–416. MR1975054

[23] M. Yamaguchi and H. Yoshida, *Nonhomogeneous string problem with periodically moving boundaries*, Operator theory and its applications (Winnipeg, MB, 1998), Fields Inst. Commun., vol. 25, Amer. Math. Soc., Providence, RI, 2000, pp. 565–574. MR1759569

UR Analysis and Control of PDEs, UR13ES64, Department of Mathematics, Faculty of Sciences of Monastir, University of Monastir, 5019 Monastir, Tunisia
Email address: kais.ammari@fsm.rnu.tn

UR Analyse Non-Linéaire et Géométrie, UR13ES32, Department of Mathematics, Faculty of Sciences of Tunis, University of Tunis El Manar, Tunisia
Email address: ahmed.bchatnia@fst.utm.tn

UR Analysis and Control of PDEs, UR13ES64, ISCAE, University of Manouba, Tunisia
Email address: karim.elmufti@iscae.rnu.tn

Contemporary Mathematics
Volume **757**, 2020
https://doi.org/10.1090/conm/757/15247

Sharp estimates of the one-dimensional boundary control cost for parabolic systems

Manuel González-Burgos and Lydia Ouaili

ABSTRACT. In this work we present new results on the cost of the boundary controllability of parabolic systems at time $T > 0$. In particular, we will study optimal estimates of the control cost at time T (T small enough) when the eigenvalues of the generator of the C_0 semigroup accumulate and do not satisfy a gap condition. The main ingredient we will use is the moment method combined with sharp estimates of the $L^2(0, T; \mathbb{C})$-norm of the elements of biorthogonal families to complex exponentials.

CONTENTS

1. Introduction

In the last years, the controllability of non scalar systems with less controls than equations has attracted the attention of many authors. In the particular case of coupled parabolic systems, one of the main problems is obtaining necessary and sufficient conditions that allow the system to be controlled with a reduced number of distributed or boundary controls (see [**1**], [**12**], [**2**], [**20**],...).

Another important problem is the study of the dependence of the so-called control cost with respect to the final observation time $T > 0$ when T is small enough and the corresponding null controllability result holds at time $T > 0$. Regarding this latter problem, we highlight the works [**10**], [**11**], [**26**], [**16**], [**17**], [**13**], [**21**], [**28**], [**8**], [**19**], etc., where the authors study an estimate of the control cost $\mathcal{K}(T)$ in

2010 *Mathematics Subject Classification.* Primary 93B05, 93B07.

The first author was supported by grant MTM2016-76990-P, Ministry of Economy and Competitiveness (Spain).

the case of scalar parabolic problems (problems that, under general assumptions, are null controllable for any $T > 0$). Most of the previous works uses the moment method in order to obtain an estimate of the control cost following the strategy:

(1) Given $\{\lambda_k\}_{k\geq 1} \subset \mathbb{C}$ the sequence of eigenvalues of the corresponding generator of the semigroup, the first objective is to prove sharp bounds on the biorthogonal family to the real or complex exponentials $\Lambda = \{e^{-\lambda_k t}\}_{k\geq 1}$ in $L^2(0,T;\mathbb{C})$. To this end, in the previous works, among other assumptions, a gap condition like

$$\inf_{k\neq n} |\lambda_k - \lambda_n| \geq c_0 > 0. \tag{1}$$

is assumed.

(2) The assumptions on the sequence Λ guarantee that the scalar parabolic system under study is null controllable at any positive time T. Combining the moment method and the estimates on the biorthogonal family, we deduce that the following property holds: there exist positive constants τ_0, C_0 and C_1 such that the control cost of the considered problem satisfies

$$\exp\left(\frac{C_0}{T}\right) \leq \mathcal{K}(T) \leq \exp\left(\frac{C_1}{T}\right), \quad \forall T \in (0,\tau_0).$$

The work [**19**] is of special relevance because in it, the author studies the cost of the controllability of the one-dimensional heat equation with a pointwise control at point x_0 and, in this framework, there might exist a positive minimal time of null-controllability $T_0 \in [0,\infty]$ (which depends on x_0 and could take any arbitrary value in $[0,\infty]$, see [**9**]). In this work, in particular, the author proves that, if $T_0 > 0$, the cost of the controllability at time $T > T_0$ when T is close to T_0, may explode in an arbitrary way.

The analysis of the control cost in the framework of the controllability of coupled parabolic systems has been addressed in [**5**]. Again, the authors use the moment method and establish sharp bounds on biorthogonal families to complex matrix exponentials associated to complex sequences $\Lambda = \{\lambda_k\}_{k\geq 1}$. As in the previous works, the authors impose appropriate assumptions on the sequence Λ which include a gap condition on the terms of the sequence. These hypotheses also assure that the system under consideration is null controllable at any time $T > 0$ and, as in the scalar case, provide an estimate for $\mathcal{K}(T)$.

However, new phenomena associated with the vectorial nature of the controllability problem arise within this framework (hyperbolic phenomena): minimal time of null controllability and dependence of the controllability result on the position of the control domain (see [**3**], [**4**], [**24**], [**22**],...). This minimal time may come from the control action itself (as in [**9**]) or from the condensation index of the sequence of eigenvalues of the generator of the semigroup (see [**3**]). Let us provide more details in the case of systems with a minimal time which comes from the condensation index of the sequence. In [**3**], the authors prove a general result of null controllability for abstract parabolic problems that develop a minimal time $T_0 \in [0,\infty]$ of controllability: the system is null-controllable at any time $T > T_0$ and not null-controllable for $T < T_0$. This minimal time is related to the Bernstein's condensation index of the sequence of eigenvalues $\Lambda = \{\lambda_k\}_{k\geq 1}$ of the generator of the semigroup (see [**3**] and [**6**] for further details). In particular, the sequence Λ does not satisfy, in general, the previous gap condition (1) and obtaining appropriate estimates of the control cost $\mathcal{K}(T)$ for $T > T_0$ is an open problem.

In this work we have two main objectives:

(1) First, we want to give a short overview on bounds on biorthogonal families to real or complex exponentials $\{e^{-\lambda_k t}\}_{k\geq 1}$ in $L^2(0,T;\mathbb{C})$ where the sequence $\Lambda = \{\lambda_k\}_{k\geq 1} \subset \mathbb{C}$ satisfies, among other hypotheses, a gap condition such as (1). Let us remark that the estimate of the norm of the elements of the biorthogonal family is crucial in order to obtain the corresponding estimate of the control cost for the system under study.

(2) We also want to advance new results which provide partial answers to the following questions: Given a complex sequence $\{\lambda_k\}_{k\geq 1}$ satisfying appropriate assumptions (see (1)–(4), (6) and (25), for an integer $q \geq 1$, below),

  (a) Is there a biorthogonal family $\{q_k\}_{k\geq 0}$ to $\{e^{-\lambda_k t}\}_{k\geq 1}$ in $L^2(0,T;\mathbb{C})$ satisfying an appropriate estimate for $\|q_k\|_{L^2(0,T;\mathbb{C})}$ (see Theorem 5.1)?

  (b) Using the moment method, is it possible to obtain estimates (from above and from below) of the control cost $\mathcal{K}(T)$, for $T > T_0$? ($T_0 \in [0,\infty]$ is the minimal time associated to the system).

  Observe that the kind of sequences under consideration satisfies a gap condition weaker than (1). Indeed, from (25) we deduce the existence of a positive constant c (depending on ρ and q) such that

$$|\lambda_k - \lambda_n| \geq c > 0, \quad \forall k,n : |k-n| \geq q. \tag{2}$$

  In particular, the sequences we are considering in this work could satisfy

$$\inf_{k\neq n} |\lambda_k - \lambda_n| = 0.$$

  As said before, when this happens, the null controllability problem for the system in study can have a minimal time T_0.

  A similar weak gap condition as (2) has been also considered in [**6**]. With this assumption the authors develop a block moment method to handle spectral condensation phenomenon in parabolic control problems.

The plan of the paper is the following: In Section 2, we will address the null controllability problem for the one-dimensional heat equation and we present the moment method. In Section 3 we analyze a boundary controllability problem for a non-scalar parabolic system with complex eigenvalues. For this system, we obtain the same estimate for the control cost $\mathcal{K}(T)$ as in the scalar case. In Section 4 we study the controllability problem for the one-dimensional phase-field system with boundary controls. This is a simple example for which the eigenvalues of the corresponding generator do not satisfy the gap condition (1). Section 5 is devoted to establishing the main result of this work: sharp estimates of the norm of biorthogonal families to complex exponentials without gap conditions. Finally, in Section 6 we apply the main result to the phase-field system and to a non-scalar parabolic system whose generator does not satisfy (1). In both cases, estimates for the control cost $\mathcal{K}(T)$ are established.

2. The one-dimensional heat equation: The moment method

Let us first present the control cost in the simplest case: the boundary null controllability problem for the one-dimensional heat equation

$$\text{(3)} \qquad \begin{cases} y_t - y_{xx} = 0 & \text{in } Q_T = (0,\pi)\times(0,T), \\ y(0,\cdot) = v, \quad y(\pi,\cdot) = 0 & \text{on } (0,T), \\ y(\cdot,0) = y_0 & \text{in } (0,\pi), \end{cases}$$

where $y_0 \in H^{-1}(0,\pi)$ is given and $v \in L^2(0,T)$ is the control function. It is well-known that system (3) is well-posed and the solution $y \in L^2(Q_T)$ (defined by transposition) depends continuously on the data y_0 and v. In fact, the solution y of (3) is related to the solutions of the corresponding adjoint problem to (3)

$$\text{(4)} \qquad \begin{cases} -\varphi_t - \varphi_{xx} = 0 & \text{in } Q_T, \\ \varphi(0,\cdot) = \varphi(\pi,\cdot) = 0 & \text{on } (0,T), \\ \varphi(\cdot,T) = \varphi_T & \text{in } (0,\pi), \end{cases}$$

as follows:

THEOREM 2.1. *Let us consider problems* (3) *and* (4). *Then, for any* $y_0 \in H^{-1}(0,\pi)$ *and* $\varphi_T \in H_0^1(0,\pi)$, *one has*

$$\text{(5)} \qquad \int_0^T \varphi_x(0,t)v(t)\,dt = \langle y(\cdot,T),\varphi_T\rangle - \langle y_0,\varphi(\cdot,0)\rangle,$$

where $y \in L^2(Q_T)\cap C^0([0,T];H^{-1}(0,\pi))$ *and*

$$\varphi \in L^2(0,T;H^2(0,\pi)\cap H_0^1(0,\pi))\cap C^0([0,T];H_0^1(0,\pi))$$

are, resp., the solutions to (3) *and* (4) *associated to* y_0 *and* v, *and* φ_T, *and* $\langle\cdot,\cdot\rangle$ *stands for the duality product between* $H^{-1}(0,\pi)$ *and* $H_0^1(0,\pi)$.

The null controllability result for system (3) was obtained in [**10**] using the moment method. This method has been successfully used to prove the boundary null controllability result for general one-dimensional scalar parabolic equations with coefficients independent of t. Let us briefly recall this method in the case of system (3).

It is well-known that the operator $-\partial_{xx}$ on $(0,\pi)$ with homogenous Dirichlet boundary conditions admits a sequence of eigenvalues and normalized eigenfunctions given by

$$\text{(6)} \qquad \lambda_k = k^2, \quad \phi_k(x) = \sqrt{\frac{2}{\pi}}\sin kx, \quad k\ge 1, \quad x\in(0,\pi)$$

which is a Hilbert basis of $L^2(0,\pi)$.

From formula (5), we deduce that, given $y_0 \in H^{-1}(0,\pi)$, there exists a control $v \in L^2(0,T)$ such that the solution y to (3) satisfies $y(\cdot,T) = 0$ in $(0,\pi)$ if and only if there exists $v \in L^2(0,T)$ satisfying

$$\int_0^T \varphi_x(0,t)v(t)\,dt = -\langle y_0,\varphi(\cdot,0)\rangle, \quad \forall\varphi_T \in H_0^1(0,\pi),$$

($\varphi \in L^2(0,T;H^2(0,\pi)\cap H_0^1(0,\pi))\cap C^0([0,T];H_0^1(0,\pi))$ is the solution of (4) associated to φ_T). Using that $\{\phi_k\}_{k\ge1}$ (see (6)) is an orthogonal basis of $H_0^1(0,\pi)$, the previous property can be rewritten as

$$\int_0^T v(t)e^{-\lambda_k(T-t)}\partial_x\phi_k(0)\,dt = -\langle y_0, e^{-\lambda_k T}\phi_k\rangle, \quad \forall k \geq 1.$$

In the previous equality λ_k and ϕ_k are given in (6). Using the Fourier decomposition of y_0, $y_0 = \sum_{k\geq 1} y_{0,k}\phi_k$, we can conclude that there exists a control $v \in L^2(0,T)$ such that the solution y to (3) satisfies $y(\cdot, T) = 0$ in $(0,\pi)$ if and only if there exists $v \in L^2(0,T)$ such that

$$k\sqrt{\frac{2}{\pi}}\int_0^T e^{-\lambda_k(T-t)}v(t)\,dt = -e^{-\lambda_k T}y_{0,k} \quad \forall k \geq 1.$$

Introducing the new function $u(t) = v(T-t)$, $t \in (0,T)$, the previous equality may be written as

$$\int_0^T e^{-\lambda_k t}u(t)\,dt = e^{-\lambda_k T}m_k, \quad \forall k \geq 1, \quad \left(m_k = -\frac{1}{k}\sqrt{\frac{\pi}{2}}y_{0,k}\right). \tag{7}$$

In conclusion, we have reformulated the null controllability problem for the heat equation (3) as a *moment problem* which involves the real exponential family $\{e^{-\lambda_k t}\}_{k\geq 1}$.

Let us now solve the moment problem (7) in $L^2(0,T)$, for every time $T > 0$. Using that the series

$$\sum_{k\geq 1}\frac{1}{\lambda_k}$$

is convergent, from the Müntz Theorem (see [**25**]), we can conclude that this family is not total (or not complete) and minimal[1] in $L^2(0,T)$. This last property also characterizes the existence of, at least, a biorthogonal family $\{q_k\}_{k\geq 1}$ to $\{e^{-\lambda_k t}\}_{k\geq 1}$ in $L^2(0,T)$ (see Lemma 5.4 in [**18**]), i.e., a family $\{q_k\}_{k\geq 1} \subset L^2(0,T)$ such that

$$\int_0^T q_k(t)e^{-\lambda_j t}\,dt = \delta_{kj}, \quad \forall k, j \in \mathbb{N}.$$

This property provides a formal solution to the moment problem (7):

$$u(t) = v(T-t) = \sum_{k\geq 1} e^{-\lambda_k T}m_k q_k(t). \tag{8}$$

The question then is whether the constructed control v belongs to $L^2(0,T)$. Of course, this strongly depends on the family $\{q_k\}_{k\geq 1}$. We have the following result:

LEMMA 2.2. *There is a positive constant C such that, for all $T > 0$ we can find a biorthogonal family $\{q_k\}_{k\geq 1} \subset L^2(0,T)$ to $\{e^{-\lambda_k t}\}_{k\geq 1}$ ($\lambda_k = k^2$) such that*

$$\|q_k\|_{L^2(0,T)} \leq Ce^{C\sqrt{\lambda_k}}e^{C/T}, \quad \forall k \geq 1. \tag{9}$$

The previous result has been proved for positive real sequences $\{\lambda_k\}_{k\geq 1}$ more general than (6); for instance, see [**10**], [**26**], [**21**], [**28**], ...

Throughout this work $C > 0$ denotes a generic positive constant that may change line to line but which does not depend on T, k, or y_0.

[1] We say that a sequence $\{x_n\}$ in a Banach space X is minimal if no vector x_m lies in the closed span of the other vectors x_n, i.e., for every $m \in \mathbb{N}$, $x_m \notin \overline{\mathrm{span}}\,\{x_n\}_{n\neq m}$.

As a consequence of Lemma 2.2 we obtain the null controllability result for system (3). Indeed, from (9) and using the inequality

$$2C\sqrt{\lambda_k} \leq \frac{C^2}{T} + \lambda_k T, \quad \forall C > 0 \text{ and } k \geq 1,$$

we deduce that the series in (8) is absolutely convergent in $L^2(0,T)$:

$$\begin{aligned}\sum_{k\geq 1} \sqrt{\frac{\pi}{2}}\frac{1}{k} |y_{0,k}| \, e^{-\lambda_k T} \|q_k\|_{L^2(0,T)} &\leq C \left(\sum_{k\geq 1} \frac{|y_{0k}|^2}{k^2} \right)^{1/2} \left(\sum_{k\geq 1} e^{-2\lambda_k T} \|q_k\|^2_{L^2(0,T)} \right)^{1/2} \\ &\leq C e^{C/T} \|y_0\|_{H^{-1}(0,\pi)} \left(\sum_{k\geq 1} e^{-2\lambda_k T} e^{2C\sqrt{\lambda_k}} \right)^{1/2} \\ &\leq C e^{C/T} \|y_0\|_{H^{-1}(0,\pi)} \left(\sum_{k\geq 1} e^{-\lambda_k T} \right)^{1/2} \leq \frac{C}{\sqrt{T}} e^{C/T} \|y_0\|_{H^{-1}(0,\pi)}.\end{aligned}$$

Therefore, the control function v given in (8) belongs to $L^2(0,T)$, satisfies

$$\|v\|_{L^2(0,T)} \leq C e^{C/T} \|y_0\|_{H^{-1}(0,\pi)}, \tag{10}$$

and solves the moment problem (7). We have then proved:

THEOREM 2.3. *For any $y_0 \in H^{-1}(0,\pi)$ and $T > 0$, there exists $v \in L^2(0,T)$ solution to the moment problem* (7)*. That is, $v(t) = u(T-t)$ is a null control for equation* (3)*. In addition, there exists a positive constant C (independent of T and y_0) such that* (10) *holds.*

Theorem 2.3 guarantees that system (3) can be driven from an initial datum $y_0 \in H^{-1}(0,\pi)$ to the equilibrium 0 at time $T > 0$, for any $T > 0$. In particular, the set

$$\mathcal{Z}_T(y_0) := \left\{ v \in L^2(0,T) : y(\cdot, T) = 0 \text{ in } (0,\pi), \text{ with } y \text{ solution of } (3) \right\},$$

is nonempty for any $y_0 \in H^{-1}(0,\pi)$ and any $T > 0$. We can then define the control cost for system (3) at time T as

$$\mathcal{K}(T) = \sup_{\|y_0\|_{H^{-1}(0,\pi)}=1} \left(\inf_{v \in \mathcal{Z}_T(y_0)} \|v\|_{L^2(0,T)} \right), \quad \forall T > 0.$$

Theorem 2.3 also provides an estimate for $\mathcal{K}(T)$ and a measure of how violent fast controls for system (3) are:

$$\mathcal{K}(T) \leq C e^{C/T}, \quad \forall T > 0.$$

The previous estimate for the control cost in $L^2(0,T)$ is optimal for problem (3). Indeed, Güichal proved in [**16**], the optimality of the previous expression with respect to the time dependence:

THEOREM 2.4 (Güichal). *Under the previous conditions, there exist positive constants C_0 and C_1 independent of $T > 0$ such that, for any $T \in (0,1]$ one has*

$$e^{C_0/T} \leq \mathcal{K}(T) \leq e^{C_1/T}, \quad \forall T \in (0,1]. \tag{11}$$

REMARK 2.1. The estimate (9) can be obtained for more general positive real sequences $\{\lambda_k\}_{k\geq 1}$. To be precise, in order to obtain (9) we have used:

(1) $\{\lambda_k\}_{k\geq 1} \subset (0, \infty)$ is an increasing sequence and there exist $K > 0$ and $\alpha \in \mathbb{R}$ such that

$$\sqrt{|\lambda_k|} = K(k+\alpha)\left[1 + O(1/k)\right]. \tag{12}$$

(2) Gap condition: There is $\rho > 0$ such that

$$|\lambda_k - \lambda_n| \geq \rho \left|k^2 - n^2\right|, \quad \forall k, n \geq 1. \tag{13}$$

In particular, Lemma 2.2, estimate (9) and Theorem 2.3 can be also proved for the corresponding sequences of eigenvalues of more general second order elliptic operators:

$$Ly = -\left(p(\cdot)y_x\right)_x + q(\cdot)y, \quad D(L) = H^2(0,\pi) \cap H_0^1(0,L),$$

with $p \in W^{1,\infty}(0,\pi)$ and $q \in L^\infty(0,\pi)$ satisfying $p(\cdot) \geq p_0 > 0$ in $(0,\pi)$.

3. A boundary controllability problem for a parabolic system with complex eigenvalues

In order to extend inequalities (9) and (11) to complex sequences or sequences that do not satisfy conditions (12) or (13), let us consider the following boundary controllability problem:

$$\begin{cases} y_t - y_{xx} + A_0 y = 0 & \text{in } Q_T = (0,\pi) \times (0,T), \\ y(0,\cdot) = Bv, \quad y(\pi,\cdot) = 0 & \text{on } (0,T), \\ y(\cdot,0) = y_0, & \text{in } (0,\pi). \end{cases} \tag{14}$$

In (14) $y = (y_1, y_2)^t$, $y_0 \in H^{-1}((0,\pi);\mathbb{R}^2)$, $v \in L^2(0,T)$ and

$$A_0 = \begin{pmatrix} 0 & 1 \\ -1 & 0 \end{pmatrix}, \quad B = \begin{pmatrix} 0 \\ 1 \end{pmatrix}.$$

As before, we want to prove a null controllability result for system (14) but, unlike system (3), we only exert a scalar control on the second equation while the first equation is indirectly controled by means of the coupling matrix A_0.

Again, system (14) is well-posed. One has:

PROPOSITION 3.1. *Assume $y_0 \in H^{-1}(0,\pi;\mathbb{R}^2)$ and $v \in L^2(0,T)$. Then, system (14) admits a unique solution (by transposition) $y \in L^2(Q_T;\mathbb{R}^2) \cap C^0([0,T]; H^{-1}(0,\pi;\mathbb{R}^2))$ which depends continuously on the data.*

We can repeat the arguments of Section 2 and prove that the null controllability of system (14) at time $T > 0$ is equivalent to an appropriate moment problem. In this case, the adjoint operator L^* is given by

$$L^*\varphi = -\varphi_{xx} + A_0^*\varphi, \quad D(L^*) = H^2(0,\pi;\mathbb{C}^2) \cap H_0^1(0,\pi;\mathbb{C}^2),$$

and admits a complex sequence of eigenvalues and normalized eigenfunctions given by

$$\Lambda = \{n^2 - i, n^2 + i\}_{n\geq 1} \quad \mathcal{B} = \left\{ \frac{1}{\sqrt{\pi}} \sin nx \begin{pmatrix} i \\ -1 \end{pmatrix}, \frac{1}{\sqrt{\pi}} \sin nx \begin{pmatrix} 1 \\ -i \end{pmatrix} \right\}_{n\geq 1}.$$

In addition, $\mathcal{B}$ is a Riesz basis of $L^2(0,\pi;\mathbb{C}^2)$.

Let us take

$$\lambda_k = \begin{cases} \left(\dfrac{k+1}{2}\right)^2 - i, & \text{if } k \text{ is odd,} \\ \left(\dfrac{k}{2}\right)^2 + i, & \text{if } k \text{ is even.} \end{cases}$$

Then, it is not difficult to see that the sequence Λ of eigenvalues of L^* is given by $\Lambda = \{\lambda_k\}_{k\geq 1} \subset \mathbb{C}$ and satisfies (12) for $K = 1/2$ and $\alpha = 0$. Nevertheless, the sequence Λ does not satisfy the gap condition (13).

Now the question is: Would it be possible to obtain Lemma 2.2 for the previous complex sequence Λ in $L^2(0,T;\mathbb{C})$? In this regard, one has the following generalization of Lemma 2.2:

LEMMA 3.2. *Let $\Lambda = \{\lambda_k\}_{k\geq 1} \subset \mathbb{C}$ be a sequence of complex numbers with the following properties:*

(1) *$\lambda_k \neq \lambda_n$ for all $k, n \in \mathbb{N}$ with $k \neq n$.*
(2) *$\Re(\lambda_k) > 0$ for every $k \geq 1$.*
(3) *For some $\beta > 0$,*

$$|\Im(\lambda_k)| \leq \beta\sqrt{\Re(\lambda_k)}, \quad \forall k \geq 1.$$

(4) *$\{\lambda_k\}_{k\geq 1}$ is non-decreasing in modulus:*

$$|\lambda_k| \leq |\lambda_{k+1}|, \quad \forall k \geq 1.$$

(5) *$\{\lambda_k\}_{k\geq 1}$ satisfies the following gap condition: for some $\rho, q > 0$,*

$$\begin{cases} |\lambda_k - \lambda_n| \geq \rho\left|k^2 - n^2\right|, & \forall k, n : |k-n| \geq q. \\ \inf\limits_{k\neq n: |k-n|<q} |\lambda_k - \lambda_n| > 0. \end{cases}$$

(6) *For some $p, \alpha > 0$,*

(15) $$\left|p\sqrt{r} - \mathcal{N}(r)\right| \leq \alpha, \quad \forall r > 0,$$

where $\mathcal{N}$ is the counting function associated with the sequence $\{\lambda_k\}_{k\geq 1}$, that is the function defined by

(16) $$\mathcal{N}(r) = \#\{k : |\lambda_k| \leq r\}, \quad \forall r > 0.$$

Then, there exists $\tau_0 > 0$ such that, for every $0 < T < \tau_0$, we can find a family of $\mathbb{C}$-valued functions

$$\{q_k\}_{k\geq 1} \subset L^2(0,T;\mathbb{C})$$

biorthogonal to $\{e^{-\lambda_k t}\}_{k\geq 1}$ in $L^2(0,T;\mathbb{C})$, with in addition

(17) $$\|q_k\|_{L^2(0,T;\mathbb{C})} \leq Ce^{C\sqrt{\Re(\Lambda_k)}+\frac{C}{T}}, \quad \forall k \geq 1.$$

The proof of this result can be found in [**5**].

We can now apply Lemma 3.2 to problem (14) in order to obtain the null controllability result for this system at time $T > 0$ and an estimate of the control cost at this time. Indeed, the sequence Λ of eigenvalues of $L^* = -\partial_{xx} + A_0$ satisfies assumptions (1)–(6) in Lemma 3.2. On the other hand, as said before, the null controllability at time $T > 0$ of system (14) is equivalent to a moment problem for the complex exponentials $\left\{e^{-\lambda_k t}\right\}_{k\geq 1}$ in $L^2(0,T;\mathbb{C})$. If we now combine inequality (17) and the reasoning of Section 2, we deduce:

THEOREM 3.3. *System* (14) *is null controllable at time* $T > 0$, *for any* $T > 0$. *In addiction, there exists a constant* $C > 0$ *such that the control cost for system* (14) *satisfies*

$$\mathcal{K}(T) \leq Ce^{C/T}, \quad \forall T > 0. \tag{18}$$

REMARK 3.1. The null controllability result for system (14) at time $T > 0$ was established in [**12**]. To be precise, in this reference the authors provided a necessary and sufficient condition for the boundary null controllability of general 2×2 parabolic systems with constant coefficients:

$$\begin{cases} y_t - y_{xx} + Ay = 0 & \text{in } Q_T = (0,\pi) \times (0,T), \\ y(0,\cdot) = Bv, \quad y(\pi,\cdot) = 0 & \text{on } (0,T), \\ y(\cdot,0) = y_0, & \text{in } (0,\pi), \end{cases} \tag{19}$$

with $A \in \mathcal{L}(\mathbb{R}^2)$, $B \in \mathbb{R}^2$, $y_0 \in H^{-1}(0,\pi;\mathbb{R}^2)$ and $v \in L^2(0,T)$. This necessary and sufficient condition for the boundary null controllability of system (19) was generalized in [**2**] for general $n \times n$ systems with the same structure as (19) with $A \in \mathcal{L}(\mathbb{R}^n)$ and $B \in \mathcal{L}(\mathbb{R}^m;\mathbb{R}^n)$ $(n, m \geq 1)$. Nevertheless, in these two references the authors did not prove the estimate of the control cost (18). This estimate was proved in [**5**]. In fact, the authors proved a generalization of Lemma 3.2 for complex exponentials $\left\{t^j e^{-\lambda_k t}\right\}_{k\geq 1, 0\leq j\leq \eta}$, with $\eta \geq 0$ a given integer, and, as a consequence, they showed (18) when system (19) satisfies the sufficient null controllability condition.

REMARK 3.2. Assumption (5) (gap condition) on the sequence Λ is crucial in the proof of the existence of a biorthogonal family $\{q_k\}_{k\geq 1}$ to $\left\{e^{-\lambda_k t}\right\}_{k\geq 1}$ in $L^2(0,T;\mathbb{C})$ that satisfies (17). However, we will see in the next section that there are some boundary controllability problems for physical systems which do not satisfy the previous gap condition. Thus, it seems pertinent to analyze the control cost for systems that do not satisfy (5).

4. The boundary controllability of a phase-field system

In this section we will study the boundary null controllability of a phase field system of Caginalp type (see [**7**]) which is a model describing the transition between the solid and liquid phases in solidification/melting processes of a material occupying the interval $(0,\pi)$. For that purpose, we consider the nonlinear system

$$\begin{cases} \theta_t - \xi\theta_{xx} + \dfrac{1}{2}\rho\xi\phi_{xx} + \dfrac{\rho}{\tau}\theta = f(\phi) & \text{in } Q_T, \\ \phi_t - \xi\phi_{xx} - \dfrac{2}{\tau}\theta = -\dfrac{2}{\rho}f(\phi) & \text{in } Q_T, \\ \theta(0,\cdot) = v, \ \phi(0,\cdot) = c, \ \theta(\pi,\cdot) = 0, \ \phi(\pi,\cdot) = c & \text{on } (0,T), \\ \theta(\cdot,0) = \theta_0, \ \phi(\cdot,0) = \phi_0 & \text{in } (0,\pi), \end{cases} \tag{20}$$

where: $\theta = \theta(x,t)$ is the temperature of the material; $\phi = \phi(x,t)$ is the phase-field function used to identify the solidification level of the material; $c \in \{-1,0,1\}$; f is the nonlinear term which comes from the derivative of the classical regular double-well potential W:

$$f(\phi) = -\frac{\rho}{4\tau}\left(\phi - \phi^3\right).$$

On the other hand, $\rho > 0$, $\tau > 0$ and $\xi > 0$ are, resp., the latent heat, a relaxation time and the thermal diffusivity. Finally, $v \in L^2(0,T)$ is the control function, and θ_0, ϕ_0 are the initial data.

The null controllability property of the nonlinear system (20) has been analyzed in [**15**] and depends on the coefficients ρ, τ and ξ. This property can be obtained from the corresponding one of the linear version of (20) (see [**15**] for more details):

$$\begin{cases} y_t - Dy_{xx} + Ay = 0 & \text{in } Q_T := (0,\pi) \times (0,T), \\ y(0,\cdot) = Bv, \quad y(\pi,\cdot) = 0 & \text{on } (0,T), \\ y(\cdot,0) = y_0 & \text{in } (0,\pi), \end{cases} \tag{21}$$

where $y = (\theta, \phi)$,

$$D = \begin{pmatrix} \xi & -\dfrac{1}{2}\rho\xi \\ 0 & \xi \end{pmatrix}, \quad A = \begin{pmatrix} \dfrac{\rho}{\tau} & -\dfrac{\rho}{2\tau} \\ -\dfrac{2}{\tau} & \dfrac{1}{\tau} \end{pmatrix}, \quad B = \begin{pmatrix} 1 \\ 0 \end{pmatrix}. \tag{22}$$

Again, our objective is to drive the solution $y = (\theta, \phi)$ of (21) to the equilibrium $(0,0)$ only acting on the first equation (the temperature) through the Dirichlet boundary condition at point $x = 0$. The phase ϕ is indirectly controlled using the coupling matrices D and A.

Let us first establish the well-posedness of system (21). One has:

PROPOSITION 4.1. *Assume $y_0 = (\theta_0, \phi_0) \in H^{-1}(0,\pi;\mathbb{R}^2)$ and $v \in L^2(0,T)$. Then, system (21) admits a unique solution (by transposition) $y = (\theta, \phi) \in L^2(Q_T;\mathbb{R}^2) \cap C^0([0,T]; H^{-1}(0,\pi;\mathbb{R}^2))$ which depends continuously on the data.*

Now, we can already analyze the controllability properties of system (21). The approximate controllability of this system is given by the next result:

THEOREM 4.2 (Approximate controllability). *Fix $T > 0$. Then, system (21) is approximately controllable in $H^{-1}(0,\pi;\mathbb{R}^2)$ at time $T > 0$ if and only if*

$$\xi^2\tau^2(\ell^2 - k^2)^2 - 2\xi\rho\tau(\ell^2 + k^2) - 2\rho - 1 \neq 0, \quad \forall k, \ell \geq 1, \quad \ell > k. \tag{23}$$

The proof of this result can be found in [**15**].

It is well-known that the approximate controllability at time T for system (21) is necessary in order to prove the null controllability result for this system at same time T. So, assuming condition (23), the next task will be to study the null controllability of the linear problem (21) at time $T > 0$ and the corresponding control cost. Again, these properties can be obtained from the properties of the sequence of eigenvalues of the operators L and L^* with $L := -D\partial_{xx} + A$ and $D(A) = H^2(0,\pi;\mathbb{R}^2) \cap H_0^1(0,\pi;\mathbb{R}^2)$.

PROPOSITION 4.3. *Let us consider the operators $L := -D\partial_{xx} + A$ and L^* (the matrices D and A are given in (22)). Then, the spectra of L and L^* are given by $\sigma(L) = \sigma(L^*) = \left\{\lambda_k^{(1)}, \lambda_k^{(2)}\right\}_{k\geq 1}$ with*

$$\lambda_k^{(1)} = \xi k^2 + \frac{\rho+1}{2\tau} - r_k, \quad \lambda_k^{(2)} = \xi k^2 + \frac{\rho+1}{2\tau} + r_k, \quad \forall k \geq 1, \tag{24}$$

where r_k is given by

$$r_k := \sqrt{\frac{\xi\rho}{\tau}k^2 + \left(\frac{\rho+1}{2\tau}\right)^2}, \quad \forall k \geq 1.$$

In addition, $\lambda_k^{(1)} \neq \lambda_n^{(2)}$ *for any integers* $k, n \geq 1$ *if and only if* (23) *holds.*

Again, the proof of this result can be found in [**15**].

Thus, the sequence of eigenvalues of the vectorial elliptic operator L is real. If condition (23) is fullfilled, it can be rearranged in such a way that $\Lambda = \{\lambda_k\}_{k\geq 1} = \left\{\lambda_k^{(1)}, \lambda_k^{(2)}\right\}_{k\geq 1}$ is an increasing sequence that satisfies (1)–(4) and (6) for appropriate $p, \alpha > 0$. However, in general

$$\inf_{k\geq 1}(\lambda_{k+1} - \lambda_k) = 0,$$

and condition (5) does not hold (see [**15**] for the details). Therefore, we have a new system where the associated sequence of eigenvalues does not satisfy the gap condition (5) and Lemma 3.2 cannot be applied. Thus, the following question arises:

QUESTION. When the gap condition (5) fails, is system (21) null controllable at time T for any $T > 0$?; in this case, can we obtain an estimate of the control cost? We will answer this two question in the next sections.

REMARK 4.1 (Minimal time). Some choices of the parameters ρ, τ and ξ make the eigenvalues of the operators $L := -D\partial_{xx} + A$ and L^* (the matrices D and A are given in (22)) concentrate and this could affect the null controllability result. To be precise, in [**3**], the authors proved that when the eigenvalues of the generator of a C_0 semigroup accumulate, the corresponding null controllability problem has a minimal time $T_0 \in [0, \infty]$ of null controllability. This hyperbolic behaviour of the parabolic system is related to the condensation index of the sequence of eigenvalues of the generator. In the case of system (21), this minimal time is $T_0 = 0$ (see [**14**]).

REMARK 4.2. The previous example shows that it is necessary to generalize Lemma 3.2 for complex sequences $\Lambda = \{\lambda_k\}_{k\geq 1} \subset \mathbb{C}$ that satisfy all the assumptions in this lemma except the gap condition:

$$\inf_{k\neq n:|k-n|<q} |\lambda_k - \lambda_n| > 0.$$

For this kind of sequences, it would also be interesting to analyze the dependence of the control cost $\mathcal{K}(T)$ with respect to time $T > 0$, when T is small.

5. The main result: Bounds on biorthogonal families to complex exponentials without gap condition

In this section we will deal with the main result of this work. To this end, let us consider a complex sequence $\Lambda = \{\lambda_k\}_{k\geq 1} \subset \mathbb{C}$ satisfying (1)–(4), (6) and

$$(25) \qquad |\lambda_k - \lambda_n| \geq \rho \left|k^2 - n^2\right|, \quad \forall k, n : |k - n| \geq q,$$

with $q \geq 1$, a given integer. Our objective is to prove a result like Lemma 3.2 for the sequence Λ but without imposing the gap condition (5).

REMARK 5.1. Observe that we do not impose assumption (5) on the sequence Λ. Therefore, the infimum

$$\inf_{k\neq n} |\lambda_k - \lambda_n|$$

could be zero. Nevertheless, it is possible to prove that the sequence Λ satisfies

$$\sum_{k\geq 1} \frac{1}{|\lambda_k|} < \infty,$$

and the exponential sequence $\left\{e^{-\lambda_k t}\right\}_{k\geq 1}$ admits a biorthogonal family $\{q_k\}_{k\geq 1}$ in $L^2(0,T;\mathbb{C})$, for any $T>0$. This fact provides a formal solution u to the moment problem

$$\int_0^T e^{-\lambda_k t}u(t)\,dt = e^{-\lambda_k T}m_k, \quad \forall k\geq 1,$$

with $\{m_k\}_{k\geq 1}\subset \ell^2$, given by

$$u(t)=\sum_{k\geq 1} e^{-\lambda_k T} m_k q_k(t).$$

In fact, the previous expression, in general, defines a function in $L^2(0,T;\mathbb{C})$ provided T is large enough. To be precise, one can prove that there exists a minimal time $T_0\in[0,\infty]$ such that the corresponding problem is null controllable at time $T>0$ if $T>T_0$. On the other hand, if $T<T_0$, the system is not null controllable at time T. For more details, see [**12**], [**2**], [**3**], [**6**] and [**14**].

Even in the case in which the minimal time is $T_0=0$, we cannot apply Lemma 3.2 to the sequence Λ and deduce inequality (18) for the corresponding control cost.

One has:

THEOREM 5.1. *Assume that the complex sequence* $\Lambda=\{\lambda_k\}_{k\geq 1}\subset\mathbb{C}$ *satisfies* (1)–(4), (6) *and* (25) *for an integer* $q\geq 1$. *Then, there exists a constant* $C>0$, *only depending on* β, q, ρ, p *and* α, *such that for any* $T>0$ *there exists a biorthogonal family* $\{q_k\}_{k\geq 1}$ *to* $\left\{e^{-\lambda_k t}\right\}_{k\geq 1}$ *in* $L^2(0,T;\mathbb{C})$ *such that*

$$\|q_k\|_{L^2(0,T;\mathbb{C})}\leq Ce^{C\sqrt{\Re(\lambda_k)}}e^{C/T}\prod_{1\leq|k-n|<q}|\lambda_k-\lambda_n|^{-1}, \quad \forall k\geq 1.$$

The proof is very technical and follows some ideas from [**5**]. It can be found in [**14**].

REMARK 5.2. Theorem (5.1) generalizes Lemma 2.2, Lemma 3.2 and some previous works. For instance:

(1) [**10**], [**26**], [**21**] and [**28**] where the authors considered real positive sequences satisfying (25) for $q=1$, i.e., for any positive integers $k,n\geq 1$, with $k\neq n$;

(2) [**17**] and [**27**] where the authors considered complex sequences that, again, satisfy a gap condition;

(3) [**8**]: in this case, the authors proved a similar result to that established in Theorem 5.1 but for positive real sequences satisfying the conditions

$$\begin{cases} \left|\lambda_k-\alpha k^2\right|\leq ck, \quad \forall k\geq 1,\\ \inf\limits_{n:n\neq k}\left|\sqrt{\lambda_k}-\sqrt{\lambda_n}\right|\geq \dfrac{r}{k}, \quad \forall k\geq 1,\end{cases}$$

with α, c and r positive constants independent of k. In particular, the previous assumptions imply that the sequence $\{\lambda_k\}_{k\geq 1}$ does not condense and satisfies the gap condition (1). Indeed, for $k\geq 2c/\alpha$ and any $n\geq 1$, one has

$$\lambda_k\geq \alpha k^2-ck\geq \frac{1}{2}\alpha k^2,$$

and

$$|\lambda_k - \lambda_n| = \left|\sqrt{\lambda_k} - \sqrt{\lambda_n}\right| \left(\sqrt{\lambda_k} + \sqrt{\lambda_n}\right) \geq \frac{r}{k}\sqrt{\lambda_k} \geq r\sqrt{\frac{\alpha}{2}} > 0.$$

On the other hand, if $k < 2c/\alpha$, inequality (1) can be easily obtained.

6. Control cost for some parabolic systems

In this section we will analyze the control cost for some parabolic systems when we exert a boundary control on the system. The estimates of the control cost are a consequence of Theorem 5.1.

6.1. The linear phase-field system. In this section we again consider system (21) with D, A and B given in (22). Now, our objective is to give a null controllability result at time $T > 0$ for this system when (23) holds and obtain a bound for the corresponding control cost. One has:

THEOREM 6.1. *Under the previous conditions, assume that condition* (23) *holds. Then, system* (21) *is exactly controllable to zero at any time* $T > 0$. *Moreover, there exists a positive constant* C *such that*

$$\mathcal{K}(T) \leq Ce^{C/T}, \quad \forall T > 0,$$

where $\mathcal{K}(T)$ *is the control cost for system* (21):

$$\mathcal{K}(T) = \sup_{\|y_0\|_{H^{-1}(0,\pi;\mathbb{R}^2)}=1} \left(\inf_{v \in \mathcal{Z}_T(y_0)} \|v\|_{L^2(0,T)} \right), \quad \forall T > 0.$$

and

$$\mathcal{Z}_T(y_0) := \left\{ v \in L^2(0,T) : y(\cdot,T) = 0 \textit{ in } (0,\pi), \textit{ with } y \textit{ solution of } (21) \right\}.$$

SKETCH OF THE PROOF. As saw in Section 2, given $T > 0$ and $y_0 \in H^{-1}(0,\pi;\mathbb{R}^2)$, there exists a control $v \in L^2(0,T)$ such that the solution of (21) satisfies $y(\cdot,T) = 0$ in $(0,\pi)$ if and only if $v \in L^2(0,T)$ solves the moment problem

$$\int_0^T e^{-\lambda_k t} u(t)\, dt = e^{-\lambda_k T} m_k, \quad \forall k \geq 1, \quad \left(m_k = m_k(y_0) \text{ is s.t. } \{m_k\}_{k\geq 1} \in \ell^2 \right),$$

with $\Lambda = \{\lambda_k\}_{k\geq 1} = \left\{\lambda_k^{(1)}, \lambda_k^{(2)}\right\}_{k\geq 1}$ and $\lambda_k^{(i)}$ given in (24). On the other hand, the real positive sequence Λ satisfies conditions (1)–(4), (6), (25) (for $q = 2$) and

$$|\lambda_k - \lambda_{k-1}|\,|\lambda_k - \lambda_{k+1}| \geq \frac{C}{k} \quad (C = C(\xi,\rho,\tau) > 0),$$

(see **[15]** for the details). We can therefore apply Theorem 5.1 to the sequence Λ and conclude the existence of a positive constant C and a biorthogonal family $\{q_k\}_{k\geq 1}$ to $\left\{e^{-\lambda_k t}\right\}_{k\geq 1}$ in $L^2(0,T)$ such that

$$\|q_k\|_{L^2(0,T)} \leq Ce^{C\sqrt{\Re(\lambda_k)}} e^{C/T} \prod_{1\leq |k-n| < 2} |\lambda_k - \lambda_n|^{-1} \leq Ce^{C\sqrt{\Re(\lambda_k)}} e^{C/T}, \quad \forall k \geq 1.$$

Combining the previous inequality and the reasoning in Section 2 we deduce the existence of a control $v \in L^2(0,T)$ which solves the previous moment problem and

$$\|v\|_{L^2(0,T)} \leq Ce^{C/T} \|y_0\|_{H^{-1}(0,\pi;\mathbb{R}^2)},$$

with $C > 0$ a positive constant only depending on ξ, ρ and τ. This provides the estimate of $\mathcal{K}(T)$ for system (21) and the proof of Theorem 6.1 $\square$

REMARK 6.1. In [**15**] the authors prove Theorem 6.1 imposing (23) and an additional condition to the parameters ξ, ρ and τ that, in particular, implies

$$\inf_{k\neq n} |\lambda_k - \lambda_n| \geq c_0 > 0$$

for a positive constant $c_0 > 0$. Therefore, Theorem 6.1generalizes the null controllability result for the linear system (21) proved in the previous reference.

6.2. A 2×2 linear coupled parabolic system. In this section we are going to give an interesting example in which the fast controls are more violent than those of the heat equation. We will see that this violent behaviour comes from the condensation of the eigenvalues of the corresponding vectorial elliptic operator.

Let us consider the problem

$$\text{(26)}\qquad \begin{cases} y_t - y_{xx} + a(x)A_1 y = 0 & \text{in } (0,\pi)\times(0,T), \\ y(0,\cdot) = Bv, \quad y(\pi,\cdot) = 0 & \text{on } (0,T), \\ y(\cdot,0) = y_0 & \text{in } (0,\pi), \end{cases}$$

where $a \in L^2(0,\pi)$ is a function such that

$$\text{(27)}\qquad \int_0^\pi a(x)\,dx = 0,$$

$y = (y_1, y_2)^t$, $y_0 \in H^{-1}(0,\pi;\mathbb{R}^2)$, $v \in L^2(0,T)$ and

$$A_1 = \begin{pmatrix} 0 & 0 \\ 0 & 1 \end{pmatrix}, \quad B = \begin{pmatrix} 1 \\ 1 \end{pmatrix}.$$

As before, our objective is to drive the system to zero at time $T > 0$ acting on the Dirichlet boundary condition at $x = 0$ with a scalar control $v \in L^2(0,T)$.

System (26) is well-posed and for any $y_0 \in H^{-1}(0,\pi;\mathbb{R}^2)$ and $v \in L^2(0,T)$, the system has a unique solution (by transposition) $y \in L^2(Q_T;\mathbb{R}^2) \cap C^0([0,T]; H^{-1}(0,\pi;\mathbb{R}^2))$ which depends continuously on the data.

The null controllability of system (26) has been studied in [**22**]. In this work, the author proves that, in general, system (26) has a minimal time of null controllability $T_0 \in [0,\infty]$ when a satisfies (27). Nevertheless, obtaining an estimate of the control cost for system (26) in $L^2(0,T)$ has not been addressed before and is an open problem when (27) holds.

As a consequence of Theorem 5.1, let us provide an estimate of this control cost in a simple case. Before, let us summarize the controllability result for system (26) proved in [**22**].

Let us introduce the vectorial operator

$$L := -\partial_{xx} + a(\cdot)A_1 \quad \text{with } D(L) = H^2(0,\pi;\mathbb{R}^2) \cap H_0^1(0,\pi;\mathbb{R}^2).$$

Under assumption (27), one has,

$$\sigma(L) = \sigma(L^*) = \Lambda = \left\{k^2, k^2 + \beta_k\right\}_{k\geq 1}, \text{with } \{\beta_k\}_{k\geq 1} \in \ell^2,$$

(see [**22**]). Since $\lim \beta_k = 0$, the sequence Λ condensates at infinity and the gap condition (5) does not hold. This property has strong implications in the controllability of the system. One has:

THEOREM 6.2. *Let us consider $a \in L^2(0,\pi)$, a function satisfying (27). Given $T > 0$, one has:*

(1) *System (26) is approximately controllable at time $T > 0$ if and only if $\Lambda_1 \cap \Lambda_2 = \emptyset$, where*

$$\Lambda_1 = \sigma(-\partial_{xx}) = \left\{k^2\right\}_{k\geq 1} \quad \textit{and} \quad \Lambda_2 = \sigma(-\partial_{xx} + a(\cdot)) = \left\{k^2 + \beta_k\right\}_{k\geq 1}.$$

(2) *Assume that $\Lambda_1 \cap \Lambda_2 = \emptyset$ and define*

$$T_0 = \limsup \frac{-\log(|\beta_k|)}{k^2} \in [0,\infty].$$

Then,

(a) *If $T > T_0$, system (26) is null controllable at time T.*
(b) *If $T < T_0$, system (26) is not null controllable at time T.*

REMARK 6.2. The proof of Theorem 6.2 can be found in [**22**]. In fact, in that reference the author analyzes the controllability of 2×2 parabolic systems more general than system (26) with distributed or boundary scalar controls.

REMARK 6.3. The study of the controllability of system (26) is easier when $a \in L^2(0,\pi)$ does not satisfy condition (27). In fact, we have the following property: system (26) is null controllable at time $T > 0$ if and only if the system is approximately controllable at this time, i.e., if and only if $\Lambda_1 \cap \Lambda_2 = \emptyset$, with Λ_1 and Λ_2 given in the first item of Theorem 6.2 (see [**22**]). In this case, we have that $T_0 = 0$ and the null controllability of the system for any $T > 0$. On the other hand, it is not difficult to check that we can apply Lemma 3.2 to the sequence Λ. As a consequence, the associated control cost $\mathcal{K}(T)$ for system (26) can be estimated as follows:

$$\exp\left(\frac{C_0}{T}\right) \leq \mathcal{K}(T) \leq \exp\left(\frac{C_1}{T}\right), \quad \forall T \in (0,\tau_0),$$

for appropriate positive constants C_0, C_1 and τ_0 independent of T.

As said before, our objective is to provide an estimate of the control cost of system (26) in a simple case (a satisfying (27)). To this end, let us state a technical result whose proof can be found in [**23**] (see also [**22**]):

LEMMA 6.3. *Given $\gamma \in (0,1)$, there exists $a \in L^2(0,\pi)$ satisfying (27) such that*

$$\sigma(L) = \Lambda = \left\{k^2, k^2 + e^{-k^{2\gamma}}\right\}_{k\geq 1}.$$

From now on, we will take $\gamma \in (0,1)$ and will work with the function a provided by the previous lemma. Observe that, in particular (see Theorem 6.2), $\Lambda_1 \cap \Lambda_2 = \emptyset$ and the minimal time T_0 for system (26) associated to a is

$$T_0 = \limsup \frac{-\log\left(e^{-k^{2\gamma}}\right)}{k^2} = \lim \frac{k^{2\gamma}}{k^2} = 0.$$

Thus, system (26) is approximately and null controllable at any time $T > 0$.

On the other hand, the real sequence Λ (Λ is given in Lemma 6.3) can be rearranged as an increasing sequence $\Lambda = \{\lambda_k\}_{k\geq 1} \subset \mathbb{R}$ as follows:

$$\lambda_k = \begin{cases} \left(\dfrac{k+1}{2}\right)^2 & \text{if } k \text{ is odd,} \\ \left(\dfrac{k}{2}\right)^2 + e^{-(k/2)^{2\gamma}} & \text{if } k \text{ is even.} \end{cases}$$

It is not difficult to check that Λ satisfies conditions (1)–(4), (6) and (25) in Theorem 5.1, with $q = 2$. Therefore, the previous result can be applied to Λ obtaining the existence of a constant $C > 0$ such that for any $T > 0$ there exists a biorthogonal family $\{q_k\}_{k\geq 1}$ to $\{e^{-\lambda_k t}\}_{k\geq 1}$ in $L^2(0,T)$ such that

$$\|q_k\|_{L^2(0,T)} \leq Ce^{C\lceil k/2\rceil} e^{C/T} e^{\lceil k/2\rceil^{2\gamma}}, \quad \forall k \geq 1,$$

where $\lceil\cdot\rceil$ is the ceiling function[2] (see [**14**] for the details).

Combining the previous estimate and the moment method, we deduce:

THEOREM 6.4. *Let us fix $\gamma \in (0,1)$ and take the function $a \in L^2(0,\pi)$ provided by Lemma 6.3. If we denote $\mathcal{K}(T)$ the control cost of system* (26) *in $L^2(0,T)$, then, there exist two positive constants τ_0 and C_1 such that*

$$\mathcal{K}(T) \leq \exp\left(\frac{C_1}{T} + \frac{C_1}{T^{\frac{\gamma}{1-\gamma}}}\right), \quad \forall T \in (0,\tau_0). \tag{28}$$

The proof of the previous result can be found in [**14**]. In fact, the previous estimate for $\mathcal{K}(T)$ is optimal in the following sense (see [**14**]):

THEOREM 6.5. *Under the assumptions of Theorem* 6.4*, there exists a positive constant C_0 such that*

$$\mathcal{K}(T) \geq \exp\left(\frac{C_0}{T} + \frac{C_0}{T^{\frac{\gamma}{1-\gamma}}}\right), \quad \forall T \in (0,\tau_0). \tag{29}$$

Again, a proof of this result can be found in [**14**].

REMARK 6.4. Observe that inequalities (28) and (29) are valid when $\gamma \in (0,1)$. In fact, we can write

(1) If $\gamma \in (0,1/2]$ one deduces the existence of positive constants C_0, C_1 such that

$$\exp\left(\frac{C_0}{T}\right) \leq \mathcal{K}(T) \leq \exp\left(\frac{C_1}{T}\right), \quad \forall T \in (0,\tau_0),$$

obtaining estimates for the control cost of system (26) similar to the heat equation (see Theorems 2.3 and 2.4).

(2) If $\gamma \in (1/2,1)$, then, for new positive constants C_0, C_1 we have

$$\exp\left(\frac{C_0}{T^{\frac{\gamma}{1-\gamma}}}\right) \leq \mathcal{K}(T) \leq \exp\left(\frac{C_1}{T^{\frac{\gamma}{1-\gamma}}}\right), \quad \forall T \in (0,\tau_0).$$

The previous expression proves that the control cost blows up when $\gamma \to 1^-$. This is natural because the minimal time for system (26) when $\gamma = 1$ is $T_0 = 1$ and the system is not null controllable at time T when $T < 1$.

[2] $\lceil x\rceil$ is the least integer greater than or equal to x.

REMARK 6.5. In this section we have analyzed the control cost for system (26) when $a \in L^2(0,\pi)$ is such that the sequence of eigenvalues of the corresponding generator is given by $\Lambda = \left\{k^2, k^2 + e^{k^{2\gamma}}\right\}_{k\geq 1}$ with $\gamma \in (0,1)$. With this choice, the minimal time of null controllability of the system is $T_0 = 0$. In [**14**], the authors analyze the general problem $a \in L^2(0,\pi)$ satisfying (27) which includes the case in which $T_0 > 0$. The estimates for $\mathcal{K}(T)$ that can be obtained when $T_0 > 0$ are very close to those obtained in [**19**] for the heat equation (see [**14**]).

References

[1] F. Ammar-Khodja, A. Benabdallah, C. Dupaix, and M. González-Burgos, *A Kalman rank condition for the localized distributed controllability of a class of linear parbolic systems*, J. Evol. Equ. **9** (2009), no. 2, 267–291, DOI 10.1007/s00028-009-0008-8. MR2511553

[2] F. Ammar-Khodja, A. Benabdallah, M. González-Burgos, and L. de Teresa, *The Kalman condition for the boundary controllability of coupled parabolic systems. Bounds on biorthogonal families to complex matrix exponentials* (English, with English and French summaries), J. Math. Pures Appl. (9) **96** (2011), no. 6, 555–590, DOI 10.1016/j.matpur.2011.06.005. MR2851682

[3] F. Ammar Khodja, A. Benabdallah, M. González-Burgos, and L. de Teresa, *Minimal time for the null controllability of parabolic systems: the effect of the condensation index of complex sequences*, J. Funct. Anal. **267** (2014), no. 7, 2077–2151, DOI 10.1016/j.jfa.2014.07.024. MR3250361

[4] F. Ammar Khodja, A. Benabdallah, M. González-Burgos, and L. de Teresa, *New phenomena for the null controllability of parabolic systems: minimal time and geometrical dependence*, J. Math. Anal. Appl. **444** (2016), no. 2, 1071–1113, DOI 10.1016/j.jmaa.2016.06.058. MR3535750

[5] A. Benabdallah, F. Boyer, M. González-Burgos, and G. Olive, *Sharp estimates of the one-dimensional boundary control cost for parabolic systems and application to the N-dimensional boundary null controllability in cylindrical domains*, SIAM J. Control Optim. **52** (2014), no. 5, 2970–3001, DOI 10.1137/130929680. MR3262588

[6] A. Benabdallah, F. Boyer, M. Morancey, *A block moment method to handle spectral condensation phenomenon in parabolic control problems*, submitted, https://hal.archives-ouvertes.fr/hal-01949391.

[7] G. Caginalp, *An analysis of a phase field model of a free boundary*, Arch. Rational Mech. Anal. **92** (1986), no. 3, 205–245, DOI 10.1007/BF00254827. MR816623

[8] N. Cîndea, S. Micu, I. Rovenţa, and M. Tucsnak, *Particle supported control of a fluid-particle system*, J. Math. Pures Appl. (9) **104** (2015), no. 2, 311–353, DOI 10.1016/j.matpur.2015.02.009. MR3365831

[9] S. Dolecki, *Observability for the one-dimensional heat equation*, Studia Math. **48** (1973), 291–305, DOI 10.4064/sm-48-3-291-305. MR333444

[10] H. O. Fattorini and D. L. Russell, *Exact controllability theorems for linear parabolic equations in one space dimension*, Arch. Rational Mech. Anal. **43** (1971), 272–292, DOI 10.1007/BF00250466. MR335014

[11] H. O. Fattorini and D. L. Russell, *Uniform bounds on biorthogonal functions for real exponentials with an application to the control theory of parabolic equations*, Quart. Appl. Math. **32** (1974/75), 45–69, DOI 10.1090/qam/510972. MR0510972

[12] E. Fernández-Cara, M. González-Burgos, and L. de Teresa, *Boundary controllability of parabolic coupled equations*, J. Funct. Anal. **259** (2010), no. 7, 1720–1758, DOI 10.1016/j.jfa.2010.06.003. MR2665408

[13] E. Fernández-Cara and E. Zuazua, *The cost of approximate controllability for heat equations: the linear case*, Adv. Differential Equations **5** (2000), no. 4-6, 465–514. MR1750109

[14] M. González-Burgos and L. Ouaili, *Sharp estimates for biorthogonal families to exponentials associated to complex sequences without gap conditions*, submitted.

[15] M. González-Burgos and G. R. Sousa-Neto, *Boundary controllability of a one-dimensional phase-field system with one control force*, J. Differential Equations **269** (2020), no. 5, 4286–4331, DOI 10.1016/j.jde.2020.03.036. MR4097937

[16] E. N. Güichal, *A lower bound of the norm of the control operator for the heat equation*, J. Math. Anal. Appl. **110** (1985), no. 2, 519–527, DOI 10.1016/0022-247X(85)90313-0. MR805273

[17] S. W. Hansen, *Bounds on functions biorthogonal to sets of complex exponentials; control of damped elastic systems*, J. Math. Anal. Appl. **158** (1991), no. 2, 487–508, DOI 10.1016/0022-247X(91)90252-U. MR1117578

[18] C. Heil, *A basis theory primer*, Expanded edition, Applied and Numerical Harmonic Analysis, Birkhäuser/Springer, New York, 2011. MR2744776

[19] P. Lissy, *The cost of the control in the case of a minimal time of control: the example of the one-dimensional heat equation*, J. Math. Anal. Appl. **451** (2017), no. 1, 497–507, DOI 10.1016/j.jmaa.2017.01.096. MR3619248

[20] P. Lissy and E. Zuazua, *Internal observability for coupled systems of linear partial differential equations*, SIAM J. Control Optim. **57** (2019), no. 2, 832–853, DOI 10.1137/17M1119160. MR3920472

[21] L. Miller, *Geometric bounds on the growth rate of null-controllability cost for the heat equation in small time*, J. Differential Equations **204** (2004), no. 1, 202–226, DOI 10.1016/j.jde.2004.05.007. MR2076164

[22] L. Ouaili, *Minimal time of null controllability of two parabolic equations*, Math. Control Relat. Fields **10** (2020), no. 1, 89–112, DOI 10.3934/mcrf.2019031. MR4063628

[23] J. Pöschel and E. Trubowitz, *Inverse spectral theory*, Pure and Applied Mathematics, vol. 130, Academic Press, Inc., Boston, MA, 1987. MR894477

[24] E. H. Samb, *Internal null-controllability of the N-dimensional heat equation in cylindrical domains* (English, with English and French summaries), C. R. Math. Acad. Sci. Paris **353** (2015), no. 10, 925–930, DOI 10.1016/j.crma.2015.04.021. MR3411223

[25] L. Schwartz, *Étude des sommes d'exponentielles réelles* (French), Actualités Sci. Ind., no. 959, Hermann et Cie., Paris, 1943. MR0014502

[26] T. I. Seidman, *Two results on exact boundary control of parabolic equations*, Appl. Math. Optim. **11** (1984), no. 2, 145–152, DOI 10.1007/BF01442174. MR743923

[27] T. I. Seidman, S. A. Avdonin, and S. A. Ivanov, *The "window problem" for series of complex exponentials*, J. Fourier Anal. Appl. **6** (2000), no. 3, 233–254, DOI 10.1007/BF02511154. MR1755142

[28] G. Tenenbaum and M. Tucsnak, *On the null-controllability of diffusion equations*, ESAIM Control Optim. Calc. Var. **17** (2011), no. 4, 1088–1100, DOI 10.1051/cocv/2010035. MR2859866

Dpto. Ecuaciones Diferenciales y Análisis Numérico and Instituto de Matemáticas de la Universidad de Sevilla (IMUS), Facultad de Matemáticas, Universidad de Sevilla, C/ Tarfia S/N, 41012 Sevilla, Spain

Email address: `manoloburgos@us.es`

Aix-Marseille Université, CNRS, Centrale Marseille, I2M, UMR 7373, 13453 Marseille, France

Email address: `ouaili.lydia@gmail.com`

Contemporary Mathematics
Volume **757**, 2020
https://doi.org/10.1090/conm/757/15243

Theoretical and numerical aspects related to the energy of some Timoshenko systems

Makram Hamouda

ABSTRACT. We recall in this note some recent results concerning the Timoshenko system which is here subject to different types of dissipation (undamped, linear damping and nonlinear damping). More precisely, the emphasis will be on the behavior, for a time large enough, of the energy associated with the solution of the damped Timoshenko system. We numerically confirm the conservation property and the theoretical results established on the decay rates of the energy associated with each type of dissipation.

CONTENTS

1. Introduction and setting of the problem

In the general theory of beams, it is common to look for the optimal damping term to stabilize the system under consideration because of the well-known phenomenon of vibrations. Hence, to reduce these harmful vibrations, several control mechanisms have been designed. In order to do that, it is first natural to study this dissipation mechanism and then model and understand the corresponding equations of these problems, see [**10, 11**].

In this context and from the theoretical point of view, it is common to derive the upper estimates of the energy for which the situation is almost understood. However, aiming to obtain an optimality result, we derive, in addition to these estimates, some lower bounds for the energy. This gives a good precision on the behavior of the energy. To the best of our knowledge, Vancostenoble and Martinez [**15**] proved optimality of the upper energy estimates for a one-dimensional (or radial in three dimensions) boundary damped wave equation and for initial data with vanishing velocities. Recently, in [**1**] and [**2**], optimality has been proved in the particular case of the one-dimensional wave equation with boundary damping.

2010 *Mathematics Subject Classification.* Primary 35B40, 74F05, 78M10, 93D15, 93D20.
Key words and phrases. Finite element methods, general decay, stability, Timoshenko system.

Now, form the numerical point of view, we need to construct a scheme which aims to preserve the most important properties of the energy. Therefore, we design a discretization scheme, based on a combination of the finite element method and the finite difference one. The obtained scheme reaches to describe the discrete energy showing the positivity, the energy conservation property and the types of the decay rates in the case when the system is damped.

For instance let us give an overview of what it is known for the 1D Timoshenko system.

Timoshenko proposed (1921) for a 1D dimensional beam of length L the following system [**14**]:

$$
\begin{cases} \rho_1\varphi_{tt} - k(\varphi_x + \psi)_x = 0, \text{in } (0, L) \times \mathbb{R}_+, \\ \rho_2\psi_{tt} - (EI\psi_x)_x + k(\varphi_x + \psi) = 0, \text{in } (0, L) \times \mathbb{R}_+, \end{cases} \tag{1.1}
$$

where t is the time variable, x is the space variable, φ is the displacement vector and ψ represents the rotation angle of the filament. The quantities ρ, E, I, k are simply positive constants which characterize the beam. Without loss of generality we set from now on $L = 1$.

Several studies were carried out on the stability of the system (1.1), namely the well-known case when the Timoshenko system is modified by a linear damping in the second equation corresponding to the rotation variation as below:

$$
\begin{cases} \rho_1\varphi_{tt} - k(\varphi_x + \psi)_x = 0, \text{ in } (0, 1) \times \mathbb{R}_+, \\ \rho_2\psi_{tt} - b\psi_{xx} + k(\varphi_x + \psi) + d\psi_t = 0, \text{in } (0, 1) \times \mathbb{R}_+, \end{cases} \tag{1.2}
$$

for which it is now widely known that the exponential stability is only available under the condition of wave speed equality when the damping term is considered in the two equations, namely

$$
\frac{k}{\rho_1} = \frac{b}{\rho_2}.
$$

We recall here that another stability number is necessary for establishing the experiential stability in our case later on, see [**3, 13**].

However, in many physical situations, the damping term is not necessarily linear and hence it is judicious to consider a nonlinear damping term instead of the linear dissipation $d\psi_t$ in $(1.2)_2$. Hence we end up with the following system:

$$
\begin{cases} \rho_1\varphi_{tt} - k(\varphi_x + \psi)_x = 0, & \text{in } (0, 1) \times \mathbb{R}_+, \\ \rho_2\psi_{tt} - b\psi_{xx} + k(\varphi_x + \psi) + g(\psi_t) = 0, & \text{in } (0, 1) \times \mathbb{R}_+, \\ \varphi(0, t) = \varphi(1, t) = \psi(0, t) = \psi(1, t) = 0, & \forall\, t \geq 0, \\ \varphi(x, 0) = \varphi_0(x),\ \varphi_t(x, 0) = \varphi_1(x), & \forall\, x \in (0, 1), \\ \psi(x, 0) = \psi_0(x),\ \psi_t(x, 0) = \psi_1(x), & \forall\, x \in (0, 1). \end{cases} \tag{1.3}
$$

The energy associated with the system (1.3) reads as follows:

$$
E_1(t) := \frac{1}{2} \int_0^1 \left(\rho_1\varphi_t^2 + \rho_2\psi_t^2 + b\psi_x^2 + k(\varphi_x + \psi)^2 \right) dx. \tag{1.4}
$$

REMARK 1. (i) We have $\frac{dE_1(t)}{dt} = 0$ which states the analytical energy conservation property.

(ii) For the numerical part, in Section 3, we will deal with the system (1.3) in addition to the undamped and linearly damped cases given, respectively, by (1.1) and (1.2), and hence prove the analogous discrete property to (i).

A more accurate model for the 1D Timoshenko system was recently considered e.g. in [**3**, **8**, **11**] where the displacement and the rotation equations are coupled with the equation of the heat flux which is now given by Cattaneo's law instead of the Fourier's law that suffers from some inconsistencies.

In this note we consider the following Timoshenko system with thermoelasticity with second sound:

(1.5)
$$\begin{cases} \rho_1\varphi_{tt} - k(\varphi_x+\psi)_x = 0, & \text{in } (0,1)\times\mathbb{R}_+,\\ \rho_2\psi_{tt} - b\psi_{xx} + k(\varphi_x+\psi) + \delta\theta_x + a(x)g(\psi_t) = 0, & \text{in } (0,1)\times\mathbb{R}_+,\\ \rho_3\theta_t + q_x + \delta\psi_{xt} = 0, & \text{in } (0,1)\times\mathbb{R}_+,\\ \tau q_t + \beta q + \theta_x = 0, & \text{in } (0,1)\times\mathbb{R}_+,\\ \varphi(0,t)=\varphi(1,t)=\psi(0,t)=\psi(1,t)=q(0,t)=q(1,t)=0, & \forall\, t\geq 0,\\ \varphi(x,0)=\varphi_0(x),\ \varphi_t(x,0)=\varphi_1(x), & \forall\, x\in(0,1),\\ \psi(x,0)=\psi_0(x),\ \psi_t(x,0)=\psi_1(x), & \forall\, x\in(0,1),\\ \theta(x,0)=\theta_0(x),\ q(x,0)=q_0(x), & \forall\, x\in(0,1), \end{cases}$$

where again here ρ_1, ρ_2, ρ_3, b, k, δ, β are positive constants, $\varphi=\varphi(x,t)$ is the displacement vector, $\psi=\psi(x,t)$ is the rotation angle of the filament, $\theta=\theta(x,t)$ is the temperature difference and $q=q(x,t)$ is the heat flux vector.

Using $(1.5)_1$, $(1.5)_3$ and the boundary conditions $(1.5)_5$, we have

$$\frac{d^2}{dt^2}\int_0^1 \varphi(x,t)dx = 0 \text{ and } \frac{d}{dt}\int_0^1 \theta(x,t)dx = 0.$$

Consequently, we obtain

$$\int_0^1 \varphi(x,t)dx = \left(\int_0^1 \varphi_1(x)dx\right)t + \int_0^1 \varphi_0(x)dx \text{ and } \int_0^1 \theta(x,t)dx = \int_0^1 \theta_0(x)dx.$$

If we set

$$\bar\varphi(x,t) = \varphi(x,t) - \left(\left(\int_0^1 \varphi_1(x)dx\right)t + \int_0^1 \varphi_0(x)dx\right),$$

and

$$\bar\theta(x,t) = \theta(x,t) - \int_0^1 \theta_0(x)dx,$$

then $(\bar\varphi,\psi,\bar\theta,q)$ satisfy also the system (1.5), and we have

$$\int_0^1 \bar\varphi(x,t)dx = 0 \text{ and } \int_0^1 \bar\theta(x,t)dx = 0.$$

From now on we deal with the new variables $(\bar\varphi,\psi,\bar\theta,q)$, but we keep the notation (φ,ψ,θ,q) for simplicity.

2. Theoretical results

In this section, we recall some results on the upper and lower estimates of the energy of the system (1.5) which is given below

$$E_2(U)(t) := \frac{1}{2}\int_0^1 (\rho_1\varphi_t^2 + \rho_2\psi_t^2 + b\psi_x^2 + k(\varphi_x+\psi)^2 + \rho_3\theta^2 + \tau q^2)dx, \tag{2.1}$$

for $U=(\varphi,\psi,\theta,q)\in\mathcal{H}$ where

$$\mathcal{H} = H_0^1(0,1)\times L^2(0,1)\times H_0^1(0,1)\times (L^2(0,1))^3.$$

The above energy satisfies $\frac{dE_2(t)}{dt}\leq 0$.

In the following we assume some hypotheses on the damping functions:

$$(H_0)\begin{cases} a \text{ is a smooth function,} \\ a(x) \geq 0, \forall\, x \in (0,1), \\ a > 0 \text{ in a nonempty subset } \Gamma \text{ of } (0,1). \end{cases}$$

$$(H_1)\begin{cases} g:\mathbb{R}\to\mathbb{R} \ \text{ is a nondecreasing } \ \mathcal{C}^0 - \text{function} \\ \text{such that for every } \ \epsilon \in (0,1), \exists\, c_1 > 0, c_2 > 0, \\ \text{and an increasing odd function } g_0 \in \mathcal{C}^1(0,+\infty), \\ \text{with } g_0(0) = 0 \text{ such that} \\ \begin{cases} g_0(|(s)|) \leq |g(s)| \leq g_0^{-1}(|(s)|), \text{ for all } |s| \leq \epsilon, \\ c_1|s| \leq |g(s)| \leq c_2|s|, \text{ for all} |s| \geq \epsilon. \end{cases} \end{cases}$$

Then, in order to apply the Dafermos' technique used in Alabau's method for the lower bound estimates, it is necessary to assume in addition the following hypotheses:

$$(HC)\begin{cases} \text{Let } (\varphi,\psi) \ \text{ be a weak solution of (1.5)} \\ \quad \text{if } \psi_t \equiv 0 \text{ on } \Gamma \text{ then } (\varphi,\psi) \equiv (0,0). \end{cases}$$

$$(H_2)\begin{cases} \exists\, r_0 > 0 \text{ such that the function } \ H:[0,r_0^2] \mapsto \mathbb{R} \\ \text{is strictly convex on } \ [0,r_0^2], \\ \text{and either } 0 < \liminf_{x\to 0}\Lambda(x) \leq \limsup_{x\to 0}\Lambda(x) < 1, \\ \text{or there exists } \mu > 0 \text{ such that} \\ 0 < \liminf_{x\to 0}\left(\frac{\Psi(\mu x)}{\mu x}\int_x^{z_1}\frac{1}{H(y)}dy\right), \ \text{ and } \ \limsup_{x\to 0}\Lambda(x) < 1, \\ \text{for some } z_1 \in (0,z_0] \text{ and for all } z_0 > 0. \end{cases}$$

We introduce

$$H(x) = \sqrt{x}g(\sqrt{x}), \quad \text{ and } \quad \Lambda(x) = \frac{H(x)}{xH'(x)}.$$

Now, the dissipation can be easily obtained and we have

$$E_2'(t) = -\beta\int_0^1 q^2dx - \int_0^1 a(x)\psi_t g(\psi_t)dx \leq 0. \tag{2.2}$$

Then, we have the following strong stability result.

THEOREM 2.1. *(New equilibrium state,* [**5**]*) Assume that* (H_0), (H_1) *and* (HC) *hold. Then, for all* $U_0 = (\varphi_0,\varphi_1,\psi_0,\psi_1,\theta_0,q_0) \in \mathcal{H}$, *the energy* E_2 *satisfies*

$$\lim_{t\to\infty} E_2(t,U) = E_\infty, \forall\, U \in \mathcal{C}([0,+\infty),\mathcal{H}), \tag{2.3}$$

where E_∞ *is the energy of* $Z \in \omega(U_0)$, *the* $\omega-$*limit of* U_0.

The energy $\mathcal{E}(t)$ *satisfies*

$$\mathcal{E}(t) \to 0, \textit{ as } t\to\infty, \tag{2.4}$$

where

$$\mathcal{E}(t) = \frac{1}{2}\int_0^1 \rho_1{\varphi_t}^2 + \rho_2{\psi_t}^2 + b{\psi_x}^2 + k(\varphi_x+\psi)^2 + \rho_3(\theta-\theta_0(0))^2 + \tau q^2dx. \tag{2.5}$$

Sketch of the proof

- $E_\star(t)$ is a non-increasing function (the first-order energy)

$$E_\star(t) := \frac{1}{2}\int_0^1 (\rho_1\varphi_{tt}^2 + \rho_2\psi_{tt}^2 + b\psi_{tx}^2 + k(\varphi_{tx}+\psi_t)^2 + \rho_3\theta_t^2 + \tau q_t^2)dx.$$

- For the initial data $U_0 = (\varphi_0, \varphi_1, \psi_0, \psi_1, \theta_0, q_0) \in D(A)$, the orbit of U_0 given by $\gamma(U_0) = \cup_{t\geq 0}\mathcal{T}(t)U_0$ is relatively compact in $\mathcal{H}$.
- Thanks to the Dafermos strong stability, [7],

$$\lim_{t\to\infty} E_2(t,U) = E_\infty, \forall\, U \in \mathcal{C}([0,+\infty),\mathcal{H}).\blacksquare$$

For the complete proof of Theorem 2.1, see [5] and for the application of the technique there, see [1, 2].

Now we state the result related to the lower bound of the energy $\mathcal{E}$ which completes hence the upper estimates proven in [3].

THEOREM 2.2. (*Lower bound estimate,* [5]) *Assume that* (H_0), (H_1) *and* (H_2) *hold. For all smooth initial data* U_0 $(\neq 0)$, $\exists$ $T_0 > 0$ *and* $T_1 > 0$, *such that the energy* $\mathcal{E}$ *satisfies the lower estimate*

$$\frac{1}{\gamma^2 C_\sigma^2}\left(H'^{-1}\left(\frac{1}{t-T_0}\right)\right)^2 \leq \mathcal{E}(t), \forall\, t \geq T_1 + T_0, \tag{2.6}$$

where σ *is a positive constant* $\sigma = \frac{\alpha_a}{\rho_2} + \frac{\beta r_0}{\tau C_1}$,

$$\alpha_a = \|a\|_{L^\infty(0,1)},$$

and

$$\gamma = \frac{4}{\rho_2}\sqrt{E_\star(0)}.$$

3. Numerical approximation of the Timoshenko system

We show in this section the numerical scheme that we construct for the energy associated with the Timoshenko systems (1.2) and (1.3). The system (1.2) is here a particular case of (1.3). However, we will start by studying the discrete energy of the undamped system (1.1), then we end up with the numerical approximation of the linearly damped Timoshenko system (1.2). We design a discretization scheme based on a combination between the finite element methods and the finite difference ones.

REMARK 2. We choose here to work on the energy to show some of its properties which should be in agreement with the theoretical results. However, it is possible to approximate directly the solutions φ and ψ, see e.g. [4, 9, 12].

For our numerical approximation, we consider the following piecewise-linear functions

$$w_i(x) = \begin{cases} \dfrac{x - x_{i-1}}{h} & x_{i-1} \leq x \leq x_i, \\ \dfrac{x_{i+1} - x}{h} & x_i \leq x \leq x_{i+1}, \\ 0 & x \in [0,1] \setminus [x_{i-1}, x_{i+1}]. \end{cases}$$

Hence, the matrix formulation for the semi-discrete problem reads

$$\begin{cases} M\dfrac{dU}{dt} = -K\Psi + S\Psi, \\ M\dfrac{dV}{dt} = -K\Psi - S\Phi - M\Psi, \\ \dfrac{d\Phi}{dt} = U(t), \\ \dfrac{d\Psi}{dt} = V(t), \end{cases} \tag{3.1}$$

where $U(t) = [u_h(x_0,t), \cdots, u_h(x_{Nx},t)]^t$ and $V(t) = [v_h(x_0,t), \cdots, v_h(x_{Nx},t)]^t$.

$M_{i,j} = (w_i, w_j)$ is the mass matrix, $K_{i,j} = (w_i', w_j')$ is the rigidity matrix and S is a matrix such that $S_{i,j} = (w_i', w_j)$.

Then, the fully-discrete scheme in finite difference methods is obtained using a leapfrog time scheme. The problem is thus formulated as follows.

Find the discrete solutions $(\Phi^n, U^n, \Psi^n, V^n)$ such that

$$\left\{\begin{array}{l} M\dfrac{U^{n+1}-U^{n-1}}{2\Delta t} = -K\Phi^n + S\Psi^n, \\ M\dfrac{V^{n+1}-V^{n-1}}{2\Delta t} = -K\Psi^n - S\Phi^n - M\Psi^n, \\ \dfrac{\Phi^{n+1}-\Phi^{n-1}}{2\Delta t} = U^n, \\ \dfrac{\Psi^{n+1}-\Psi^{n-1}}{2\Delta t} = V^n. \end{array}\right. \tag{3.2}$$

The initial data are evaluated using the explicit Euler approximation and we have

$$\begin{aligned} \Phi^0 &= \phi_0(x), \\ \Psi^0 &= \psi_0(x), \\ U^0 &= \phi_1(x), \\ V^0 &= \psi_1(x), \\ \Phi^1 &= \Phi^0 + \Delta t U^0, \\ \Psi^1 &= \Psi^0 + \Delta t V^0, \\ MU^1 &= MU^0 + \Delta t(-K\Phi^0 + S\Psi^0), \\ MV^1 &= MV^0 + \Delta t(-K\Psi^0 + S^T\Phi^0 - M\Psi^0), \end{aligned}$$

where S is an antisymmetric Matrix, namely $S^T = -S$ (S^T is the transpose matrix).

3.1. The discrete energy. Let $\Delta t > 0$. We define below the discrete energy

$$\begin{aligned} E^{n+\frac{1}{2}} \;:=\; & \left(M\frac{\Phi^{n+2}-\Phi^n}{2\Delta t}, \frac{\Phi^{n+2}-\Phi^n}{2\Delta t}\right) \\ & + \left(M\frac{\Psi^{n+2}-\Psi^n}{2\Delta t}, \frac{\Psi^{n+2}-\Psi^n}{2\Delta t}\right) \\ & + (K\Phi^n, \Phi^{n+2}) + ((K+M)\Psi^n, \Psi^{n+2}) \\ & + (S\Psi^n, \Phi^{n+2}) - (S\Phi^n, \Psi^{n+2}), \end{aligned} \tag{3.3}$$

which corresponds to the systems (1.1), (1.2) and (1.3).

PROPOSITION 3.1. *Let $\Delta x > 0$, $\Delta t > 0$ and (Φ^n, Ψ^n) be the solution of the finite difference scheme* (3.2). *Then, for all $n \in \{0, \cdots, N_t\}$, we have*

$$E^{n+\frac{1}{2}} = E^{n-\frac{1}{2}}. \tag{3.4}$$

PROOF. We rewrite the problem (3.2) in terms of Φ and Ψ as:

$$M\left(\frac{\Phi^{n+1} - 2\Phi^n + \Phi^{n-1}}{\Delta t^2}\right) = -K\Phi^n + S\Psi^n, \tag{3.5}$$

$$M\left(\frac{\Psi^{n+1} - 2\Psi^n + \Psi^{n-1}}{\Delta t^2}\right) = -K\Psi^n - S\Phi^n - M\Psi^n. \tag{3.6}$$

Multiplying (3.5) by $\frac{\Phi^{n+1}-\Phi^{n-1}}{2\Delta t}$ and (3.6) by $\frac{\Psi^{n+1}-\Psi^{n-1}}{2\Delta t}$, we obtain

$$\left(M\frac{\Phi^{n+1}-2\Phi^n+\Phi^{n-1}}{\Delta t^2}+K\Phi^n-S\Psi^n,\frac{\Phi^{n+1}-\Phi^{n-1}}{2\Delta t}\right)$$

$$=\frac{1}{2\Delta t}\left(\left(M\frac{\Phi^{n+1}-\Phi^n}{\Delta t},\frac{\Phi^{n+1}-\Phi^n}{\Delta t}\right)-\left(M\frac{\Phi^n-\Phi^{n-1}}{\Delta t},\frac{\Phi^n-\Phi^{n-1}}{\Delta t}\right)\right)$$

$$+\frac{1}{2\Delta t}\left((K\Phi^n,\Phi^{n+1})-(K\Phi^n,\Phi^{n-1})\right)-\frac{1}{2\Delta t}\left((S\Psi^n,\Phi^{n+1})-(S\Psi^n,\Phi^{n-1})\right)=0, \tag{3.7}$$

and

$$\left(M\frac{\Psi^{n+1}-2\Psi^n+\Psi^{n-1}}{\Delta t^2}+K\Psi^n+S\Phi^n+M\Psi^n,\frac{\Psi^{n+1}-\Psi^{n-1}}{2\Delta t}\right)$$

$$=\frac{1}{2\Delta t}\left(\left(M\frac{\Psi^{n+1}-\Psi^n}{\Delta t},\frac{\Psi^{n+1}-\Psi^n}{\Delta t}\right)-\left(M\frac{\Psi^n-\Psi^{n-1}}{\Delta t},\frac{\Psi^n-\Psi^{n-1}}{\Delta t}\right)\right)$$

$$+\frac{1}{2\Delta t}\left((K\Psi^n,\Psi^{n+1})-(K\Psi^n,\Psi^{n-1})\right)+\frac{1}{2\Delta t}\left((S\Phi^n,\Psi^{n+1})-(S\Phi^n,\Psi^{n-1})\right)$$

$$+\frac{1}{2\Delta t}\left((M\Psi^n,\Psi^{n+1})-(M\Psi^n,\Psi^{n-1})\right)=0. \tag{3.8}$$

Taking into account the fact that K, M are symmetric definite positive matrices and S is a skew symmetric one, we observe that

$$E^{n+\frac{1}{2}} := \left(M\frac{\Phi^{n+1}-\Phi^n}{\Delta t},\frac{\Phi^{n+1}-\Phi^n}{\Delta t}\right)+\left(M\frac{\Psi^{n+1}-\Psi^n}{\Delta t},\frac{\Psi^{n+1}-\Psi^n}{\Delta t}\right)$$

$$+\left((K\Phi^n,\Phi^{n+1})+(K\Psi^n,\Psi^{n+1})\right) \tag{3.9}$$

$$+\left((S\Psi^n,\Phi^{n+1})-(S\Phi^n,\Psi^{n+1})+(M\Psi^n,\Psi^{n+1})\right). \tag{3.10}$$

Then, we infer that

$$\frac{E^{n+\frac{1}{2}}-E^{n-\frac{1}{2}}}{2\Delta t}=0. \tag{3.11}$$

Therefore, $E^{n+\frac{1}{2}}$ is conserved along the discrete time. □

THEOREM 3.1. (*The damped system*) *For all $\Delta t>0$, the discrete energy $E^{n+\frac{1}{2}}$ associated with* (1.3) *verifies*

$$\frac{1}{2\Delta t}\left(E^{n+\frac{1}{2}}-E^{n-\frac{1}{2}}\right)=-\mu.\|\frac{\Psi^{n+1}-\Psi^{n-1}}{2\Delta t}\|_M^2\leq 0,\forall n\in 0,\cdots,N_t, \tag{3.12}$$

where we denote here by $\|.\|_M$ the norm $\|u\|_M^2=(Mu,u)$.

PROOF. The proof is very similar to the one of Proposition 3.1. The only difference is the damping term that we approximate by:

$$\psi_t\approx\frac{\Psi^{n+1}-\Psi^{n-1}}{2\Delta t}.$$

Then, it suffices to perform the same computations as in (3.5)–(3.8). □

REMARK 3. More details about the discrete energy associated with the systems (1.1), (1.2) and (1.3) and the corresponding numerical results and their proofs can be found in [**6**].

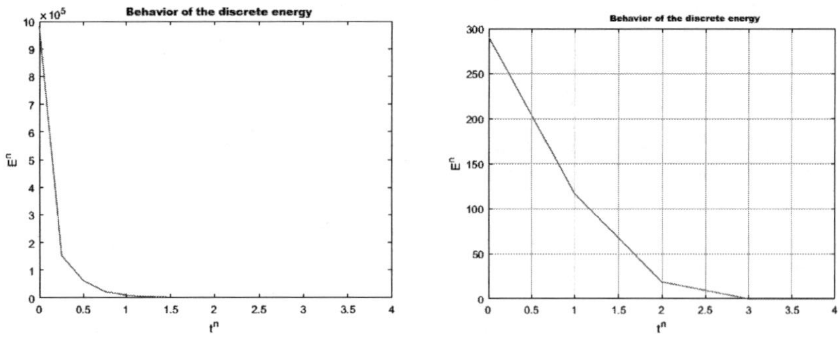

FIGURE 1. The damped case: the discrete energy expressed as a function of t_n (exponential decay with different initial data).

3.2. Numerical experiments. We present in this section some numerical tests that we perform using the scheme defined in (3.2).

Figure 1 represents $log(E^n)$ using $P1-$linear approximation.

Figure 2 represents $log(E^n)$ using $P1-$linear approximation.

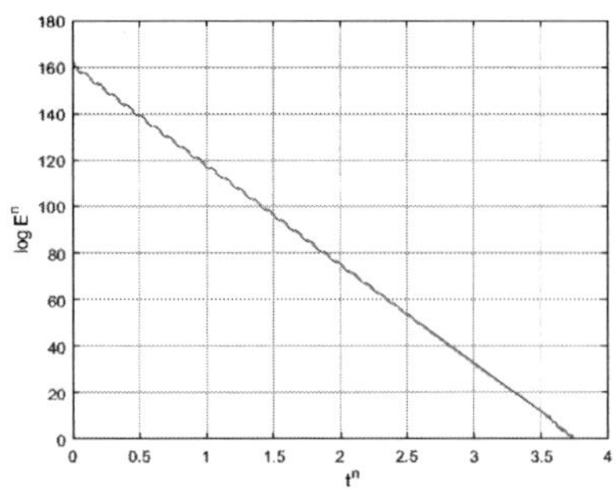

FIGURE 2. Exponential decay rate of discrete energy

The energy E^n decays like an **exponential function** $\exp(-\mu t^n)$ for $\mu > 0$. In the linearly damped case, the discrete energy of the Timoshenko system is exponentially stable.

$$E^n = exp(b)\exp(-\mu t^n), \forall\ n = 1, \cdots, Nt.$$

References

[1] F. Alabau-Boussouira, *Convexity and weighted integral inequalities for energy decay rates of nonlinear dissipative hyperbolic systems*, Appl. Math. Optim. **51** (2005), no. 1, 61–105, DOI 10.1007/s00245. MR2101382

[2] F. Alabau-Boussouira, *Strong lower energy estimates for nonlinearly damped Timoshenko beams and Petrowsky equations*, NoDEA Nonlinear Differential Equations Appl. **18** (2011), no. 5, 571–597, DOI 10.1007/s00030-011-0108-3. MR2846920

[3] M. A. Ayadi, A. Bchatnia, M. Hamouda, and S. Messaoudi, *General decay in a Timoshenko-type system with thermoelasticity with second sound*, Adv. Nonlinear Anal. **4** (2015), no. 4, 263–284, DOI 10.1515/anona-2015-0038. MR3420319

[4] M. A. Ayadi, A. Bchatnia, and M. Hamouda, *Numerical Solutions For A Timoshenko-Type System With Thermoelasticity With Second Sound.* In preparation.

[5] A. Bchatnia, S. Chebbi, M. Hamouda, and A. Soufyane, *Lower bound and optimality for a nonlinearly damped Timoshenko system with thermoelasticity*, Asymptot. Anal. **114** (2019), no. 1-2, 73–91. MR3998021

[6] S. Chebbi and M. Hamouda, *Discrete energy behavior of a damped Timoshenko system*, Comput. Appl. Math. **39** (2020), no. 1, Paper No. 4, 19, DOI 10.1007/s40314-019-0982-6. MR4036536

[7] C. M. Dafermos, *Asymptotic behavior of solutions of evolution equations*, Nonlinear evolution equations (Proc. Sympos., Univ. Wisconsin, Madison, Wis., 1977), Publ. Math. Res. Center Univ. Wisconsin, vol. 40, Academic Press, New York-London, 1978, pp. 103–123. MR513814

[8] H. D. Fernández Sare and R. Racke, *On the stability of damped Timoshenko systems: Cattaneo versus Fourier law*, Arch. Ration. Mech. Anal. **194** (2009), no. 1, 221–251, DOI 10.1007/s00205-009-0220-2. MR2533927

[9] R. D. Krieg, *On the behavior of a numerical approximation to the rotatory inertia and transverse shear plate.* Journal of Applied Mechanics, vol. **40**, no. 4, pp. 977–982, (1973).

[10] S. A. Messaoudi and M. I. Mustafa, *On the stabilization of the Timoshenko system by a weak nonlinear dissipation*, Math. Methods Appl. Sci. **32** (2009), no. 4, 454–469, DOI 10.1002/mma.1047. MR2493590

[11] S. A. Messaoudi, M. Pokojovy, and B. Said-Houari, *Nonlinear damped Timoshenko systems with second sound—global existence and exponential stability*, Math. Methods Appl. Sci. **32** (2009), no. 5, 505–534, DOI 10.1002/mma.1049. MR2500690

[12] G. Papukashvili, *On a numerical algorithm for a Timoshenko type nonlinear beam equation*, Rep. Enlarged Sess. Semin. I. Vekua Appl. Math. **31** (2017), 115–118. MR3746924

[13] M. L. Santos, D. S. Almeida Júnior, and J. E. Muñoz Rivera, *The stability number of the Timoshenko system with second sound*, J. Differential Equations **253** (2012), no. 9, 2715–2733, DOI 10.1016/j.jde.2012.07.012. MR2959386

[14] Timoshenko S P. *On the correction for shear of the differential equation for transverse vibrations of prismatic bars.* Philosophical Magazine Series, 6 (41), 245 (1921), pp. 744–746.

[15] J. Vancostenoble and P. Martinez, *Optimality of energy estimates for the wave equation with nonlinear boundary velocity feedbacks*, SIAM J. Control Optim. **39** (2000), no. 3, 776–797, DOI 10.1137/S0363012999354211. MR1786329

Imam Abdulrahman Bin Faisal University, Deanship of Preparatory Year and Supporting Studies, Department of Basic Sciences, Dammam, Kingdom of Saudi Arabia

Email address: `mmhamouda@iau.edu.sa`

Contemporary Mathematics
Volume **757**, 2020
https://doi.org/10.1090/conm/757/15244

On the finite-time stabilization of some hyperbolic control systems by boundary feedback laws: Lyapunov approach

Chaker Jammazi and Ghada Ben Belgacem

ABSTRACT. This paper discusses the finite-time stabilization of some linear hyperbolic partial differential equations by persistently exciting boundary feedbacks. The construction of stabilizing feedbacks is based on Lyapunov approach. A particular attention is paid to transport and wave control systems. The obtained results are used to solve two problems: the first one is finite-time stability of the SMB chromatography while the second application is the finite-time controllers of two strings connected by point mass.

CONTENTS

1. Introduction

Since remarkable papers and books,(e.g. [**1**, **3**–**5**, **15**, **19**, **22**, **28**, **29**, **45**–**47**, **49**, **53**, **55**, **58**, **60**], and the references therein), problems of controllability and stabilizability of controlled partial differential equations become one of challenging topics in control theory.

Among of new problems in this area, are the stabilization of some partial differential equations with conservation laws. It is worth mentioning that many results exist in the literature around this kind of conservative systems; authors in [**32**, **62**] developed the asymptotic stability for the case of second-order systems of conservation laws. Moreover, a generalization to $n \times n$ systems was given later in [**48**]. Various strategies used a strict Lyapunov function whose time derivative can be made strictly negative definite by an appropriate choice of the boundary controls and leads to rapid stabilization, was presented in [**6**, **21**–**23**].

2010 *Mathematics Subject Classification.* Primary 93C20.

Key words and phrases. Transport equation, wave equation, strings, point mass, boundary feedback laws, finite-time stabilization.

Recently, a special and interesting case of stabilizability of some partial differential equations, called "finite-time stabilization" has spread. This type of vanishing solutions of dynamical systems seems a natural phenomenon, since all biologic living finished by disappears in finite-time. In control theory, this problem was investigated by several authors the last years, see, for instance; [**1, 10–12, 18–20, 33, 35–42, 53, 54, 56–58**], and references therein.

Basically, for controlled dynamical systems, the finite-time stabilization means the design of feedback controller which renders the closed loop system Lyapunov stable and all trajectories converge to the origin in finite-time. In the most papers cited below, e.g. [**10, 11, 33, 35–39, 57**], the finite-time stability required the construction of Lyapunov function V satisfying the differential inequality

$$\dot{V} + cV^{\alpha} \leq 0, \, \alpha \in (0, 1), \, c > 0. \tag{1}$$

For partial differential equation, the above inequality renders the stabilization task more difficult, and it is used by fewer authors [**14, 24, 59**] for the problems of Schrodinger equation, and for the wave and the heat equation respectively. In order to avoid the construction of Lyapunov function, a new idea developed in the remarkable papers [**1, 25, 58**] consists in apply boundary feedbacks that are solutions of finite-time stable ordinary differential equation. This makes to stabilize in finite-time some hyperbolic systems. As applications, in [**1, 58**], the authors has treated examples of networks as well as the regulation problem of water flow in channels, and networks of strings.

The common approach used for the stabilization of some hyperbolic systems is through the explicit evolution of the Riemann invariants along the characteristics [**5, 8, 27, 60**]. These invariants are solutions of transport equations which are widely considered in literature, predominate in the modeling of traffic flow [**2**], heat exchangers [**13**], open channel flow [**26**], SMB chromatography [**7**], and in many other fields. However, this approach is limited in higher dimension and becomes a difficult task when characteristics intersect. Moreover, the finite-time stabilization is limited and just occurs in a definite domain which is attentively built in order to avoid shocks or discontinuities.

In recent years, the stabilization of some evolution equations by persistently exciting feedback laws has attracted many authors [**16, 17, 34, 43, 50, 52**].

For example, in finite dimension, this notion of persistently exciting signal is known (see, for example, [**16, 50, 52**], and references therein), and seems a good alternative for the rapid stabilization problems (e.g. exponential stabilization).

For infinite dimensional, this notion is again used by many authors, for example, in [**17**] the authors showed that exponential stabilization of a system of transport equations on a network of N circles intersecting at a single point O is possible by using persistently exciting signals.

It is remarked in [**34**], that the notion of persistently exciting signals only is not sufficient for the stabilization problems in infinite dimension. The authors in [**34**] showed that the persistently exciting condition coupled with a generalized observability inequality lead to exponential stability in Hilbert space.

In contrast, in this work, we prove that the uniform persistently exciting condition is crucial to get the finite-time stabilization of some evolution control systems. Hence, in order to give a physical meaning to our state feedback laws, it is possible, to assume that, the square of the right of the boundary conditions are "uniformly persistently exciting" this punctual condition means, for example, for the transport equation, the following property $(\mathcal{P})$: there exists $c > 0$ such that for $h > 0$, $\frac{1}{h}\int_t^{t+h} |y(s, L)|^2 \, ds \geq c$. This property forces the solution to reaches the equilibrium point in finite-time.

Based on the above observation, our aim is to give a robust condition for finite-time stabilization of some examples of hyperbolic systems by avoiding the characteristic lines and the explicit solutions. Our approach consists on the construction of suitable Lyapunov function together with appropriate boundary feedback laws leading to the finite-time stabilization of controlled transport and wave equations.

Our study is illustrated by two examples: The first one is the SMB chromatography model and the second is the system of strings connected by point mass.

For the best knowledge, this two hybrid hyperbolic systems are attractive and studied by many authors. For the SMB chromatography where the system's parameters and boundary conditions switch periodically in time, the issue of exponential stability was established by Coron et al. [**7**] based on so-called multiple Lyapunov function.

Regarding the problem of vibrating strings connected by point mass one can cite e.g. [**4, 8, 15, 29, 49, 55**]. Controllability and asymptotic stabilizability issues are largely discussed in the famous paper of Hansen and Zuazua [**29**]. In particular, necessary conditions for boundary controllability are given and stabilizing feedback laws are established to put the system in the equilibrium configuration. Briefly, the analysis requires the energy estimate in the good asymmetric space, which leads to decreasing exponentially the energy function provided that the velocity feedbacks are applied at both ends.

This work expanded quickly in the control community. For example, in [**49**], authors have improved feedbacks obtained in [**29**], by showing that the energy to the right of the mass decays like C/t provided that the initial data belong to certain invariant subspace of semigroup of operators that generates the solution of the system.

In [**9**], the problem of finite-time stabilization of two strings linked by point mass is discussed. The analysis shows that with Bhat-Bernstein feedbacks in the boundary, the stabilization is partial (i.e. weak). To handled the problem, we have suggested a feedback in the charge.

In this work we show how feedbacks given by Hansan and Zuazua in [**29**] lead not only to the asymptotic stability of strings, but provide again the finite-time stability by excited boundary feedbacks.

The paper is structured as follows. The next section deals with definition and technical results on linear transport equation. The main results are presented in sections 3 and 4; in particular, (a) the problem of finite-time stabilization of some hyperbolic equations by boundary feedback laws is proved by using Lyapunov approach, (b) as applications, we have studied two problems:(1) the controllability and finite-time stabilizability of SMB chromatography, and (2) the finite-time stabilization of strings connected by point mass.

2. Preliminary on transport equation

2.1. The transport equation revisited. Let $T > 0$ and $L > 0$, we consider the linear control system describing the transport equation

$$y_t + v(y(t, x))\, y_x = 0, \qquad 0 < x < L,\ 0 < t < T, \tag{2}$$

$$y(t, 0) = u(t), \qquad 0 < t < T, \tag{3}$$

$$y(0, x) = y_0(x), \qquad 0 < x < L, \tag{4}$$

where, at time t, the control is $u(t) \in \mathbf{R}$ and the state is $y(t, .) : (0, L) \to \mathbf{R}$. We assume that the velocity of the propagation denoted by v is such that $v \geq c$, where c denotes some positif constant.

In the case $v = 1$, it was proved in [**19**, section 2.1] that for every $y_0 \in L^2(0, L)$ and $u \in L^2(0, T)$ the Cauchy problem (2)-(4) admits a unique weak solution in $L^2((0, L) \times (0, T))$. This result is recalled in the following theorem

THEOREM 2.1. ([**19**, Theorem 2.4, pp. 27]) *For every $y_0 \in L^2(0, L)$ and $u \in L^2(0, T)$, the Cauchy problem* (2)-4 *has unique solution. Moreover, this solution satisfies*

$$||y(t, .)||_{L^2(0,\, L)} \leq ||y^0||_{L^2(0,\, L)} + ||u||_{L^2(0,\, T)}. \tag{5}$$

In addition, it is known, see e.g. [**31**] that this solution can be found explicitly thanks to the method of characteristics which reduces the study of the transport equation as ordinary differential equation along the characteristics of the transport operator. Hence the solution of (2)-(4) can be expressed as follows

$$y(t, x) = \begin{cases} y_0(x - tv) & \text{if } x - tv \geq 0, \\ u(t - \frac{x}{v}) & \text{if } x - tv < 0. \end{cases} \tag{6}$$

Before to present our work, we need to recall the notion of finite-time stabilization of control systems. Let be the dynamical controlled system

$$\begin{aligned} &\dot{x} = f(x, u), \qquad x \in \mathcal{X},\ u \in \mathcal{U}, \\ &f(0, 0) = 0, \end{aligned} \tag{7}$$

where $f : \mathcal{X} \times \mathcal{U} \to \mathcal{X}$ is a continuous function, with $\mathcal{X}$ and $\mathcal{U}$ are Hilbert spaces denotes respectively the state space and the control space.

DEFINITION 2.1. We assume that the control system (7) is well-posed. System (7) is said to be finite-time stabilizable if there exists an admissible control $x \mapsto u(x) \in \mathcal{U}$ with $u(0) = 0$ and the closed loop system $\dot{x} = f(x, u(x))$ is finite-time stable. More precisely:

(a) The origin of $\dot{x} = f(x, u(x))$ is Lyapunov stable.
(b) there exists $r > 0$ and $T = T(x(0)) > 0$, called the settling-time function such that if $|x(0)| < r$, then $x(t) = 0$ for every $t \geq T$.

2.2. Finite-time stabilization of transport equation by explicit solution. In this part, we will explain the idea proposed in [**58**] for the finite-time stabilization of the transport equation.

Since the velocity of the transport equation depends uniformly on the solution, then the main strategy developed in [**58**] is to authorized feedbacks of the form

$$u(y(t, 0)) = -k\ sgn(y(t, 0))|y(t, 0)|^{\alpha}, \tag{8}$$

with $(k,\alpha) \in (0,+\infty) \times (0,1)$, and "sgn" is the usually sign function defined by

$$sgn(x) := \begin{cases} \dfrac{x}{|x|} & \text{if } x \neq 0, \\ 0 & \text{if } x = 0, \end{cases}$$

and the boundary condition satisfies the ODE:

$$\dot{y}(t,\,0) = -k\; sgn(y(t,\,0))|y(t,\,0)|^{\alpha}. \tag{9}$$

Clearly, $y(t,\,0)$ vanishes in finite-time, and therefore, by the explicit solution (6), the state $y(t,\,x)$ solution of (2)-(9) again vanish in finite-time.

3. Finite-time stabilization by Lyapunov approach

The main goal of this part is to present another alternative to stabilize in finite-time the control system (2)-(4) by using Lyapunov approach. The stabilizing feedback laws are boundary. For sake of simplicity, we assume that $v = 1$.

Before to give the main result of this subsection, we need some useful lemmas.

LEMMA 3.1. [**10**] *Assume that* V *is a nonnegative function defined on* $[0,\,+\infty)$, *and*

$$\dot{V} \leq -c\,V^{\alpha}, \tag{10}$$

where $c > 0$ *and* $0 < \alpha < 1$ *are two constants. Then, there exists a settling time* $T(V(0))$ *depending on initial condition such that*

$$V = 0,\, \forall\, t \geq T, \text{ and } T \leq \frac{V^{1-\alpha}(0)}{c(1-\alpha)}.$$

Sometimes, in the stabilization problem of evolution equation by boundary feedback laws, we need the following version.

LEMMA 3.2. *Let* $E(t)$ *be a nonnegative function defined on* $[0,\,+\infty)$ *satisfying the differential inequality*

$$\dot{E} \leq -c\,f(t)\,E^{\beta}, \tag{11}$$

where $\beta \in (0,\,1)$, $f(.)$ *is nonnegative continuous function and satisfies on* $[0,\,+\infty)$ *the "growth condition"* $\displaystyle\int_0^t f(s)ds \geq \mu\, t^m > 0$ *with* $\mu,\, m > 0$.

Then, there exists a settling time T^* *such that*

$$E(t) = 0,\, \forall\, t \geq T^*.$$

PROOF. The proof is based on comparison principle [**44**, page 110]. Let V be positive function solves the differential equation

$$\dot{V} = -c\,f(t)\,V^{\beta},\; c > 0,\; \beta \in (0,\,1), \tag{12}$$

with the initial condition $V(0) := V_0$. This solution is given by

$$V(t) = \begin{cases} [V_0^{1-\beta} - c(1-\beta)\int_0^t f(s)ds]^{1/(1-\beta)},\; 0 \leq t \leq T^* \textit{ and } V_0 \neq 0, \\ 0,\; if\; t \geq T^* \textit{ and } V_0 \neq 0, \\ 0,\; if\; V_0 = 0. \end{cases} \tag{13}$$

where the time T^* will be defined later. Clearly for $V_0 \neq 0$, we have

$$(14) \qquad V_0^{1-\beta} - c(1-\beta)\int_0^t f(s)ds \leq V_0^{1-\beta} - c(1-\beta)\mu t^m.$$

Since, the first case holds provided that $V_0^{1-\beta} - c(1-\beta)\int_0^t f(s)ds \geq 0$, then, if $V_0^{1-\beta} - c(1-\beta)\mu t^m \geq 0 \Leftrightarrow t \leq \big(\frac{V_0^{1-\beta}}{c\mu(1-\beta)}\big)^{1/m}$, then one takes in this case

$$T^* := \Big(\frac{V_0^{1-\beta}}{c\mu(1-\beta)}\Big)^{1/m}$$

and then for $t \geq T^*$, we have immediately $V = 0$ □

Now, we consider the initial condition $E(0) := E_0$ is such that $E_0 \leq V_0$, then by comparison principle [**44**, page 110], we get

$$E(t) \leq V(t), \forall t \geq 0.$$

Therefore, the energy E vanish in finite-time.

LEMMA 3.3. *We consider the assumptions of the Lemma 3.2 and we assume that the function f satisfies the "uniformly persistently exciting condition"* [**50**].
($\mathcal{P}$): There exists $c > 0$ such that for every $t \geq 0$ and $h > 0$ we have

$$\frac{1}{h}\int_t^{t+h} f(s)ds \geq c.$$

Then, the energy E vanish in finite-time.

The above Lemma is a particular case of the next proposition.

PROPOSITION 3.4. *Let $0 < r < 1$, and we consider $u(t)$ a locally Lipschitz positive function defined on $[0, +\infty)$. Let f a nonnegative continuous function satisfies the property ($\mathcal{P}$) and k be a positive real number, and we assume that we have the inequality*

$$(15) \qquad u(t) \leq u(t_0) - k\int_{t_0}^t f(s)u^r(s)ds,\ t \geq t_0 \geq 0.$$

Then u vanishes in finite-time.

PROOF. Since u is locally Lipschitz, then the time derivative of u is defined in Dini sense by $D^+u(t) := \limsup_{h\to 0^+} \dfrac{u(t+h)-u(t)}{h} < \infty$. Hence, by the constraint (15) we get $D^+u(t) \leq -\limsup_{h\to 0^+} k\displaystyle\int_t^{t+h} \frac{f(s)u^r(s)}{h}ds \leq 0$, and therefore u is a decreasing function over $[t_0, +\infty)$. Thus for $t \leq s \leq t+h$, we have $-u^r(s) \leq -u^r(t+h)$. Taking into account of continuity of u and the growth condition ($\mathcal{P}$), we get

$$D^+u(t) \leq -c\,k\,u^r(t).$$

By Lemma 3.1 we conclude that u vanish in finite-time. This ends the proof. □

For example, the scalar control system $\dot{x} = u$, where, $u \in \mathbf{R}$ is the control and $x \in \mathbf{R}$ is the state can be stabilized in finite-time by $u(t, x) = -\alpha(t)x^{1/3}$ with $\alpha(t) > 0$ is the persistently exciting function satisfying the constraint $\int_t^{t+h}\alpha(s)ds \geq c\,h$.

Indeed, the candidate Lyapunov function $V = \frac{1}{2}x^2$ satisfies along the system $\dot{V} \leq -\alpha(t)V^{2/3}$. Lemma 3.3 allows to conclude.

Now, we turn to prove the finite-time stabilizability of (2)-(4) by using Lyapunov function.

3.1. The case of transport equation. In this part, by using Lyapunov approach, we will prove the finite-time stabilization of the transport equation by boundary feedback laws provided that the property of uniformly persistently exciting is satisfied. Before to present the stabilizing feedback law, we remark that if $y(t, L)$ vanish for all $t \geq 0$, then the finite-time stabilizability of (2)-(4) is excluded in force.

LEMMA 3.5. *If $y(t, L) = 0$ for all $t \geq 0$, then (2)-(4) cannot be finite-time stabilizable.*

PROOF. Assume for the control system (2)-(4) that $y(t, L) = 0$ for all $t \geq 0$. Let E be the energy function defined by $E := \int_0^L |y(t, x)|^2 dx$. By a simple calculation, we get $\dot{E} = -(y^2(t, L) - u^2(t))$ which turn implies with the assumption that $\dot{E} = u^2 \geq 0$. Hence for $u = 0$ we have a system with conservation laws and then the finite-time stabilization is impossible. Again, for $u > 0$ for almost every where, we conclude by Lyapunov first method that (2)-(4) is unstable [**44**]. □

In the next, we will assume that $y(t, L) \neq 0$ for almost every $t \geq 0$, and to handle the problem it me possible to excite uniformly the boundary condition $y(t, L)$. The notion of persistently exciting condition is significative and seems an attractive idea to stabilize many control systems, see for instance [**16, 17, 34, 50**] and references therein. For control systems in finite dimension, this idea is used by Loria and its collaborators [**50, 52**]. For PDE's, the notion of persistence is used in [**17, 34**]. Then we have the following finite-time stabilization result.

THEOREM 3.6. *Consider the system (2)-(4). If there exists a nonnegative constant c such that for every $t \geq 0$ and $h > 0$ is small enough, we have*

$$\int_t^{t+h} y^2(s, L)ds \geq ch \tag{16}$$

Then, for every $y_0 \in L^2(0, L)$, the boundary feedback law

$$u(y) = k\,y(t, L),\ k \in (0, 1) \tag{17}$$

stabilizes in finite-time (2)-(4). More precisely, there exists a settling time T^* depending on initial condition such that $y(t, x) = 0$, $\forall t \geq T^*$ and for almost every $x \in [0, L]$.

PROOF. For $y_0 \in L^2(0, L)$, the Cauchy system

$$y_t + y_x = 0, \qquad x \in (0, L), t > 0, \tag{18}$$

$$y(t, 0) = ky(t, L), \tag{19}$$

$$y(0, x) = y_0(x), \qquad x \in (0, L), \tag{20}$$

admits a unique week solution [**19**, Proposition 2.12]. Let us consider the Lyapunov function defined by

$$V = \left(\int_0^L |y(t, x)|^2 dx\right)^\alpha, \text{ with } \alpha > 1. \tag{21}$$

The time derivative of V along the system (2)-4 is given by

$$\begin{aligned}\dot{V} &= \alpha\Big(\int_0^L |y(t,\,x)|^2dx\Big)^{\alpha-1}\Big(2\int_0^L y(t,\,x)y_t(t,\,x)dx\Big)\\ &= -\alpha\,V^{\frac{\alpha-1}{\alpha}}\Big(y^2(t,\,L)-u^2(t)\Big).\end{aligned} \tag{22}$$

Hence, under the feedback law $u(y)=k\,y(t,\,L)$, with $0<k<1$ we get

$$\dot{V}=-\alpha(1-k^2)\,y^2(t,\,L)\,V^{1-\frac{1}{\alpha}}\leq -c\,f(t)\,V^{\beta}, \tag{23}$$

where $c:=\alpha(1-k^2)>0$ and $\beta:=1-\frac{1}{\alpha}\in(0,\,1)$ and $f(t)=y^2(t,\,L)$ is nonnegative and continuous (the continuity of the solution in the boundary is due to hidden regularity, see e. g. [**19**]), and satisfies the property $(\mathcal{P})$ of the Prop 3.4. Then by Lemma 3.3, there exists a settling time T^* such that $V(t)=0,\,\forall\,t\geq T^*$. Then for every $x\in[0,\,L]$, the state $y(t,\,x)$ vanishes in finite-time. □

COROLLARY 3.7. *We consider the control system*

$$\begin{aligned} &y_t-y_x=0, && (t,\,x)\in(0,\,\infty)\times(0,\,L),\\ &y(t,\,L)=u(t), && t>0,\\ &y(0,x)=y_0(x), && x\in(0,\,L).\end{aligned}$$

Then, for every $y_0\in L^2(0,\,L)$ and for every $k\in\mathbf{R}$ such that $0<|k|<1$, this control system is finite-time stabilizable under the excited boundary feedback law

$$u(y)=k\,y(t,\,0),\ \textit{with},\ \int_t^{t+h}y^2(s,\,0)ds\geq c\,h. \tag{24}$$

3.2. The problem of wave equation. Consider the control system modeled by means of a partial differential equation describing the motion of the wave equation excited by one boundary control

$$y_{tt}-y_{xx}=0, \qquad x\in(0,1),\,t\geq 0, \tag{25}$$

$$y(t,0)=0, \qquad t\geq 0, \tag{26}$$

$$y_x(t,1)=u(t), \qquad t\geq 0, \tag{27}$$

$$y(0,\,x)=\alpha^0(x),\,y_t(0,\,x)=\beta^0(x),\,x\in(0,\,1). \tag{28}$$

For every $\alpha^0\in H^1_{(0)}(0,\,1)$[1] and $\beta^0\in L^2(0,\,1)$, the above Cauchy problem is well posed [**19**, Section 2.4]. More precisely, the solution lives in

$$\mathcal{C}^0([0,\,\infty);H^1_{(0)}(0,\,1))\bigcap\mathcal{C}^1([0,\,\infty);L^2(0,\,1)).$$

This vector space is equipped with the scalar product

$$\langle f,\,g\rangle_{H^1_{(0)}(0,\,1)}:=\int_0^1 f_x\,g_x\,dx.$$

A basic tool to study the asymptotic stability is the Lyapunov function which constitutes a generalization of the energy. We will show that under a suitable choice of Lyapunov function and feedback law, the assumptions of Lemma 3.2 are satisfied and therefore the finite-time stabilization of the wave equation is established.

Before to present the main result of this section, we give a boundary obstruction to finite-time stabilizability of the wave equation.

[1] $H^1_{(0)}(0,\,1):=\{\alpha\in H^1(0,\,1):\alpha(0)=0\}$.

LEMMA 3.8. *If $y_t(t,\,1) = 0$, then the control wave system* (25)-(27) *cannot be finite-time stabilizable.*

PROOF. We consider system (25)-(27) and we assume that $y_t(t,\,1) = 0$. Let E be the energy function defined by $E := \int_0^1 (|y_t|^2 + |y_x|^2)dx$, then by a simple calculation we get along the system that

$$\dot{E} = 0,$$

and therefore the system is with conservation laws. Then it cannot be finite-time stable. □

In the sequel, we will consider the case that $y_t(t,\,1) \neq 0$ for almost every $t > 0$.

THEOREM 3.9. *Let k be a non-negative real number, and we assume that $y_t(t,\,1)$ satisfies the persistently exciting condition property:*

($\mathcal{P}$) there exists a nonnegative constant c such that for every $t \geq 0$ and $h > 0$ is small enough, we have

$$\int_t^{t+h} y_t^2(s,\,1)ds \geq c\,h. \tag{29}$$

Then for every $\alpha^0 \in H^1_{(0)}(0,\,1)$, $\beta^0 \in L^2(0,\,1)$ and under the boundary feedback law

$$u(y) = -k\,y_t(t,\,1), \tag{30}$$

the wave control system (25) *is finite-time stable. More precisely, there exists a settling time T^* depending on initial condition such that for every almost $x \in [0,\,L], y(t,\,x) = 0, \forall\, t \geq T^*$.*

PROOF. Let $\alpha^0 \in H^1_0(0,\,1)$ and $\beta^0 \in L^2(0,\,1)$. For the control system (25)-(27) we will choose the following control Lyapunov function

$$E = \left(\int_0^1 (|y_t|^2 + |y_x|^2)dx\right)^{\alpha},\ \alpha > 1. \tag{31}$$

Using integration by parts, the time derivative of E along the wave control system is given by

$$\begin{aligned}\dot{E} &= \alpha\left(\int_0^1 (|y_t|^2 + |y_x|^2)dx\right)^{\alpha-1}\left(\int_0^1 (2y_t y_{tt} + 2y_x y_{tx})dx\right)\\ &= 2\alpha\, E^{\frac{\alpha-1}{\alpha}}\, u(t) y_t(t,1).\end{aligned}$$

Once if we take $u(t) = -k\,y_t(t,1)$ with $k > 0$ a positive constant, then we have

$$\dot{E} = -2\alpha\, k\, E^{\frac{\alpha-1}{\alpha}}\, y_t^2(t,1) \leq -c\, f(t)\, E^{\beta}, \tag{32}$$

where $c := 2\alpha\, k, \beta := \frac{\alpha-1}{\alpha} \in (0,\,1)$, and $f(t) := y_t^2(t,1)$. Thanks to (31) and (32) one deduces that E admits a finite limit as $t \to \infty$ and therefore $y_t(t,\,.) \in L^2(0,\,1)$. In particular due to the regularity property of the wave equation, we have again $y_t(t,1) \in \mathcal{C}^0(0,\infty)$ this implies that $t \mapsto f(t) := y_t^2(t,1)$ is nonnegative and continuous function. Hence, by the persistence condition (29), one conclude by Lemma 3.3 that E vanishes in finite-time. □

Let T^* be the settling time of E, then we have; for every $x \in [0,\,1]$, $y_x(t,\,x) = 0$ for $t \geq T^*$. Since the Poincaré Inequality is applicable in $H^1_{(0)}(0,\,1)$ then we have

$$||y||_{H^1_{(0)}(0,\,1)} \leq c'||y_x||_{L^2(0,\,1)},\ c' > 0, \tag{33}$$

which implies again for every $x \in [0,\, 1]$

$$y(t,x) = 0, \quad \forall t \geq T^*.$$

COROLLARY 3.10. *Let a be non-negative real number and we consider the system*

$$y_{tt} - y_{xx} + ay = 0, \qquad x \in (0,1),\, t \geq 0, \tag{34}$$

with the boundary and initial conditions (26)-(28)*. Then, under the boundary feedback law $u(y) = -k\, y_t(t,\, 1)$ subject to persistence excitation* (29)*, and $k > 0$, the control system (*(34)-(26)-(28)*) is finite-time stable.*

PROOF. For $\alpha > 1$, let the candidate Lyapunov function V defined by

$$V := \Big(\int_0^1 (|y_t|^2 + |y_x|^2 + a|y|^2)dx\Big)^{\alpha}.$$

A simple calculation yields to

$$\begin{aligned} \dot{V} : &= \alpha\, V^{(\alpha-1)/\alpha} u(y) y_t(t,\, 1) \\ &\leq -c\, f(t)\, V^{\beta}. \end{aligned}$$

By the same argument, as in the proof of Theorem 3.9, one conclude that f is continuous and nonnegative function. Then the conclusion of the corollary follows from Lemma 3.3. □

The aim of the next result is to prove the finite-time stabilization by boundary feedback law if the term y is affected by a perturbation a function on t and x. We assume that $a \in L^{\infty}([0,\, +\infty) \times [0,\, 1])$ and satisfies the condition $||a||_{\infty} \leq r$.

THEOREM 3.11. *Let the perturbed wave equation*

$$y_{tt} - y_{xx} + a(t,\, x)y = 0, \qquad x \in (0,1),\, t \geq 0, \tag{35}$$

with the boundary and initial conditions (26)-(27)*. We assume that $y(0,\, 1) = \alpha^0(1) \neq y_t(t,\, 1)$ then under the boundary feedback law*

$$u(y) = -\frac{r||y||_{L^2[0,\,1]}||y_t||_{L^2[0,\,1]}}{y_t(t,\, 1)} - k\, y_t(t,\, 1), \tag{36}$$

where the persistence exciting condition (29) *holds and $k > 0$, the control system* (26)-(28)- (35) *is finite-time stable. Moreover, if $y_t(t,\, 1)$ vanish in finite-time, the feedback $u(y)$ is well defined and cannot blow up in finite-time.*

PROOF. For $\alpha > 1$, we consider the candidate Lyapunov function defined by

$$V := \Big(\int_0^1 (|y_t|^2 + |y_x|^2)dx\Big)^{\alpha}.$$

Using integrations by parts, in closed loop, the time derivative of V along the new system leads to

$$\dot{V} = \alpha\, V^{(\alpha-1)/\alpha}\Big(-\int_0^1 ayy_t\, dx + u\, y_t(t,\, 1)\Big), \tag{37}$$

by Cauchy-Schwartz inequality we have

$$\int_0^1 |ayy_t|\, dx \leq r||y||_{L^2[0,\,1]}||y_t||_{L^2[0,\,1]}, \tag{38}$$

then, from (38), and under the proposed feedback law u in (36), we get

$$\dot{V} \leq -\alpha\, k\, y_t^2(t,\, 1)V^{\beta},\ where\, \beta = \frac{\alpha - 1}{\alpha} \in (0,\, 1). \tag{39}$$

Clearly, by Lemma (3.3) combined with Poincaré inequality on $H^1_{(0)}(0,\, 1)$, the state $y(t,\, x)$ vanishes in finite-time for all $x \in [0,\, 1]$. Let T^* be the settling time of $y(t,\, x)$.

If $y_t(t,\, 1)$ vanish in finite-time, we show that $||y||_{L^2[0,\, 1]}||y_t||_{L^2[0,\, 1]}$ vanishes in finite-time faster than $y_t(t,\, 1)$.

We have, for all $t \geq 0$,

$$\begin{aligned} y(t,\, 1) &= y(0,\, 1) + \int_0^t y_s(s,\, 1)ds && (40)\\ &= \alpha^0(1) + \int_0^t y_s(s,\, 1)ds, && (41) \end{aligned}$$

since $y_t(t,\, x) = 0$ for every $x \in [0,\, 1]$ and $t \geq T^*$, we get

$$\begin{aligned} |y(t,\, 1) - \alpha^0(1)| &\leq \int_0^t |y_s(s,\, 1)|ds && (42)\\ &\leq \int_0^{T^*} |y_s(s,\, 1)|ds + \int_{T^*}^t |y_s(s,\, 1)|ds && (43)\\ &\leq T^*||y_t(t,\, 1)||_{L^\infty[0,\, T^*]}. && (44) \end{aligned}$$

Then

$$\frac{||y||_{L^2[0,\, 1]}||y_t||_{L^2[0,\, 1]}}{|y_t(t,\, 1)|} \leq \frac{T^*||y||_{L^2[0,\, 1]}||y_t||_{L^2[0,\, 1]}}{|y(t,\, 1) - \alpha^0(1)|},$$

since $\alpha^0(1) \neq y(t,\, 1)$ vanishes of $t \geq T^*$. Then for $t \geq T^*$, we have

$$\frac{T^*||y||_{L^2[0,\, 1]}||y_t||_{L^2[0,\, 1]}}{|y(t,\, 1) - \alpha^0(1)|} = 0,$$

and $u(y) = 0$ for $t \geq T^*$. This ends the proof. □

4. Applications

In this section, we will apply our results to some control systems governed by partial differential equations. The first one is the SMB chromatography where we prove some controllability result around periodic trajectory and we establish finite-time boundary stabilizing feedback laws. The second example treated in this paper is the vibrating strings connected by point mass, again we construct boundary feedback laws making the system in closed loop finite-time stable. Various feedback laws are discussed.

4.1. Finite-time stabilization of SMB chromotography.

4.1.1. *Physical presentation.* The description of the SMB chromatograph is given in [**5**,**7**,**63**]. As defined in [**5**], the Simulated moving bed (SMB) chromatography is a process used by chemists and engineers to separate two (or more) chemical species in a fluid phase that would be difficult or impossible to resolve otherwise. It is a technology where several interconnected chromatographic columns are switched periodically against the fluid flow. The use of many columns allows for a continuous separation with a better performance than the discontinuous single-column chromatography. A standard SMB process is represented in Fig.1.

The existence of two inlets flows: feed mixture which contains the components A and B to be separated and the solvent which is the carrier that moves the analyst through the chromatograph, and of two outlets flows: extract and raffinate, divide the system in four zones each containing one chromatographic column. The four columns are labeled 1,2,3,4. The four operating zones are labeled I, II, III, IV. Pumps connected at each port determine the liquid phase flow rates in all the zones which is the part used to store the eluant and injecting it under pressure into the column. The feed mixture is injected between zone II and III. The less adsorbed component (A) is collected at the raffinate port and the more adsorbed component (B) at the extract port. The separation of the two species is performed in zones II and III, whereas zones I and IV dedicated to adsorbent regeneration and solvent recycling.

A liquid-solid counter-current movement is obtained by a periodic circular switching of the columns in the opposite direction to the liquid phase flow as shown in Fig.2.

4.1.2. *Mathematical Modeling.* Let us introduce the following notations:

- The switching time period is T and the column length is L.
- $C_i^\ell(t,x) \geq 0,\ 0 \leq x \leq L,\ t \geq 0$, is the concentration of species $\ell \in \{A, B\}$ in the column $i \in \{1,2,3,4\}$.
- V_1 is the fluid velocity in the columns located in zones I and III, V_2 is the fluid velocity in the columns located in zones II and IV, with $V_1 > V_2 > 0$.
- $h^\ell > 0$ is the Henry coefficient.
- $V_F > 0$ is the constant fluid velocity while C_F^A, $C_F^B > 0$ are the constant species concentrations in the input flow.

We introduce also the following vector and matrix notations, where $'$ denotes the symbol of transposition.

$$C^\ell(t,x) = \left(C_1^\ell(t,x), C_2^\ell(t,x), C_3^\ell(t,x), C_4^\ell(t,x)\right)',$$

$$U^\ell = \left((V_F/V_1)C_F^\ell, 0, 0, 0\right)',$$

$$\gamma = diag\{V_1, V_2, V_1, V_2\},$$

$$K = \begin{pmatrix} 0 & 0 & 0 & V_2/V_1 \\ 1 & 0 & 0 & 0 \\ 0 & V_2/V_1 & 0 & 0 \\ 0 & 0 & 1 & 0 \end{pmatrix},$$

$$P = \begin{pmatrix} 0 & 0 & 0 & 1 \\ 1 & 0 & 0 & 0 \\ 0 & 1 & 0 & 0 \\ 0 & 0 & 1 & 0 \end{pmatrix} \quad \text{is the permutation matrix.}$$

To simplify the notations, we write C for C^ℓ, U for U^ℓ and h rather than h^ℓ.

The general form for the hyperbolic system of conservation laws describing the periodic SMB chromatography is given for $mT \leq t < (m+1)T, \quad m = 0, 1, 2, 3, 4, 5, ...$ by

$$(1+h)\partial_t C + (P^m)\gamma(P^m)\partial_x C = 0, \tag{45}$$

$$C(t,0) = P^m K(P^m)'C(t,L) + (P^m)U, \tag{46}$$

$$C(0,x) = C_0(x). \tag{47}$$

The posedness of the problem (45)-(47) is discussed in [**5**, Appendix A], but the controllability problem of SMB around periodic trajectories has not been established.

Let us recall the following result given by the authors in [**7**] without proof.

PROPOSITION 4.1. *The system* (45) *has a single quasi-periodic time solution* $C^* = (C^{*A}, C^{*B})$ *such that* $C^*(t,x) = C^*(t+4T,x)$, $x \in [0,L]$, $t \geq 0$ *provided that* $v_F C_F^A$ *and* $v_F C_F^B$ *are sufficiently small.*

PROOF.

STEP 1 (Uniqueness). We assume that (45) admits two solutions C^* and $\tilde{C^*}$ such that

$$\begin{aligned} C^*(0,x) &= \tilde{C^*}(0,x), \\ &\textit{and} \\ C^*(t,L) &= \tilde{C^*}(t,L). \end{aligned}$$

Let $\widehat{C} = C^* - \tilde{C^*}$. Then, we have

$$(1+h)\partial_t \widehat{C} + (P^m)\Upsilon(P^m)^T\partial_x \widehat{C} = 0, \tag{48}$$

$$\widehat{C}(0,x) = 0, \tag{49}$$

$$\widehat{C}(t,0) = \widehat{C}(t,L) = 0. \tag{50}$$

We multiply (48) by $2\widehat{C}$, we integrate in $(0,t) \times (0,L)$, we have

$$\int_0^t \int_0^L (1+h)\partial_s \widehat{C}.2\widehat{C}dxds + \int_0^t \int_0^L (P^m)\Upsilon(P^m)^T\partial_x \widehat{C}.2\widehat{C}dxds = 0$$

then, with integration by parts, and using (49) and (50) one has

$$\|\widehat{C}(t,x)\|^2_{L^2([0,t]\times[0,L])} = 0$$

STEP 2 (Periodicity). In this step, we study the problem of periodicity for the case when $V_f C_f$ is small enough.

We consider the system (45) with the following conditions

$$\begin{aligned} C(t,\,0) &= C(t+4T,\,0), \\ C(0,\,x) &= C(4T,\,x). \end{aligned}$$

Let $z(t,\,x) = C(t,\,x) - C(t+4T,\,x)$ where $C(t,\,x)$ is a solution of (45).

We have

$$\begin{aligned} z(t,0) &= P^m K(P^m)^T\big(C(t,L) - C(t+4T,L)\big), \\ &= P^m K(P^m)^T z(t,L), \\ &= 0. \end{aligned}$$

Is not hard to see again, that for $mT \leq t < (m+1)T$, $m = 0,1,2,...$

$$(1+h)\partial_t z + (P^m)\Upsilon(P^m)^T\partial_x z = 0, \tag{51}$$

$$z(t,\,0) = P^m K(P^m)^T z(t,L) = 0, \tag{52}$$

$$z(0,\,x) = 0. \tag{53}$$

By the first step, it is clear that $z \equiv 0$ is solution of the new problem (51)-(53), and therefore by the uniqueness of solution for (45) "generalized" transport equation, we have

$$z \equiv 0 \qquad (t, x) \in [mT, (m+1)T[\times[0, L].$$

Which proves

$$C(t, x) = C(t + 4T, x), \qquad \text{for } mT \leq t < (m+1)T, \text{ and } m = 0, 1, 2, ...$$

This ends the proof. □

4.1.3. *Controllability around quasi-periodic trajectories.* In this section we analyze the controllability for the SMB chromatography system in the presence of switching controls. The switching design is based only on the past dynamics, regardless of what the future dynamics is. Note that there are few papers regarding the notion of switching control [**64**], [**51**]. In this part of paper, we inspire from the construction proposed in [**51**] to prove the controllability of SMB around periodic trajectories.

For all $t \in [mT, (m+1)T[$, $m = 0, 1, 2, ...$ and for $C \in L^2([mT, (m+1)T[\times[0, L])$ solution of (45)-(47), let

$$z(t, x) = (z_1, z_2, z_3, z_4)'$$

such that

$$z(t,\, x) = C(t,\, x) - C(t + 4T,\, x)$$

be the solution for the following problem

$$\partial_t z + A_m \partial_x z = 0, \tag{54}$$

$$z(t,\, 0) = u_m(t), \tag{55}$$

$$z(0,\, x) = 0,\;\; z_t(0, x) = z^1(x). \tag{56}$$

where u_m is the control applied at the boundary $x = 0$, and $A_m = \dfrac{(P^m)\Upsilon(P^m)'}{(1+h)}$, is a diagonalizable matrix with eigenvalues given by $\mu = diag(\mu_1, \mu_2, \mu_3, \mu_4)$ such that, for all $i = 1, 2, 3, 4$, $\mu = \lambda_1$ or λ_2, with

$$\lambda_1 = \frac{V_1}{1+h}, \quad \lambda_2 = \frac{V_2}{1+h}.$$

Let define the Riemann invariants of (54)-(56), see e.g. [**30**] and [**61**], by

$$w_i(t, x) = L_i z(t, x),$$

where L_i is the left eigenvector satisfying

$$L_i . A_m = \mu_i L_i,$$

and let

$$W(t, x) = (w_1(t, x), w_2(t, x), w_3(t, x), w_4(t, x))'.$$

In these Riemann coordinates, the linear system can be written in this characteristic form

$$\partial_t W + \Lambda_m \partial_x W = 0, \tag{57}$$

$$W(t, 0) = u_m(t), \tag{58}$$

$$W(0, x) = W^0(x) = 0\;, W_t(0, x) = W^1. \tag{59}$$

Set

$$\Lambda_m = \begin{cases} diag\{\lambda_1, \lambda_2, \lambda_1, \lambda_2\} & \text{if m even,} \\ diag\{\lambda_2, \lambda_1, \lambda_2, \lambda_1\} & \text{if m odd,} \end{cases}$$

define

$$u_m(t) = \begin{cases} u^1(t) & \text{if m even,} \\ u^2(t) & \text{if m odd,} \end{cases}$$

with

$$u^1(t) = \begin{pmatrix} v_1(t) \\ v_2(t) \\ v_3(t) \\ v_4(t) \end{pmatrix} = \begin{pmatrix} V_2/V_1 R^4(t,L) \\ R^1(t,L) \\ V_2/V_1 R^2(t,L) \\ R^3(t,L) \end{pmatrix}$$

and

$$u^2(t) = \begin{pmatrix} \overline{v}_1(t) \\ \overline{v}_2(t) \\ \overline{v}_3(t) \\ \overline{v}_4(t) \end{pmatrix} = \begin{pmatrix} R^4(t,L) \\ V_2/V_1 R^1(t,L) \\ R^2(t,L) \\ V_2/V_1 R^3(t,L) \end{pmatrix}.$$

Let $\gamma_v^m(t),\ \gamma_{\overline{v}}^m(t) : \mathbf{R} \to \{0,1\}$ be two measurable functions with values assumed to switch between 0 and 1 so that, the control activates alternating in a manner that, in each time t, only one control is active.

$$\gamma_v^m(t) = \begin{cases} 1 & \text{if m even,} \\ 0 & \text{if m odd,} \end{cases}$$

$$\gamma_{\overline{v}}^m(t) = \begin{cases} 0 & \text{if m even,} \\ 1 & \text{if m odd.} \end{cases}$$

Then the control can be written as follows

$$u_m(t) = u^1(t)\gamma_v^m(t) + u^2(t)\gamma_{\overline{v}}^m(t).$$

Consequently, each Riemann invariant $w_i(t,x)$ is a solution of the scalar advection problem

$$\partial_t w_i + \mu_i \partial_x w_i = 0, \tag{60}$$

$$w_i(t,0) = v_i(t)\gamma_v^m(t) + \overline{v}_i(t)\gamma_{\overline{v}}^m(t), \tag{61}$$

$$w_i(0,x) = w_i^0(x) = L_i z_i^0(x) = 0, \quad \partial_t w_i(x,0) = w_i^1(x) = L_i z_i^1(x). \tag{62}$$

We are now able to give the explicit solution of the Cauchy problem (60)-(62) according to the parity of m for every $t \in [mT,(m+1)T[$.

In fact, if m is even, the solution is

$$w_i(t,x) = \begin{cases} w_i^0(x-\mu_i t) & \text{if } x-\mu_i t \geq 0, \\ v_i(t-\frac{x}{\mu_i}) & \text{if } x-\mu_i t < 0, \end{cases} \tag{63}$$

and in the case when m is odd the solution expression is

$$w_i(t,x) = \begin{cases} w_i^0(x-\mu_i t) & \text{if } x-\mu_i t \geq 0, \\ \overline{v}_i(t-\frac{x}{\mu_i}) & \text{if } x-\mu_i t < 0, \end{cases}$$

where

$$\mu_i = \lambda_1 \gamma_{i,m}^1(t) + \lambda_2 \gamma_{i,m}^2(t),$$

and

$$\gamma_{i,m}^1(t) = \begin{cases} 1 & \text{if } t \in [mT,(m+1)T[\ \chi_{\{S_1 \cup S_4\}}, \\ 0 & \text{if } t \in [mT,(m+1)T[\ \chi_{\{S_2 \cup S_3\}}, \end{cases}$$

$$\gamma_{i,m}^2(t) = \begin{cases} 0 & \text{if } t \in [mT,(m+1)T[\ \chi_{\{S_1 \cup S_4\}}, \\ 1 & \text{if } t \in [mT,(m+1)T[\ \chi_{\{S_2 \cup S_3\}}, \end{cases}$$

with $S_1 = \{m \text{ even }, i \text{ odd}\}$, $S_2 = \{m \text{ even }, i \text{ even}\}$, $S_3 = \{m \text{ odd }, i \text{ odd }\}$ and $S_4 = \{m \text{ odd }, i \text{ even }\}$.

Now, we are already to state our main result.

THEOREM 4.2. *The switching control system* (54)-(56) *is controllable in time* T *if and only if*

$$T \geq \frac{L}{4\lambda_2}.$$

PROOF. We begin with the case $m = 0$, in particular we have $0 \leq t < T$.

Since m is even, then the explicit solution of w_i, for $i = 1, 2, 3, 4$ is given by (63). Based on the Coron's proof [**19**, Theorem 2.6, pp. 30], we can easily show that for every w_i^0, $w_i^1 \in L^2(0, L)$, the control v_i steers the solution in time T from w_i^0 to w_i^1, more precisely,

$$w_i(T, x) = v_i(T - \frac{x}{\mu_i}) = w_i^1(x) \qquad \text{for} \quad T \geq \max_{j=1,2}\left(\frac{L}{\lambda_j}\right).$$

Recall now that

$$\lambda_j = \frac{V_j}{1+h}, \qquad j = 1, 2.$$

Since $V_1 > V_2$, then $\lambda_1 > \lambda_2$,

hence

$$T \geq \frac{L}{\lambda_2}.$$

Now, we treat the case $m = 1$, i.e. $T \leq t < 2T$. By the same aspect we have

$$w_i(T, x) = \overline{v}_i(T - \frac{x}{\mu_i}) = w_i^1(x) \qquad \text{for} \quad T \geq \frac{L}{2\lambda_2}.$$

It is then clear that, the time interval $[mT, (m+1)T[$ is decomposed into sequence of subintervals which depends in the parity of m, and where the control is built to switch in an iterative manner so that the system is controllable for $T \geq \dfrac{L}{(m+1)\lambda_2}$, but here the periodicity of the solution intervenes to reduce the time of controllability.

In fact, since

$$w_i(t, x) \quad = \quad w_i(t + 4T, x), \qquad \text{for } 1 \leq i \leq 4$$

we deduce that

$$T \geq \frac{L}{4\lambda_2}.$$

This ends the proof. □

4.1.4. *Finite-time stability of the SMB chromatography.* In this part, we keep with the finite-time stability of the SMB by means of boundary feedback laws. To this, we return to the general form for the hyperbolic system (45)-(47) describing the quasi-periodic SMB chromatography. Recall that this system has a unique quasi-periodic time solution C^* of period $4T$. In order to show the finite-time stability, we define the Riemann coordinates as follows

$$\xi_i = (1 + h)(C_i - C_i^*), \qquad 1 \leq i \leq 4.$$

In these Riemann coordinates, the quasi-periodic linear system (45)-(47) is written

$$\partial_t \xi + \Lambda_m \partial_x \xi = 0, \tag{64}$$

$$\xi(t, 0) = K_m \xi(t, L), \tag{65}$$

$$\xi(0, x) = \xi^0(x). \tag{66}$$

With the following notations:

$$\xi(t,x) = (\xi_1(t,x), \xi_2(t,x), \xi_3(t,x), \xi_4(t,x))',$$

$$\Lambda_m = (P^m)\Lambda(P^m)',\ K_m = (P^m)K(P^m)',$$

$$\Lambda = diag\{\lambda_1, \lambda_2, \lambda_1, \lambda_2\}.$$

In order to get the finite-time stability we will consider (64)-(66) in its scalar form. In fact, we are faced two cases that depend on the parity of m:

When m is even, (64) and (66) are equivalent to

$$\partial_t\xi_1 + \lambda_1\partial_x\xi_1 = 0, \tag{67}$$

$$\partial_t\xi_2 + \lambda_2\partial_x\xi_2 = 0, \tag{68}$$

$$\partial_t\xi_3 + \lambda_1\partial_x\xi_3 = 0, \tag{69}$$

$$\partial_t\xi_4 + \lambda_2\partial_x\xi_4 = 0, \tag{70}$$

$$\xi_i(0,x) = \xi_i^0(x); \quad 1 \le i \le 4, \tag{71}$$

and the boundary condition (65) is given in its scalar form as follows

$$\xi_1(t,0) = \frac{V_2}{V_1}\xi_4(t,L), \tag{72}$$

$$\xi_2(t,0) = \xi_1(t,L), \tag{73}$$

$$\xi_3(t,0) = \frac{V_2}{V_1}\xi_2(t,L), \tag{74}$$

$$\xi_4(t,0) = \xi_3(t,L). \tag{75}$$

For the case where m is odd, (64) and (66) are equivalent to

$$\begin{aligned} &\partial_t\xi_1 + \lambda_2\partial_x\xi_1 = 0,\\ &\partial_t\xi_2 + \lambda_1\partial_x\xi_2 = 0,\\ &\partial_t\xi_3 + \lambda_2\partial_x\xi_3 = 0,\\ &\partial_t\xi_4 + \lambda_1\partial_x\xi_4 = 0,\\ &\xi_i(0,x) = \xi_i^0(x); \quad 1 \le i \le 4, \end{aligned} \tag{76}$$

while the boundary condition (65) is given as follows

$$\begin{aligned} \xi_1(t,0) &= \xi_4(t,L),\\ \xi_2(t,0) &= \frac{V_2}{V_1}\xi_1(t,L),\\ \xi_3(t,0) &= \xi_2(t,L),\\ \xi_4(t,0) &= \frac{V_2}{V_1}\xi_3(t,L). \end{aligned}$$

We assume that one of the boundary conditions satisfies the persistently exciting condition, for example

$$\int_t^{t+h} \xi_4^2(s,L)ds \ge c\,h. \tag{77}$$

We will treat the case where m is even, the other case is treated by the same manner. So, let us consider the following Lyapunov function

$$E(t) = \sum_{i=1}^{4} \Big(\int_0^L |\xi_i(t,x)|^2 dx \Big)^{\alpha}, \quad \alpha > 1 \tag{78}$$

Calculate the derivative of (78) along the trajectories of (67)-(75)

$$\begin{aligned} \dot{E}(t) &= \alpha E^{\frac{\alpha-1}{\alpha}} \int_0^L \Big(2\xi_1\xi_{1,t} + 2\xi_2\xi_{2,t} + 2\xi_3\xi_{3,t} + 2\xi_4\xi_{4,t} \Big) dx \\ &= \alpha E^{\frac{\alpha-1}{\alpha}} \int_0^L \Big(-2\lambda_1\xi_1\xi_{1,x} - 2\lambda_2\xi_2\xi_{2,x} - 2\lambda_1\xi_3\xi_{3,x} - 2\lambda_2\xi_4\xi_{4,x} \Big) dx \\ &= \alpha E^{\frac{\alpha-1}{\alpha}} \Big((\xi_1^2(t,0) + \xi_3^2(t,0))(\lambda_1 - \lambda_2(\frac{V_1}{V_2})^2) + (\xi_2^2(t,0) + \xi_4^2(t,0))(\lambda_2 - \lambda_1) \Big). \end{aligned}$$

Recall that $\lambda_1 = \dfrac{V_1}{1+h}$ and $\lambda_2 = \dfrac{V_2}{1+h}$, since we have $V_1 > V_2 > 0$, it is simple to verify that

$$\lambda_2 - \lambda_1 < 0,$$

and

$$\lambda_1 - \lambda_2(\frac{V_1}{V_2})^2 < 0.$$

Taking the continuous function f defined by

$$f(t) = -(\lambda_2 - \lambda_1)(\xi_2^2(t,0) + \xi_4^2(t,0)) - \alpha(\lambda_1 - \lambda_2(\frac{V_1}{V_2})^2)(\xi_1^2(t,0) + \xi_3^2(t,0)) \geq 0$$

and

$$\beta = \frac{\alpha - 1}{\alpha},$$

Clearly, the continuous function satisfies the persistently exciting condition (77). Hence one gets

$$\dot{E} \leq -\alpha f(t) E^{\beta}, \ \beta \in (0,1).$$

Then $E = 0$ in finite-time and for almost every $x \in [0,\ L]$, the components ξ_1, ξ_2, ξ_3 and ξ_4 vanish in finite-time.

Thus, the SMB system (45)-(47) is finite-time stable.

4.2. Finite-time stabilization of vibrating strings connected by point mass. The vibrating string system under study can be modeled as a hybrid system. In fact, it can be considered as two separate strings in which one end of each string is attached to a common point mass. The first string occupies $\Omega_1 = (-\ell_1, 0) \subset \mathbf{R}$ and the second one $\Omega_2 = (0, \ell_2) \subset \mathbf{R}$, where ℓ_1 and ℓ_2 are positive.

We suppose that both strings are homogeneous. The deformation of the first and second string at abscissa x and time t will be described respectively by the functions u and v where each one is solution of a one dimensional wave equation associated to each interval between the boundary and the point mass;

$$\rho_1 u_{tt} = \sigma_1 u_{xx}, \qquad x \in \Omega_1,\ t > 0, \tag{79}$$

$$\rho_2 v_{tt} = \sigma_2 v_{xx}, \qquad x \in \Omega_2,\ t > 0. \tag{80}$$

The physical parameters $\rho_i > 0$ and $\sigma_i > 0$ ($i = 1, 2$) represent, respectively, the density and the tension of each string.

The position $z(t)$ of the mass M, attached to the string at $x=0$, satisfies the continuity relation

$$u(0,t)=v(0,t)=z(t), \quad t>0, \tag{81}$$

and the second order linear ordinary differential equation describing the motion of the mass

$$Mz_{tt}(t)+\sigma_1 u_x(0,t)-\sigma_2 v_x(0,t)=0, \quad t>0. \tag{82}$$

The boundary control $p(t)$ and $q(t)$ are applied as follows

$$u_x(-\ell_1,t)=p(t), \quad t>0, \tag{83}$$

$$v_x(\ell_2,t)=q(t), \quad t>0. \tag{84}$$

The initial conditions are

$$(u(x,0),u_t(x,0))=(u^0(x),u^1(x)), \qquad x\in\Omega_1, \tag{85}$$

$$(v(x,0),v_t(x,0))=(v^0(x),v^1(x)), \qquad x\in\Omega_2, \tag{86}$$

$$(z(0),z_t(0))=(z^0,z^1). \tag{87}$$

In the control system (79)-(87), the state is $y=:(u,\ v,\ z,\ u_t,\ v_t,\ z_t)'$ lives in $\mathcal{H}:=H^1(\Omega_1)\times H^1(\Omega_2)\times\mathbf{R}\times L^2(\Omega_1)\times L^2(\Omega_2)\times\mathbf{R}$ and p, q the two controllers.

It is proved in [**29**] that if $y^0=(u^0,\ v^0,\ z^0,\ u^1,\ v^1,\ z^1)'\in\mathcal{H}$, then system (79)-(87) is well posed. More precisely, $(u,\ v,\ z)'\in\mathcal{C}^0([0,\ \infty);H^1(\Omega_1)\times H^1(\Omega_2)\times\mathbf{R})$ and $(u_t,\ v_t,\ z_t)\in\mathcal{C}([0,\ \infty);L^2(\Omega_1)\times L^2(\Omega_2)\times\mathbf{R}$.

In this section, we study the finite-time stabilization by boundary condition in three configurations. The first one is the action by one feedback at the extreme-point ℓ_2. The second is the action by boundary feedbacks at both extremes $-\ell_1$ and ℓ_2 while the third idea is the action by boundary feedbacks at the point $x=0$. Along of our study, we show how feedbacks proposed by the authors in [**29**] lead to finite-time stability of the closed loop system.

4.2.1. *Action to extreme-point ℓ_2.* For this action to extreme point we have the following stability result.

THEOREM 4.3. *For every $y^0\in\mathcal{H}$, and with the action of the excited boundary feedback laws*

$$\begin{cases} u(-\ell_1,t)=0, \\ q(t)=-\frac{k}{\sigma_2}v_t(\ell_2,t),\ k>0, \\ \int_t^{t+h} v_s^2(\ell_2,\ s)ds\geq c\,h,\ \text{where } c,\ h>0. \end{cases}$$

the control system (79)-(84) *is finite-time stable.*

PROOF. For every $y^0\in\mathcal{H}$, the problem (79)-(84) admits a unique solution in $\mathcal{H}$. Let $\alpha>1$ and we consider the following Lyapunov function which is merely a generalization of its energy function

$$\begin{aligned} 2V^{1/\alpha} &= \int_{-\ell_1}^0 [\rho_1|u_t(x,t)|^2+\sigma_1|u_x(x,t)|^2]\,dx+\int_0^{\ell_2}[\rho_2|v_t(x,t)|^2+\sigma_2|v_x(x,t)|^2]\,dx \\ (88) \quad &+ M|z_t(t)|^2 \end{aligned}$$

Using an integration by parts, the time derivative of V along the system (79)-(84) is given by

$$\dot{V}(t)=\alpha V^{\frac{\alpha-1}{\alpha}}\left[\int_{-\ell_1}^{0}(\rho_1 u_t u_{tt}+\sigma_1 u_x u_{xt})dx+\int_{0}^{\ell_2}(\rho_2 v_t v_{tt}+\sigma_2 v_x v_{xt})dx+Mz_t z_{tt}\right]$$

$$=\alpha V^{\frac{\alpha-1}{\alpha}}\Big(\sigma_2 v_x(\ell_2,t)v_t(\ell_2,t)-\sigma_1 u_x(-\ell_1,t)u_t(-\ell_1,t)\Big) \tag{89}$$

$$=\alpha V^{\frac{\alpha-1}{\alpha}}\Big(\sigma_2 v_t(\ell_2,t)q(t)-\sigma_1 u_t(-\ell_1,t)p(t)\Big). \tag{90}$$

Hence, with the boundary condition $u(-\ell_1,t)=0$ ($\Rightarrow u_t(-\ell_1,\,t)=0$) and under the feedback law

$$q(t)=-\frac{k}{\sigma_2}v_t(\ell_2,t),\; k>0$$

we get

$$\dot{V}(t)=-k\alpha V^{\frac{\alpha-1}{\alpha}}v_t^2(\ell_2,t)\leq 0. \tag{91}$$

Since $\beta:=\frac{\alpha-1}{\alpha}\in(0,\,1)$ and $f(t):=v_t^2(\ell_2,t)\in\mathcal{C}^0(0,\,+\infty)$ then, Lemma 3.3 leads to V vanish in finite-time. Hence

$$u_t(x,t)=u_x(x,t)=v_t(x,t)=v_x(x,t)=z_t(t)=0\quad\text{in finite-time.}$$

By the fact $u_t(x,\,t)=0$ in finite-time for all $x\in[-\ell_1,\,0]$, we deduce that $u(x,\,t)$ is a constant c_1 for t is large enough and for $x\in[-\ell_1,\,0]$. Since $u(-\ell_1,\,t)=0$ then $c_1=0$ and then, $\forall x\in[-\ell_1,\,0]$ $u(x,\,t)=0$ for t is is large enough. Consequently $z(t)=u(0,\,t)=0$ vanishes again in finite-time and leads to vanishing the state $v(x,\,t)$ in finite-time. This ends the proof. □

4.2.2. *Action at both extreme-points $-\ell_1$ and ℓ_2.* In this case, we will assume that only one extreme is excited, for example,

$$\int_t^{t+h}u_s^2(-\ell_1,s)ds\geq c\,h. \tag{92}$$

Then we have.

PROPOSITION 4.4. *For every $y^0\in\mathcal{H}$, we consider the boundary feedback laws*

$$\begin{cases}p(u)=\frac{k}{\sigma_1}u_t(-\ell_1,t),\\[2ex] q(v)=-\frac{k}{\sigma_2}v_t(\ell_2,t)\end{cases}$$

then system (79)-(84) *subject to persistence excitation condition* (92) *is finite-time partially stable. More precisely, there exists a constant k depending on initial conditions such that $u(x,\,t)=v(x,\,t)=z(t)=k$ for t is large enough.*

PROOF. Under our boundary feedback laws, the time derivative of the Lyapunov function V becomes

$$\dot{V}=-\alpha kV^{\frac{\alpha-1}{\alpha}}\Big(\sigma_2 v_t^2(\ell_2,t)+\sigma_1 u_t^2(-\ell_1,t)\Big)\leq 0$$

Taking $f(t)=\alpha k(\sigma_2 v_t^2(\ell_2,t)+\sigma_1 u_t^2(-\ell_1,t))$ which is a positive and continuous function and satisfies the persistence excitation condition of Lemma 3.3, and $\beta=\frac{\alpha-1}{\alpha}\in(0,\,1)$, we deduce that

$$\dot{V}\leq -f(t)V^{\beta},$$

more precisely, the Lyapunov energy function V equal to zero in finite-time, hence

$$u_t(x,t) = u_x(x,t) = v_t(x,t) = v_x(x,t) = z_t(t) = 0 \quad \text{in finite-time.}$$

Thus, the partial state $(u,\ v,\ z)(t) = k$ constant for t is large enough. □

Acknowledgment

The authors would like to thank Prof. Kaïs Ammari for his valuable comments and discussions.

References

[1] F. Alabau-Boussouira, V. Perrollaz, and L. Rosier, *Finite-time stabilization of a network of strings*, Math. Control Relat. Fields **5** (2015), no. 4, 721–742, DOI 10.3934/mcrf.2015.5.721. MR3485747

[2] S. Amin, F. M. Hante, and A. M. Bayen, *On stability of switched linear hyperbolic conservation laws with reflecting boundaries*, Hybrid systems: computation and control, Lecture Notes in Comput. Sci., vol. 4981, Springer, Berlin, 2008, pp. 602–605, DOI 10.1007/978-3-540-78929-1_44. MR2728909

[3] K. Ammari and E. Crépeau, *Feedback stabilization and boundary controllability of the Korteweg–de Vries equation on a star-shaped network*, SIAM J. Control Optim. **56** (2018), no. 3, 1620–1639, DOI 10.1137/17M113959X. MR3799048

[4] S. A. Avdonin and W. Moran, *Simultaneous control problems for systems of elastic strings and beams*, Systems Control Lett. **44** (2001), no. 2, 147–155, DOI 10.1016/S0167-6911(01)00137-2. MR2007587

[5] G. Bastin and J.-M. Coron, *Stability and boundary stabilization of 1-D hyperbolic systems*, Progress in Nonlinear Differential Equations and their Applications, vol. 88, Birkhäuser/Springer, [Cham], 2016. Subseries in Control. MR3561145

[6] G. Bastin, J.-M. Coron, and B. d'Andréa-Novel, *On Lyapunov stability of linearised Saint-Venant equations for a sloping channel*, Netw. Heterog. Media **4** (2009), no. 2, 177–187, DOI 10.3934/nhm.2009.4.177. MR2525202

[7] G. Bastin, J.-M. Coron, B. d'Andréa Novel, P. Suvarov, A.-V. Wouwer, and A. Kienle. Stability analysis of switching hyperbolic systems: the example of SMB chromatography. pages 2153–2157, Strasbourg, France, June 24-27 2014. European Control Conference.

[8] G. Ben Belgacem and C. Jammazi. On the finite-time stabilization of strings connected by point mass. In Hitay Ozbay Catherine Bonnet, Bozenna Pasik-Duncan and Qing Zhang, editors, *SIAM Conference on Control and its Applications*, pages 212–219, Paris, July 8-10 2015.

[9] G. Ben Belgacem and C. Jammazi, *On the finite-time Bhat-Bernstein feedbacks for the strings connected by point mass*, Discrete Contin. Dyn. Syst. Ser. B **24** (2019), no. 4, 1653–1675. MR3927473

[10] S. P. Bhat and D. S. Bernstein, *Continuous finite-time stabilization of the translational and rotational double integrators*, IEEE Trans. Automat. Control **43** (1998), no. 5, 678–682, DOI 10.1109/9.668834. MR1618028

[11] S. P. Bhat and D. S. Bernstein, *Finite-time stability of continuous autonomous systems*, SIAM J. Control Optim. **38** (2000), no. 3, 751–766, DOI 10.1137/S0363012997321358. MR1756893

[12] S. P. Bhat and D. S. Bernstein, *Geometric homogeneity with applications to finite-time stability*, Math. Control Signals Systems **17** (2005), no. 2, 101–127, DOI 10.1007/s00498-005-0151-x. MR2150956

[13] C.-Z. Xu and G. Sallet, *Exponential stability and transfer functions of processes governed by symmetric hyperbolic systems*, ESAIM Control Optim. Calc. Var. **7** (2002), 421–442, DOI 10.1051/cocv:2002062. MR1925036

[14] R. Carles and C. Gallo, *Finite time extinction by nonlinear damping for the Schrödinger equation*, Comm. Partial Differential Equations **36** (2011), no. 6, 961–975, DOI 10.1080/03605302.2010.531074. MR2765425

[15] C. Castro and E. Zuazua, *High frequency asymptotic analysis of a string with rapidly oscillating density*, European J. Appl. Math. **11** (2000), no. 6, 595–622, DOI 10.1017/S0956792500004307. MR1811309

[16] A. Chaillet, Y. Chitour, A. Loría, and M. Sigalotti, *Uniform stabilization for linear systems with persistency of excitation: the neutrally stable and the double integrator cases*, Math. Control Signals Systems **20** (2008), no. 2, 135–156, DOI 10.1007/s00498-008-0024-1. MR2411416

[17] Y. Chitour, G. Mazanti, and M. Sigalotti, *Persistently damped transport on a network of circles*, Trans. Amer. Math. Soc. **369** (2017), no. 6, 3841–3881, DOI 10.1090/tran/6778. MR3624395

[18] J.-M. Coron, *On the stabilization in finite time of locally controllable systems by means of continuous time-varying feedback law*, SIAM J. Control Optim. **33** (1995), no. 3, 804–833, DOI 10.1137/S0363012992240497. MR1327239

[19] J.-M. Coron, *Control and nonlinearity*, Mathematical Surveys and Monographs, vol. 136, American Mathematical Society, Providence, RI, 2007. MR2302744

[20] J.-M. Coron, *Phantom tracking method, homogeneity and rapid stabilization*, Math. Control Relat. Fields **3** (2013), no. 3, 303–322, DOI 10.3934/mcrf.2013.3.303. MR3110065

[21] J.-M. Coron and G. Bastin, *Dissipative boundary conditions for one-dimensional quasi-linear hyperbolic systems: Lyapunov stability for the C^1-norm*, SIAM J. Control Optim. **53** (2015), no. 3, 1464–1483, DOI 10.1137/14097080X. MR3354994

[22] J.-M. Coron, G. Bastin, and B. d'Andréa-Novel, *Dissipative boundary conditions for one-dimensional nonlinear hyperbolic systems*, SIAM J. Control Optim. **47** (2008), no. 3, 1460–1498, DOI 10.1137/070706847. MR2407024

[23] J.-M. Coron, B. d'Andréa-Novel, and G. Bastin, *A strict Lyapunov function for boundary control of hyperbolic systems of conservation laws*, IEEE Trans. Automat. Control **52** (2007), no. 1, 2–11, DOI 10.1109/TAC.2006.887903. MR2286756

[24] H. Cui, G. Xu, and Z. Han, *Distributed tracking control of the Schrödinger equation with internal disturbance*, IMA J. Math. Control Inform. **34** (2017), no. 3, 831–850, DOI 10.1093/imamci/dnv071. MR3746078

[25] B. d'Andréa-Novel, I. Moyano, and L. Rosier, *Finite-time stabilization of an overhead crane with a flexible cable*, Math. Control Signals Systems **31** (2019), no. 2, Art. 6, 19, DOI 10.1007/s00498-019-0235-7. MR3953136

[26] J. de Halleux, C. Prieur, J.-M. Coron, B. d'Andréa-Novel, and G. Bastin, *Boundary feedback control in networks of open channels*, Automatica J. IFAC **39** (2003), no. 8, 1365–1376, DOI 10.1016/S0005-1098(03)00109-2. MR2141681

[27] A. Diagne, G. Bastin, and J.-M. Coron, *Lyapunov exponential stability of 1-D linear hyperbolic systems of balance laws*, Automatica J. IFAC **48** (2012), no. 1, 109–114, DOI 10.1016/j.automatica.2011.09.030. MR2879417

[28] M. Dick, M. Gugat, and G. Leugering, *Classical solutions and feedback stabilization for the gas flow in a sequence of pipes*, Netw. Heterog. Media **5** (2010), no. 4, 691–709, DOI 10.3934/nhm.2010.5.691. MR2740529

[29] S. Hansen and E. Zuazua, *Exact controllability and stabilization of a vibrating string with an interior point mass*, SIAM J. Control Optim. **33** (1995), no. 5, 1357–1391, DOI 10.1137/S0363012993248347. MR1348113

[30] L. C. Evans, *Partial differential equations*, Graduate Studies in Mathematics, vol. 19, American Mathematical Society, Providence, RI, 1998. MR1625845

[31] E. Godlewski and P.-A. Raviart, *Numerical approximation of hyperbolic systems of conservation laws*, Applied Mathematical Sciences, vol. 118, Springer-Verlag, New York, 1996. MR1410987

[32] J. M. Greenberg and T. T. Li, *The effect of boundary damping for the quasilinear wave equation*, J. Differential Equations **52** (1984), no. 1, 66–75, DOI 10.1016/0022-0396(84)90135-9. MR737964

[33] V. T. Haimo, *Finite time controllers*, SIAM J. Control Optim. **24** (1986), no. 4, 760–770, DOI 10.1137/0324047. MR846381

[34] F. M. Hante, M. Sigalotti, and M. Tucsnak, *On conditions for asymptotic stability of dissipative infinite-dimensional systems with intermittent damping*, J. Differential Equations **252** (2012), no. 10, 5569–5593, DOI 10.1016/j.jde.2012.01.037. MR2902127

[35] Y. Hong and Z.-P. Jiang, *Finite-time stabilization of nonlinear systems with parametric and dynamic uncertainties*, IEEE Trans. Automat. Control **51** (2006), no. 12, 1950–1956, DOI 10.1109/TAC.2006.886515. MR2284421

[36] Y. Hong, G. Yang, D. Cheng, and S. Spurgeon, *Finite time convergent control using terminal sliding mode*, J. Control Theory Appl. **2** (2004), no. 1, 69–74, DOI 10.1007/s11768-004-0026-6. MR2093814

[37] Y. Hong, *Finite-time stabilization and stabilizability of a class of controllable systems*, Systems Control Lett. **46** (2002), no. 4, 231–236, DOI 10.1016/S0167-6911(02)00119-6. MR2010240

[38] Y. Hong, J. Huang, and Y. Xu, *On an output feedback finite-time stabilization problem*, IEEE Trans. Automat. Control **46** (2001), no. 2, 305–309, DOI 10.1109/9.905699. MR1814578

[39] X. Huang, W. Lin, and B. Yang, *Global finite-time stabilization of a class of uncertain nonlinear systems*, Automatica J. IFAC **41** (2005), no. 5, 881–888, DOI 10.1016/j.automatica.2004.11.036. MR2157720

[40] C. Jammazi, *A discussion on the Hölder and robust finite-time partial stabilizability of Brockett's integrator*, ESAIM Control Optim. Calc. Var. **18** (2012), no. 2, 360–382, DOI 10.1051/cocv/2010101. MR2954630

[41] C. Jammazi, *Continuous and discontinuous homogeneous feedbacks finite-time partially stabilizing controllable multichained systems*, SIAM J. Control Optim. **52** (2014), no. 1, 520–544, DOI 10.1137/110856393. MR3164120

[42] C. Jammazi, *Some results on finite-time stabilizability: application to triangular control systems*, IMA J. Math. Control Inform. **35** (2018), no. 3, 877–899, DOI 10.1093/imamci/dnw076. MR3858295

[43] C. Jammazi, M. Zaghdoudi, and M. Boutayeb, *On the global polynomial stabilization of nonlinear dynamical systems*, Nonlinear Anal. Real World Appl. **46** (2019), 29–42, DOI 10.1016/j.nonrwa.2018.07.020. MR3887118

[44] H. K. Khalil. *Nonlinear Systms*. Prentice Hall, third edition, 2002.

[45] V. Komornik and P. Loreti, *Fourier series in control theory*, Springer Monographs in Mathematics, Springer-Verlag, New York, 2005. MR2114325

[46] E. B. Lee and Y. C. You, *Stabilization of a vibrating string system linked by point masses*, Control of boundaries and stabilization (Clermont-Ferrand, 1988), Lect. Notes Control Inf. Sci., vol. 125, Springer, Berlin, 1989, pp. 177–198, DOI 10.1007/BFb0043360. MR1015970

[47] G. Leugering and E. J. P. G. Schmidt, *On the modelling and stabilization of flows in networks of open canals*, SIAM J. Control Optim. **41** (2002), no. 1, 164–180, DOI 10.1137/S0363012900375664. MR1920161

[48] T. T. Li, *Global classical solutions for quasilinear hyperbolic systems*, RAM: Research in Applied Mathematics, vol. 32, Masson, Paris; John Wiley & Sons, Ltd., Chichester, 1994. MR1291392

[49] W. Littman and S. W. Taylor, *Boundary feedback stabilization of a vibrating string with an interior point mass*, Nonlinear problems in mathematical physics and related topics, I, Int. Math. Ser. (N. Y.), vol. 1, Kluwer/Plenum, New York, 2002, pp. 271–287, DOI 10.1007/978-1-4615-0777-2_16. MR1970616

[50] A. Loria, E. Panteley, and K. Melhem. Ugas of skew-symmetric time-varying systems: Applications to stabilization of chained systems. *European Journal of Control (EJC)*, 8:33–43, 2002.

[51] Q. Lü and E. Zuazua, *Robust null controllability for heat equations with unknown switching control mode*, Discrete Contin. Dyn. Syst. **34** (2014), no. 10, 4183–4210, DOI 10.3934/dcds.2014.34.4183. MR3195365

[52] M. Maghenem, A. Loria, and E. Panteley, *A robust δ-persistently exciting controller for leader-follower tracking-agreement of multiple vehicles*, Eur. J. Control **40** (2018), 1–12, DOI 10.1016/j.ejcon.2017.09.001. MR3767576

[53] A. Majda, *Disappearing solutions for the dissipative wave equation*, Indiana Univ. Math. J. **24** (1974/75), no. 12, 1119–1133, DOI 10.1512/iumj.1975.24.24093. MR385330

[54] E. Moulay, *Stability and stabilization of homogeneous systems depending on a parameter*, IEEE Trans. Automat. Control **54** (2009), no. 6, 1382–1385, DOI 10.1109/TAC.2009.2015560. MR2532634

[55] H. Mounier, J. Rudolph, M. Fliess, and P. Rouchon, *Tracking control of a vibrating string with an interior mass viewed as delay system*, ESAIM Control Optim. Calc. Var. **3** (1998), 315–321, DOI 10.1051/cocv:1998112. MR1644431

[56] Y. Orlov, *Finite time stability and robust control synthesis of uncertain switched systems*, SIAM J. Control Optim. **43** (2004/05), no. 4, 1253–1271, DOI 10.1137/S0363012903425593. MR2124272

[57] Y. V. Orlov, *Discontinuous systems: Lyapunov analysis and robust synthesis under uncertainty conditions*, Communications and Control Engineering Series, Springer-Verlag London, Ltd., London, 2009. MR2731468

[58] V. Perrollaz and L. Rosier, *Finite-time stabilization of systems of conservation laws on networks*, SIAM Contr. and Optimization, 52(1):143–163, 2014.

[59] A. Pisano, Y. Orlov, and E. Usai, *Tracking control of the uncertain heat and wave equation via power-fractional and sliding-mode techniques*, SIAM J. Control Optim. **49** (2011), no. 2, 363–382, DOI 10.1137/090781140. MR2784692

[60] C. Prieur, *Control of systems of conservation laws with boundary errors*, Netw. Heterog. Media **4** (2009), no. 2, 393–407, DOI 10.3934/nhm.2009.4.393. MR2525213

[61] D. Serre, *Systems of conservation laws. 2*, Cambridge University Press, Cambridge, 2000. Geometric structures, oscillations, and initial-boundary value problems; Translated from the 1996 French original by I. N. Sneddon. MR1775057

[62] M. Slemrod, *Boundary feedback stabilization for a quasilinear wave equation*, Control theory for distributed parameter systems and applications (Vorau, 1982), Lect. Notes Control Inf. Sci., vol. 54, Springer, Berlin, 1983, pp. 221–237, DOI 10.1007/BFb0043951. MR793047

[63] P. Suvarov, A. Vande Wouwer, and A. Kienle. A simple robust control for simulated moving bed chromatography separation. *In Proceedings IFAC Symposium on Advanced Control of Chemical processes*, pages 137–142, 2012.

[64] E. Zuazua, *Switching control*, J. Eur. Math. Soc. (JEMS) **13** (2011), no. 1, 85–117, DOI 10.4171/JEMS/245. MR2735077

Faculté des Sciences de Bizerte and Laboratoire d'Ingénierie Mathématique, Université de Carthage, Tunisia

Email address: chaker.jammazi@ept.rnu.tn

Faculté des Sciences de Bizerte and Laboratoire d'Ingénierie Mathématique, Université de Carthage, Tunisia

Email address: benbelgacemghada13@gmail.com

Contemporary Mathematics
Volume **757**, 2020
https://doi.org/10.1090/conm/757/15245

A stochastic sub-grid viscosity model and wavelet based method for images assimilation

Souleymane Kadri Harouna and Etienne Mémin

Abstract. In this work, we present a wavelet based numerical scheme to discretize the Navier-Stokes equations with location uncertainties [Geophys. Astrophys. Fluid Dyn. 108 (2014), no. 2, pp. 119–146]. This model relies on the decomposition of the flow in terms of a smooth in time resolved component and a highly fluctuating random component. This modeling enables to constitute large-scale dynamical models in which the effect of the unresolved random components is taken into account through the constitution of an anisotropic sub-grid viscosity model without needing any eddy-viscosity assumption. Image data assimilation is used to estimate and optimize the contribution of this stress tensor, directly related to the variance tensor of the unresolved small-scale flow components. This leads to an explicit treatment of the sub-grid viscosity model that enforced the time step limit. Numerical Examples are provided in two space dimension in the case of an inhomogeneous turbulent flow.

Contents

1. Introduction

The large scale analysis of complex fluid flows in domains ranging from climate sciences to engineering sciences requires to constitute efficient dynamical models incorporating properly various physical contributions that are difficult to specify precisely at the envisaged resolution scale. This includes for instance a small-scale physical forcing or the action of boundary layers, but also partially known inlet

2010 *Mathematics Subject Classification.* Primary 93C20, 93B07.
Key words and phrases. Navier-Stokes, uncertainty modeling, wavelet numerical scheme, sub-grid modeling, image assimilation.

or boundary conditions, and eventually numerical errors arising from truncation policy and scale coarsening procedures. An accurate deterministic modeling of the effects of these very different contributions is obviously hardly achievable in complex situations and we advocate instead the use of a stochastic modeling. Within this prospect, we aim at describing these missing contributions as random variables that will be referred to as flow uncertainties in the following. The modeling of such uncertainties but also of their evolution along time is of the utmost importance in ensemble methods used in geophysics, either for data assimilation or forecasting issues. In both cases, a modeling of the flow dynamics errors enables to rank the quality of some flow configurations. This authorizes ideally to set up efficient filtering strategies in which high errors are corrected by more meaningful data.

These errors or uncertainties are meant to represent principally small-scale physical processes ignored at the resolution scale. They include small-scales eddies, topographic forcing or boundary layers turbulence for instance and are responsible both of an energy dissipation but also of local backscatters of energy. The introduction of random variables constitutes an appealing mechanism to model inverse energy cascade mechanism [**23, 26, 35**], in so far as they enable a phenomenological modeling of the uncertainties involved. Recently those models have regained a great interest for the modeling of geophysical flows [**3**] in climate sciences and for data assimilation issue. In fluid mechanics, Large eddies simulations (LES) and Reynolds average simulations (RANS) face also the very same question: how to model the action on the resolved component of the small scales of the flow. Numerous turbulence modeling used in the aforementioned methodologies rely on eddies viscosity concept to model the energy dissipation due to unresolved scales. Eddy viscosity concept dates back to the work of Boussinesq [**4**] and Prandl mixing length [**30**]. It relies on the hypothesis that the energy transfer from the resolved scales to the subgrid scales can be described in a similar way as the molecular viscosity mechanism. It is therefore not at all related to any uncertainty or error quantities. In models dealing explicitly with a probabilistic modeling of the small scales there is thus some incoherency in representing the dissipative mechanism attached to random component through an eddy viscosity assumption. In this work we will not make use on such hypothesis. Instead, we will rely on a a rigorous expression of the subgrid stress tensor ensuing directly from our stoichastic modeling and that explicitly depends on the uncertainty variance.

More precisely, this subgrid model is properly derived from a general stochastic model of the fluid motion in which the fluid parcels displacement is decomposed in two components: a smooth differentiable function and an uncertainty function uncorrelated in time but correlated in space [**28**]. The whole displacement field is defined as an Eulerian description of the form:

$$\boldsymbol{U}(\boldsymbol{x},t) = \boldsymbol{w}(\boldsymbol{x},t)dt + \boldsymbol{\sigma}(\boldsymbol{x},t)d\tilde{\boldsymbol{B}}_t. \tag{1}$$

In this expression, $\boldsymbol{w} = (w_1, w_2, w_3)$, corresponds to the smooth resolved velocity component of the flow referred to as the *drift* or resolved component. In this work it will be assumed to be a deterministic differentiable function (and more precisely a finite variation process). This assumption will allows to build deterministic evolution laws in which the stochasticity of the small-scale component is taken into account. Nevertheless, more general models can be obtained when the resolved velocity component is a semi-martingale [**31–33**]. The second component is a random component encoding the uncertainties we have on the flow. This uncertainty component, which is not differentiable in time, involves a diffusion tensor that has to be properly specified. Compared to the smooth drift, this component lives at a much smaller time scale. It is nevertheless defined at all the spatial scales.

This diffusion tensor can be specified explicitly as temporal or spatial correlation of the resolved component, see [**13**]. Another approach explored in this work consists to use the beforehand knowledge on the velocity field to optimally set up the effects of the unresolved uncertainty component through a data assimilation procedure.

The random field $\boldsymbol{U}(\boldsymbol{x},t)$, is assumed to follow a stochastic linear momentum conservation principle that is mainly derived from a stochastic version of the Reynolds transport theorem. This modified transport theorem is presented in the following section.

The paper is organized as follow. After a brief presentation of the stochastic Navier-Stokes system devised in [**28**] in section 2, we describe a particular numerical scheme based on divergence free wavelets [**14, 15**]. This particular type of wavelets method together with the semi-implicit Euler scheme used to set up the discrete scheme are briefly presented in section 3. Therefore, we analyse the stability of this numerical scheme and give detail for the associated spatial discretization. Section 4 recalls the principle of variational data assimilation method for the discret model and numerical results in two space dimension in the case of an inhomogeneous turbulent flow are provided in section 6.

2. Navier-Stokes equation under uncertainty

As the system of Navier-Stokes equations we consider incorporates random uncertainties to describe unknown forcing or scale coarsening approximations, they constitute in essence a stochastic evolution system. Numerous methodological choices can be envisaged to constitute such a system. It is possible for instance to consider additional random forcing to a system of equations whose structure ensues from a deterministic formalization. This is the choice that has been the most often done since the work of Benssoussan [**2**]. Another choice, in the wake of Kraichnan's work [**16**], consists to close the large-scale representation in the Fourier space by relying on a Langevin stochastic representation [**18, 20, 22**]. Obviously the frontiers between these two methodologies are sometimes fuzzy, and numerous works rely on both of these strategies in order to setup the shape that should take the random variables evolution law [**20, 35**].

In this work, we shall rely on a different strategy that consists in directly defining the uncertainty as a location uncertainty on the resolution grid. A fluid particle location will be here defined only up to a Gaussian random field. The Navier-Stokes equations are then deduced from acceleration and second Newton's principle. In a similar way as for the deterministic case, the main methodological tool used here is a stochastic version of the Reynolds transport theorem.

2.1. Stochastic Reynolds transport theorem. The Reynolds transport theorem provides the expression of the rate of change of a scalar function q within a material volume $\mathcal{V}(t)$ transported by the flow. Its extension to stochastic flows is given as [**28**]:

$$(2)\quad d\int_{\mathcal{V}(t)} q(\boldsymbol{x},t)d\boldsymbol{x} = \int_{\mathcal{V}(t)} \{dq_t + [\boldsymbol{\nabla}\cdot(q\boldsymbol{w}) + \frac{1}{2}\|\boldsymbol{\nabla}\cdot\boldsymbol{\sigma}\|^2 q - \sum_{i,j}\frac{1}{2}\frac{\partial^2}{\partial x_i\partial x_j}(a_{ij}q)_{|\boldsymbol{\nabla}\cdot\boldsymbol{\sigma}=0}]dt + \boldsymbol{\nabla}\cdot(q\boldsymbol{\sigma}d\tilde{\boldsymbol{B}}_t)\}d\boldsymbol{x}.$$

In this expression the first term is a time increment at fixed coordinates, $\boldsymbol{x}$, of the scalar quantity q. Let us note that this function is random and is not differentiable in time. The randomness of function q is an essential point in the derivation of this theorem as it requires to use a generalized form of Ito calculus – the Ito-Wentzell formula – adapted to the differentiation of the composition of two stochastic processes [**19**]. The last random term is built from a tempered representation of a Brownian map noted in a formal way through a convolution product $d\tilde{\boldsymbol{B}}_t = d\boldsymbol{B}_t \star \varphi_\nu$ and, $\boldsymbol{\sigma}_t$, a linear deterministic symmetric operator with null value outside the domain interior referred to as the diffusion tensor in the following. The random oscillating component is denoted:

$$\boldsymbol{\sigma}(\boldsymbol{x},t)d\tilde{\boldsymbol{B}}_t = \int \boldsymbol{\sigma}_t(\boldsymbol{x},\boldsymbol{y})d\tilde{\boldsymbol{B}}_t(\boldsymbol{y})d\boldsymbol{y}.$$

It is important to outline that an incompressibility condition on this component requires necessarily a divergence free diffusion tensor. The third term is related to the compression strength of the uncertainty field. The fourth term must be computed considering the diffusion tensor is divergence free. The tensor $\mathbf{a}(\boldsymbol{x})$ involved in this term is a matrix function associated to the diagonal elements of the covariance tensor. It corresponds hence to the uncertainty variance and is defined as:

$$(3)\qquad a_{ij}(\boldsymbol{x},t) = \sum_k \sigma^\nu_{ik}(\boldsymbol{x},t)\sigma^\nu_{kj}(\boldsymbol{x},t),$$

where $\boldsymbol{\sigma}^\nu(\boldsymbol{x},\boldsymbol{y},t) = \boldsymbol{\sigma}(\boldsymbol{x},\bullet,t)\star\varphi_\nu(\boldsymbol{y})$ denotes a filtered version of the diffusion tensor along its second component. This rate of change formula is obtained from Ito-Wentzell differentiation of a function tending to the material volume characteristic function. The differentiation of the product of this characteristic function with the scalar function of interest, followed by a formal integration by part gives us the sought expression [**28**]. This relation allows us stating a mass conservation principle that accounts for the considered uncertainty on the fluid flow. Applying the previous transport theorem to the fluid density $\rho(\boldsymbol{x},t)$ and canceling this expression for arbitrary volumes, we get the following mass conservation constraint:

$$(4)\quad d\rho_t + \boldsymbol{\nabla}\cdot(\rho\boldsymbol{w})dt = \frac{1}{2}(\sum_{i,j}\frac{\partial^2}{\partial x_i\partial x_j}(a_{ij}\rho)_{|\boldsymbol{\nabla}\cdot\boldsymbol{\sigma}=0} - \frac{1}{2}\|\boldsymbol{\nabla}\cdot\boldsymbol{\sigma}\|^2\rho)dt - \boldsymbol{\nabla}\cdot(\rho\boldsymbol{\sigma}d\tilde{\boldsymbol{B}}_t).$$

For an incompressible fluid with constant density, canceling separately the slow deterministic terms and the rapid oscillating stochastic terms, and imposing to the whole deformation field (1) to be volume preserving, this system simplifies in a set of incompressibility relations:

$$(5)\qquad \boldsymbol{\nabla}\cdot(\boldsymbol{\sigma}d\tilde{\boldsymbol{B}}_t) = 0,\quad \boldsymbol{\nabla}\cdot\boldsymbol{w} = 0,\quad \boldsymbol{\nabla}\cdot(\boldsymbol{\nabla}\cdot\mathbf{a}) = 0,$$

composed of two standard volume preserving constraints accompanied with a less intuitive additional volume preserving constraint on the divergence of the uncertainty variance tensor. For divergence free homogeneous isotropic random fields such as the Kraichnan random field [**17**] this last constraint is naturally satisfied as those fields are associated to a constant diagonal variance tensor. In this case the system reduces hence to the standard divergence free constraint.

For isochoric flow with varying density we get a mass conservation constraint of the form:

$$d\rho_t + \boldsymbol{\nabla}\rho\boldsymbol{w}dt - \frac{1}{2}\sum_{i,j}\frac{\partial^2}{\partial x_i x_j}(\rho a_{ij})dt = \boldsymbol{\nabla}\rho\sigma d\tilde{\boldsymbol{B}}_t. \tag{6}$$

In the case of the Kraichnan model the density variation involves a Laplacian diffusion and the density conditional expectation with respect to a given initial condition evolution comes to an intuitive advection diffusion equation. Let us note that the same kind of deterministic advection equation with an anisotropic diffusion is also obtained if the noise lies in the tangent plane of isodensity surfaces. This type of diffusion for the transport of a given scalar is often considered in geophysics to encode large scale mixing of stratified fluids. They are called isopycnal or isoneutral diffusion in this context [**11**].

2.2. Linear momentum conservation. The mass conservation constraint and the stochastic version of the Reynolds theorem allows us expressing the balance between the momentum and the forces:

$$d\int_{\mathcal{V}(t)}\rho(\boldsymbol{w}(\boldsymbol{x},t)dt + \boldsymbol{\sigma}(\boldsymbol{x},t)d\tilde{\boldsymbol{B}}_t)d\boldsymbol{x} = \int_{\mathcal{V}(t)}\boldsymbol{F}(\boldsymbol{x},t)d\boldsymbol{x}.$$

In this momentum equation, the differentiation in left hand term must be interpreted in a distribution sense (since the random term is non-differentiable). As for the forces in the right hand term, they are composed of standard deterministic forces such as the gravity force and forces acting on both the random and deterministic velocity components. For instance, the surface forces may be defined, as a direct extension of the deterministic, as:

$$\boldsymbol{\Sigma} = \int_{\mathcal{V}} -\boldsymbol{\nabla}(pdt + d\tilde{p}_t) + \mu(\Delta\boldsymbol{U} + \frac{1}{3}\boldsymbol{\nabla}(\boldsymbol{\nabla}\cdot\boldsymbol{U})).$$

In this expression μ is the dynamic viscosity, $p(\boldsymbol{x},t)$ denotes the deterministic contribution of the pressure and $d\tilde{p}_t$ is a zero mean stochastic pressure fluctuation attached to the random component of the velocity. The Navier-Stokes equations under uncertainty are obtained incorporating the mass preservation principle (4) and separating the contributions of the finite variation and martingale terms [**28**]. For a constant density and for a general divergence free (incompressible) uncertainty component, they read:

$$(\partial_t\boldsymbol{w}+\boldsymbol{w}\boldsymbol{\nabla}^T\boldsymbol{w}-\frac{1}{2}\sum_{i,j}\frac{\partial^2}{\partial x_i x_j}(a_{ij}\boldsymbol{w}))\rho=\rho\boldsymbol{g}-\boldsymbol{\nabla}p+\mu\Delta\boldsymbol{w}, \tag{7a}$$

$$\boldsymbol{\nabla}d\tilde{p}_t = -\boldsymbol{w}\boldsymbol{\nabla}^T\rho\boldsymbol{\sigma}d\tilde{\boldsymbol{B}}_t + \mu\Delta(\boldsymbol{\sigma}d\tilde{\boldsymbol{B}}_t), \tag{7b}$$

$$\boldsymbol{\nabla}\cdot(\boldsymbol{\sigma}d\tilde{\boldsymbol{B}}_t) = 0,\;\; \boldsymbol{\nabla}\cdot\boldsymbol{w} = 0,\;\; \boldsymbol{\nabla}\cdot(\boldsymbol{\nabla}\cdot\mathbf{a}) = 0. \tag{7c}$$

This system involves a first equation describing the evolution of the deterministic resolved component. It corresponds to the dynamics of the slow differentiable velocity component. Compared to the original Navier-Stokes formulation and similarly to the classical Reynolds decomposition, it includes an additional stress term that depends here on the resolved velocity component and on the uncertainty variance. The subsequent equations of this system denote a stochastic balance on the diffusion tensor and a mass conservation constraint respectively. Let us note that relaxing the finite variation assumption on $\boldsymbol{w}$, and letting this component be a semi-martingale woul lead to a full stochastic system [**31**–**33**]. In that case no separation between $\boldsymbol{w}$ and the martingale terms is possible.

In the simplest case of a divergence free isotropic model with constant density, we get a Navier-Stokes formulation with a constant eddy viscosity diffusivity coefficient :

$$(\frac{\partial \boldsymbol{w}}{\partial t}+\boldsymbol{w}\boldsymbol{\nabla}^{T}\boldsymbol{w}-\gamma\frac{1}{2}\Delta\boldsymbol{w})\rho=\rho\boldsymbol{g}-\boldsymbol{\nabla}p+\mu\Delta\boldsymbol{w},\ \boldsymbol{\nabla}\cdot\boldsymbol{w}=0, \tag{8}$$

where the diffusion is augmented by the noise variance. Note that in the previous non homogeneous model the diffusion term attached to the uncertainties cannot be directly related to the Boussinesq eddy viscosity formulation. However, it can be checked that for divergence free random field this term is globally dissipative as its energy is

$$\int_{\Omega}\boldsymbol{w}^{T}\sum_{i,j}\frac{\partial^2}{\partial x_i\partial x_j}(a_{ij}\boldsymbol{w})d\boldsymbol{x}=-\int_{\Omega}\|\nabla\boldsymbol{w}\|^2_{\mathbf{a}}d\boldsymbol{x}.$$

The subgrid stress tensor constitutes an anisotropic diffusion whose preferential directions of diffusion are given by the uncertainty variance. Setting the uncertainty diffusion tensor, $\boldsymbol{\sigma}$, or its variance tensor allows defining directly the subgrid diffusion term that has to be incorporated in the resolved drift component. For instance, considering uncertainties along iso-density surfaces provides immediately a clear justification of the isopycnal diffusion employed in oceanic circulation models [**28**]. The use also of constant eddy viscosity is also justified as the direct consequence of an isotropic homogeneous uncertainty component. Such an approach opens new perspectives for flow modeling that goes from *a priori* uncertainty specification to data based strategies. This framework, which does not rely neither on Reynolds averaging nor on spatial filtering concept, might be of great interest when uncertainties are prevalent as it is the case in geophysical flows or climate modeling. As another practical consequence, if one consider velocity fields supplied by particle image velocimetry methods, which are related to the true flow kinematics only up to a Gaussian uncertainty, then those measurements does not follow exactly the actual flow dynamics. Their physical interpretation should then be carried out with some care.

In the following we will experiment simple choices related in spirit to the scale similarity principles used to define subgrid stress tensor [**1**]. Before presenting the variance tensors we used, we present the numerical scheme we devised for the numerical simulation of the drift component.

3. A divergence-free wavelet numerical scheme for the drift component

The objective here is to construct a wavelet-based numerical discretization for the drift component equations of the stochastic system (7). Precisely, we consider the following system of an incompressible fluid:

$$\begin{cases} \frac{\partial \boldsymbol{w}}{\partial t}-\nu\Delta\boldsymbol{w}+\boldsymbol{w}\boldsymbol{\nabla}^{T}\boldsymbol{w}-\frac{1}{2}\sum_{i,j}\frac{\partial^2}{\partial x_i x_j}(a_{ij}\boldsymbol{w})+\boldsymbol{\nabla}p=\boldsymbol{f},\ (\boldsymbol{x},t)\in\Omega\times]0,T[,\\ \boldsymbol{\nabla}\cdot\boldsymbol{w}=0,\ \boldsymbol{x}\in\Omega,\\ \boldsymbol{w}(\boldsymbol{x},0)=\boldsymbol{w}_0,\ \boldsymbol{x}\in\Omega, \end{cases} \tag{9}$$

where $\Omega \subset \mathbb{R}^d$ is a *regular* open bounded subset satisfying periodic boundary conditions and $\nu = \mu/\rho$ denotes the kinematic viscosity. Since the matrix $\boldsymbol{a}(\boldsymbol{x}, t)$ represents the covariance of the uncertainties on the flow, it is assumed to satisfy:

$$\sum_{i,j} a_{ij}\xi_i\xi_j \geq \gamma|\xi|^2, \ \forall\, \xi \in \mathbb{R}^d, \tag{10}$$

with $\gamma > 0$ a positive constant. Let us suppose $\boldsymbol{w}$ to be a *regular* solution. Then, using integration by part and Poincaré's inequality, we get:

$$\frac{1}{2}\frac{d}{dt}\int_\Omega |\boldsymbol{w}|^2 + \frac{\nu}{2}\int_\Omega |\boldsymbol{\nabla w}|^2 + \frac{1}{2}\int_\Omega \|\nabla \boldsymbol{w}\|^2_{\mathbf{a}} \leq C\|\boldsymbol{f}\|^2_{H^{-1}(\Omega)^d}. \tag{11}$$

Again, using Poincaré for $\alpha > 0$, we have:

$$\begin{aligned} \frac{d}{dt}\left(e^{\alpha t}\|\boldsymbol{w}\|^2_{L^2(\Omega)^d}\right) &= e^{\alpha t}\left(\alpha\|\boldsymbol{w}\|^2_{L^2(\Omega)^d} + \frac{d}{dt}\|\boldsymbol{w}\|^2_{L^2(\Omega)^d}\right) \\ &\leq e^{\alpha t}\left(\frac{d}{dt}\|\boldsymbol{w}\|^2_{L^2(\Omega)^d} + \alpha C\int_\Omega |\boldsymbol{\nabla w}|^2\right). \end{aligned} \tag{12}$$

Thus, appropriately setting the value of α, we obtained:

$$\|\boldsymbol{w}\|^2_{L^2(\Omega)^d} \leq e^{-\alpha t}\|\boldsymbol{w}_0\|^2_{L^2(\Omega)^d} + \int_0^t e^{\alpha(s-t)}\|\boldsymbol{f}\|_{H^{-1}(\Omega)^d}ds. \tag{13}$$

If there is no forcing term ($\boldsymbol{f} = 0$), from (11) and (13), we deduce that a *regular* solution $\boldsymbol{w}$ is uniformly bounded. Moreover, the energy of the solution decays exponentially in time:

$$\|\boldsymbol{w}\|^2_{L^2(\Omega)^d} \leq C(\boldsymbol{w}_0) \text{ and } \|\boldsymbol{w}\|^2_{L^2(\Omega)^d} \leq e^{-\alpha t}C(\boldsymbol{w}_0). \tag{14}$$

To keep this stability condition, it is important for any numerical discretization associated to (9) to satisfy condition similar to (13) or (14). The objective of the forthcoming sections is to provide a wavelet-based numerical scheme of this type for the spatial discretization. The temporal discretization will rely on a finite difference method. Moreover, we will show that a semi-implicit Euler scheme combined to an explicit computation of the non linear term $\boldsymbol{w}\boldsymbol{\nabla}^T\boldsymbol{w}$ and the diffusion term $\frac{1}{2}\sum_{i,j}\frac{\partial^2}{\partial x_i x_j}(a_{ij}\boldsymbol{w})$ is stable under a particular CFL type condition. For convenience, we will first describ the spatial discretization and then, we will present the temporal scheme.

3.1. Spatial discretization. Galerkin type method is used to discretize in space the system (9). Since the velocity $\boldsymbol{w}$ is incompressible: $\nabla \cdot \boldsymbol{w} = 0$, to incorporate this constraint, a divergence-free wavelet basis is considered [**8**, **14**]. Then, we suppose that the spatial domain is the hypercube $\Omega = [0,1]^d$. At a fixed spatial resolution $j > 0$, the approximated solution $\boldsymbol{w}_j$ is defined by its wavelet series:

$$\boldsymbol{w}_j(\boldsymbol{x}, t) = \sum_{|\boldsymbol{j}|\leq j, \boldsymbol{k}\in\Lambda_j} \boldsymbol{d}_{\boldsymbol{j},\boldsymbol{k}}(t)\Psi^{\text{div}}_{\boldsymbol{j},\boldsymbol{k}}(\boldsymbol{x}),\ \boldsymbol{\nabla}\cdot\Psi^{\text{div}}_{\boldsymbol{j},\boldsymbol{k}} = 0,\ j \in \mathbb{N} \text{ and } \Lambda_j = [0, 2^j - 1]^d, \tag{15}$$

where $\{\Psi^{\text{div}}_{\boldsymbol{j},\boldsymbol{k}}\}_{\boldsymbol{j},\boldsymbol{k}\in\mathbb{Z}^d}$ is the divergence-free wavelet basis for the velocity function space. For details on the construction of a such wavelet basis, we refer the reader to [**8**, **14**] and references therein.

Now, let us denote by $\mathbf{V}_j$ the multiresolution analysis spaces associated to the divergence-free wavelets basis $\{\Psi^{\text{div}}_{\boldsymbol{j},\boldsymbol{k}}\}_{\boldsymbol{j},\boldsymbol{k}\in\mathbb{Z}^d}$. It is known that supplemented with boundary conditions, $\mathbf{V}_j$ is a finite dimensional space [**8**], thus all the norm on $\mathbf{V}_j$ are equivalent. Particularly, we have:

$$\lim_{j\to+\infty} \|\boldsymbol{w} - \boldsymbol{w}_j\|_{L^2(\Omega)^d} = 0 \text{ and } \|\boldsymbol{w}_j\|_{L^2(\Omega)^d} \lesssim \|\boldsymbol{w}\|_{L^2(\Omega)^d}.$$

One advantage of the wavelet basis is that we have a direct and an inverse estimation on the solutions $\boldsymbol{w}_j$ and $\boldsymbol{w}$. These estimations are very important in the stability proof of the numerical scheme. More precisely, we have Bernstein and Jackson type of inequalities:

$$\|\boldsymbol{w}_j\|_{H^s(\Omega)^d} \lesssim 2^{js}\|\boldsymbol{w}_j\|_{L^2(\Omega)^d} \text{ and } \inf_{\boldsymbol{w}_j\in\mathbf{V}_j} \|\boldsymbol{w} - \boldsymbol{w}_j\|_{L^2(\Omega)^d} \lesssim 2^{-js}\|\boldsymbol{w}\|_{H^s(\Omega)^d}. \tag{16}$$

In comparison to finite element method, the mesh size is $\delta x = 2^{-j}$, where $j > 0$ denotes the spatial resolution. We emphasise that the decomposition (15) provides a time and space scale separation and since the wavelet basis $\{\Psi^{\text{div}}_{\boldsymbol{j},\boldsymbol{k}}\}_{\boldsymbol{j},\boldsymbol{k}\in\mathbb{Z}^d}$ is explicitly defined, the unknowns are the wavelet coefficients $\boldsymbol{d_{j,k}}(t)$. Then, to compute the coefficients $\boldsymbol{d_{j,k}}(t)$ we must invert the mass and stiffness matrices of the wavelet basis $\{\Psi^{\text{div}}_{\boldsymbol{j},\boldsymbol{k}}\}_{\boldsymbol{j},\boldsymbol{k}\in\mathbb{Z}^d}$. This step will be detailed in section 3.4 for the 2D case, the generalization to higher dimension is analogous.

3.2. Temporal discretization. The time discretization of (9) uses similar methods such as those proposed by [**8**]. For the sake of simplicity, we assume that $\boldsymbol{f} = 0$. Then, the projection of (7a) onto the divergence-free function space leads to:

$$\frac{\partial \boldsymbol{w}}{\partial t} - \nu\Delta\boldsymbol{w} = \mathbb{P}[-\boldsymbol{w}\boldsymbol{\nabla}^T\boldsymbol{w} + \frac{1}{2}\sum_{i,j}\frac{\partial^2}{\partial x_i x_j}(a_{ij}\boldsymbol{w})], \tag{17}$$

where $\mathbb{P}$ denotes the Leray projector: the orthogonal projector from $L^2(\Omega)^d$ onto the divergence-free function space $\mathcal{H}_{div}(\Omega) = \{\boldsymbol{u} \in L^2(\Omega)^d \ : \ \boldsymbol{\nabla}\cdot\boldsymbol{u} = 0\}$. The projector $\mathbb{P}$ is in general explicitly defined in Fourier domain. In the present work, the computation of this projector is done though the projection onto the divergence-free wavelet basis $\{\Psi^{\text{div}}_{\boldsymbol{j},\boldsymbol{k}}\}$, see [**8, 14**].

Equation (17) can be seen as a heat equation with source term given by:

$$\mathbb{P}[-\boldsymbol{w}\boldsymbol{\nabla}^T\boldsymbol{w} + \frac{1}{2}\sum_{i,j}\frac{\partial^2}{\partial x_i x_j}(a_{ij}\boldsymbol{w})].$$

Hence, classical numerical schemes for the time discretization of the heat kernel can be used. Sticking to an implicit Euler scheme with a time step δt and setting $\boldsymbol{w}^n(\boldsymbol{x}) \simeq \boldsymbol{w}_j(n\delta t, \boldsymbol{x})$ leads to:

$$(I - \nu\delta t\Delta)\boldsymbol{w}^{n+1} = \boldsymbol{w}^n - \delta t\mathbb{P}[\boldsymbol{w}^n\boldsymbol{\nabla}^T\boldsymbol{w}^n - \frac{1}{2}\sum_{i,j}\frac{\partial^2}{\partial x_i x_j}(a^n_{ij}\boldsymbol{w}^n)]. \tag{18}$$

In this case, the pressure p is recovered from the Helmholtz-Hodge decomposition:

$$\boldsymbol{w}\boldsymbol{\nabla}^T\boldsymbol{w} - \frac{1}{2}\sum_{i,j}\frac{\partial^2}{\partial x_i x_j}(a_{ij}\boldsymbol{w}) = \mathbb{P}[\boldsymbol{w}\boldsymbol{\nabla}^T\boldsymbol{w} - \frac{1}{2}\sum_{i,j}\frac{\partial^2}{\partial x_i x_j}(a_{ij}\boldsymbol{w})] - \nabla p.$$

The convergence and the computational efficiency of the discrete scheme (18) highly depends on the time step value. In the following section we establish a necessary stability condition for the method convergence. This condition will allow us to fix a maximal value for the time step in practice.

3.3. Stability of the semi-implicit Euler scheme. In this section, we will show that the semi-implicit Euler scheme (18) is stable in a specified meaning, if the matrix $\boldsymbol{a}(\boldsymbol{x},t)$ is fixed and bounded on $\Omega \times [0,T]$:

$$\|\boldsymbol{a}\|_{L^\infty(\Omega\times[0,T])} \le C \text{ and } \|\boldsymbol{\nabla}\cdot\boldsymbol{a}\|_{L^\infty(\Omega\times[0,T])} \le C. \tag{19}$$

Condition (19) is always satisfied for an isotropic model with constant density [**13, 28**]: in this case we have $\boldsymbol{\nabla}\cdot\boldsymbol{a} = 0$. Then, if (19) is satisfied, we have the following proposition:

PROPOSITION 1. *If the kinematic viscosity ν, the anisotropic diffusion coefficients a_{ij}, the time step δt and the mesh size $\delta x = 2^{-j}$ satisfy:*

$$\nu \ge \frac{1}{\sqrt{2}}\sqrt{\|\boldsymbol{a}\|^2_{L^\infty} + C(\Omega)^2\|\boldsymbol{\nabla}\cdot\boldsymbol{a}\|^2_{L^\infty}} \qquad \text{and} \qquad \frac{\delta t}{\delta x^d} \le C(\boldsymbol{w}^0,\boldsymbol{a},\nu), \tag{20}$$

then there exists $\lambda(\nu,\Omega) > 0$, a positive constant independent of δt and δx, such that the solution $\boldsymbol{w}^n$ of (18) is stable:

$$\|\boldsymbol{w}^{n+1}\|^2_{L^2(\Omega)^d} \le (1+\delta t\lambda(\nu,\Omega))^{-n}\|\boldsymbol{w}^0\|^2_{L^2(\Omega)^d} + C_0\sum_{k=0}^{n-1}(1+\delta t\lambda(\nu,\Omega))^{k-n}, \tag{21}$$

where C_0 is a constant depending on ν, $\boldsymbol{w}^0$ and a_{ij}.

We infer that relation (21) is a discret version of (13). Then, a uniform bound for $\|\boldsymbol{w}^n\|^2_{L^2(\Omega)^d}$ can be deduced.

The proof of proposition 1 relies on the following intermediate lemma:

LEMMA 1. *If the parameters ν, $\boldsymbol{a}$, δt and $\delta x = 2^{-j}$ satisfy the condition of (20), the solution $\boldsymbol{w}^n$ given by (18) remains bounded:*

$$\|\boldsymbol{w}^n\|^2_{L^2(\Omega)^d} \le C(\boldsymbol{w}^0,\boldsymbol{a}), \;\; n = 0,\cdots,N \tag{22}$$

$$\delta t\sum_{k=1}^{N}\|\nabla\boldsymbol{w}^k\|^2_{L^2(\Omega)^{d\times d}} \le C(\boldsymbol{w}^0,\boldsymbol{a}). \tag{23}$$

where $C(\boldsymbol{w}^0,\boldsymbol{a}) > 0$ is a constant depending on the initial data.

Lemma 1 is analogous to lemma 5.3 of [**37**]. Very similar arguments can be used for its proof, this proof is provided in A. Let us now prove Proposition 1.

PROOF. Taking $2\boldsymbol{w}^{n+1}$ as a test function in (18), for each term we obtain:

$$2\int_\Omega (1-\nu\delta t\Delta)\boldsymbol{w}^{n+1}\cdot\boldsymbol{w}^{n+1} - 2\int_\Omega \boldsymbol{w}^n\cdot\boldsymbol{w}^{n+1} =$$
$$\|\boldsymbol{w}^{n+1}\|^2_{L^2(\Omega)^d} - \|\boldsymbol{w}^n\|^2_{L^2(\Omega)^d} + 2\delta t\nu\|\nabla\boldsymbol{w}^{n+1}\|^2_{L^2(\Omega)^{d\times d}} + \|\boldsymbol{w}^{n+1}-\boldsymbol{w}^n\|^2_{L^2(\Omega)^d},$$

and due to the incompressibility constraint $\boldsymbol{\nabla}\cdot(\boldsymbol{\nabla}\cdot\boldsymbol{a}) = 0$, we have:

$$\frac{\partial^2}{\partial x_i\partial x_j}(a_{ij}\boldsymbol{w}^n) = \frac{\partial a_{ij}}{\partial x_j}\frac{\partial \boldsymbol{w}^n}{\partial x_i} + \frac{\partial}{\partial x_i}(a_{ij}\frac{\partial \boldsymbol{w}^n}{\partial x_j}).$$

Thus:

$$\begin{aligned}\int_\Omega \mathbb{P}[\frac{\partial^2}{\partial x_i \partial x_j}(a_{ij}\boldsymbol{w}^n)]\cdot \boldsymbol{w}^{n+1} &\leq \|\boldsymbol{a}\|_{L^\infty(\Omega\times[0,T])}\|\boldsymbol{\nabla}\boldsymbol{w}^{n+1}\|_{L^2(\Omega)^{d\times d}}\|\boldsymbol{\nabla}\boldsymbol{w}^n\|_{L^2(\Omega)^{d\times d}}\\ &+ \|\boldsymbol{\nabla}\cdot\boldsymbol{a}\|_{L^\infty(\Omega\times[0,T])}\|\boldsymbol{w}^n\|_{L^2(\Omega)^d}\|\boldsymbol{\nabla}\boldsymbol{w}^{n+1}\|_{L^2(\Omega)^{d\times d}}.\end{aligned}$$

For the non linear term, since

$$\int_\Omega (\boldsymbol{w}^n\cdot\nabla)\boldsymbol{w}^n\cdot\boldsymbol{w}^n = -\frac{1}{2}\int_\Omega(\nabla\cdot\boldsymbol{w}^n)|\boldsymbol{w}^n|^2 = 0,$$

one obtains:

$$2\int_\Omega (\boldsymbol{w}^n\cdot\nabla)\boldsymbol{w}^n\cdot(\boldsymbol{w}^{n+1}-\boldsymbol{w}^n) \leq 2\|\boldsymbol{w}^n\|_{L^\infty(\Omega)^d}\|\nabla\boldsymbol{w}^n\|_{L^2(\Omega)^{d\times d}}\|\boldsymbol{w}^{n+1}-\boldsymbol{w}^n\|_{L^2(\Omega)^d}.$$

As we are in a finite dimensional space, in addition to (16), we have:

$$\|\boldsymbol{w}^n\|_{L^\infty(\Omega)^d} \leq 2^{jd/2}\|\boldsymbol{w}^n\|_{L^2(\Omega)^d}, \tag{24}$$

then, we get:

$$\begin{aligned}2\|\boldsymbol{w}^n\|_{L^\infty}\|\nabla\boldsymbol{w}^n\|_{L^2(\Omega)^d}\|\boldsymbol{w}^{n+1}-\boldsymbol{w}^n\|_{L^2(\Omega)^d} &\leq 2.2^{jd/2}\|\boldsymbol{w}^n\|_{L^2(\Omega)^d}\|\nabla\boldsymbol{w}^n\|_{L^2(\Omega)^{d\times d}}\\ &\quad \|\boldsymbol{w}^{n+1}-\boldsymbol{w}^n\|_{L^2(\Omega)^d} \quad .\end{aligned}$$

Gathering these estimations, we have:

$$\begin{aligned}\|\boldsymbol{w}^{n+1}\|^2_{L^2(\Omega)^d} &-\|\boldsymbol{w}^n\|^2_{L^2(\Omega)^d} + 2\delta t\nu\|\nabla\boldsymbol{w}^{n+1}\|^2_{L^2(\Omega)^{d\times d}} + \|\boldsymbol{w}^{n+1}-\boldsymbol{w}^n\|^2_{L^2(\Omega)^d}\\ \leq\ & 2\delta t 2^{\frac{jd}{2}}\|\boldsymbol{w}^n\|_{L^2(\Omega)^d}\|\nabla\boldsymbol{w}^n\|_{L^2(\Omega)^{d\times d}}\|\boldsymbol{w}^{n+1}-\boldsymbol{w}^n\|_{L^2(\Omega)^d} +\\ +\ & \delta t\|\boldsymbol{a}\|_{L^\infty(\Omega\times[0,T])}\|\nabla\boldsymbol{w}^{n+1}\|_{L^2(\Omega)^{d\times d}}\|\nabla\boldsymbol{w}^n\|_{L^2(\Omega)^{d\times d}} +\\ +\ & \delta t\|\boldsymbol{\nabla}\cdot\boldsymbol{a}\|_{L^\infty(\Omega\times[0,T])}\|\boldsymbol{w}^n\|_{L^2(\Omega)^d}\|\nabla\boldsymbol{w}^{n+1}\|_{L^2(\Omega)^{d\times d}}.\end{aligned}$$

Young's inequality leads to:

$$\begin{aligned}\|\boldsymbol{w}^{n+1}\|^2_{L^2(\Omega)^d} - \|\boldsymbol{w}^n\|^2_{L^2(\Omega)^d} &+ \delta t\nu\|\nabla\boldsymbol{w}^{n+1}\|^2_{L^2(\Omega)^{d\times d}}\\ &\leq \delta t^2 2^{jd}\|\boldsymbol{w}^n\|^2_{L^2(\Omega)^d}\|\nabla\boldsymbol{w}^n\|^2_{L^2(\Omega)^{d\times d}}\\ &+ \frac{\delta t\|\boldsymbol{a}\|^2_{L^\infty(\Omega\times[0,T])}}{2\nu}\|\nabla\boldsymbol{w}^n\|^2_{L^2(\Omega)^{d\times d}}\\ &+ \frac{\delta t\|\boldsymbol{\nabla}\cdot\boldsymbol{a}\|^2_{L^\infty(\Omega\times[0,T])}}{2\nu}\|\boldsymbol{w}^n\|^2_{L^2(\Omega)^d} \quad .\end{aligned}\tag{25}$$

Using Poincaré's inequality for the last term of (25), we deduce that:

$$\begin{aligned}\|\boldsymbol{w}^{n+1}\|^2_{L^2(\Omega)^d} - \|\boldsymbol{w}^n\|^2_{L^2(\Omega)^d} &+ \delta t\nu\|\nabla\boldsymbol{w}^{n+1}\|^2_{L^2(\Omega)^{d\times d}}\\ &\leq \delta t^2 2^{jd}\|\boldsymbol{w}^n\|^2_{L^2(\Omega)^d}\|\nabla\boldsymbol{w}^n\|^2_{L^2(\Omega)^{d\times d}}\\ &+ \frac{\delta t\|\boldsymbol{a}\|^2_{L^\infty(\Omega\times[0,T])}}{2\nu}\|\nabla\boldsymbol{w}^n\|^2_{L^2(\Omega)^{d\times d}}\\ &+ \frac{\delta t C(\Omega)^2\|\boldsymbol{\nabla}\cdot\boldsymbol{a}\|^2_{L^\infty(\Omega\times[0,T])}}{2\nu}\|\nabla\boldsymbol{w}^n\|^2_{L^2(\Omega)^d} \quad .\end{aligned}\tag{26}$$

Now let us define the constant $K(\boldsymbol{a},\nu)$ by:

$$K(\boldsymbol{a},\nu) = \frac{\|\boldsymbol{a}\|^2_{L^\infty(\Omega\times[0,T])} + C(\Omega)^2\|\boldsymbol{\nabla}\cdot\boldsymbol{a}\|^2_{L^\infty(\Omega\times[0,T])}}{2\nu}, \tag{27}$$

and fixed δt small enough such that:

$$\delta t 2^{jd} C(\boldsymbol{w}^0, \boldsymbol{a}) \leq \frac{1}{2}(\nu - K(\boldsymbol{a}, \nu)) \quad .$$

So we obtain :

$$\frac{\delta t}{\delta x^d} \leq \frac{(\nu - K(\boldsymbol{a}, \nu))}{2C(\boldsymbol{w}^0, \boldsymbol{a})} = C(\boldsymbol{w}^0, \boldsymbol{a}, \nu). \tag{28}$$

Since $\|\boldsymbol{w}^n\|^2_{L^2(\Omega)^d} \leq C(\boldsymbol{w}^0, \boldsymbol{a})$, by Lemma 1 and from (26) we deduce that:

$$\begin{aligned} \|\boldsymbol{w}^{n+1}\|^2_{L^2(\Omega)^d} - \|\boldsymbol{w}^n\|^2_{L^2(\Omega)^d} \quad &+ \quad \delta t \nu \|\nabla \boldsymbol{w}^{n+1}\|^2_{L^2(\Omega)^{d\times d}} \\ &\leq \quad \frac{\delta t}{2}(\nu + K(\boldsymbol{a}, \nu))\|\nabla \boldsymbol{w}^n\|^2_{L^2(\Omega)^{d\times d}} \quad . \end{aligned} \tag{29}$$

Again, using Poincaré's inequality, we get:

$$\begin{aligned} \|\boldsymbol{w}^{n+1}\|^2_{L^2(\Omega)^d} - \|\boldsymbol{w}^n\|^2_{L^2(\Omega)^d} \quad &+ \quad \frac{\delta t \nu}{C(\Omega)^2}\|\boldsymbol{w}^{n+1}\|^2_{L^2(\Omega)^d} \\ &\leq \quad \frac{\delta t}{2}(\nu + K(\boldsymbol{a}, \nu))\|\nabla \boldsymbol{w}^n\|^2_{L^2(\Omega)^{d\times d}} \quad . \end{aligned} \tag{30}$$

We set $\lambda(\nu, \Omega) = \frac{\nu}{C(\Omega)^2}$. Then, the sequence A_n defined by:

$$A_n = (1 + \delta t \lambda(\nu, \Omega))^n \|\boldsymbol{w}^n\|^2_{L^2(\Omega)^{d\times d}}, \tag{31}$$

satisfies:

$$\begin{aligned} A_{n+1} - A_n \quad &= \quad (1 + \delta t \lambda(\nu, \Omega))^n \\ &\quad \left(\|\boldsymbol{w}^{n+1}\|^2_{L^2(\Omega)^d} - \|\boldsymbol{w}^n\|^2_{L^2(\Omega)^d} + \delta t \lambda(\nu, \Omega)\|\boldsymbol{w}^{n+1}\|^2_{L^2(\Omega)^d}\right) \\ &\leq \quad (1 + \delta t \lambda(\nu, \Omega))^n \frac{\delta t}{2}(\nu + K(\boldsymbol{a}, \nu))\|\nabla \boldsymbol{w}^n\|^2_{L^2(\Omega)^{d\times d}}. \end{aligned}$$

Summation over n leads to:

$$A_n - A_0 \leq \sum_{k=0}^{n-1}(1 + \delta t \lambda(\nu, \Omega))^k \frac{\delta t}{2}(\nu + K(\boldsymbol{a}, \nu))\|\nabla \boldsymbol{w}^k\|^2_{L^2(\Omega)^{d\times d}}.$$

Again, from Lemma 1 we have:

$$\delta t \|\nabla \boldsymbol{w}^n\|^2_{L^2(\Omega)^{d\times d}} \leq C(\boldsymbol{w}^0, \boldsymbol{a}), \ \forall\, n = 0, 1, \cdots,$$

thus:

$$A_n - A_0 \leq C_0 \sum_{k=0}^{n-1}(1 + \delta t \lambda(\nu, \Omega))^k, \tag{32}$$

with $C_0 = \frac{1}{2}(\nu + K(\boldsymbol{a}, \nu))C(\boldsymbol{w}^0, \boldsymbol{a})$. Replacing A_n by its expression (31) ends the proof. □

To complete this section, we analyze now the behavior of the discretization error as δt goes to zero and $j > 0$ goes to infinity:

$$\boldsymbol{e}^n = \boldsymbol{w}(\boldsymbol{x}, t_n) - \boldsymbol{w}^n,$$

where $t_n = n\delta t$ denotes the corresponding discrete time and $\boldsymbol{w}(t,x)$ is the exact regular solution of (7a), which we suppose at least that $\boldsymbol{w} \in H^2(0,T;L^2(\Omega)^d) \cap \mathcal{C}^0(0,T;W^{1,+\infty}(\Omega)^d \cap H^1(\Omega)^d)$. Then, replacing $\boldsymbol{w}^n$ by $\boldsymbol{w}(\boldsymbol{x},t_n)$ in (18), we get:

$$\begin{aligned}\boldsymbol{w}(\boldsymbol{x},t_{n+1})-\boldsymbol{w}(\boldsymbol{x},t_n) \quad - \quad & \nu\delta t\Delta\boldsymbol{w}(\boldsymbol{x},t_{n+1})\delta t\mathbb{P}[(\boldsymbol{w}(\boldsymbol{x},t_n)\cdot\nabla)\boldsymbol{w}(\boldsymbol{x},t_n) - \\ & \frac{1}{2}\sum_{ij}\frac{\partial^2}{\partial x_i\partial x_j}(a_{ij}\boldsymbol{w}(\boldsymbol{x},t_n))] = \delta t\boldsymbol{\epsilon}^{n+1}\end{aligned} \tag{33}$$

where $\boldsymbol{\epsilon}^n$ denotes the consistency error. From (33), we infer that: $\boldsymbol{\epsilon}^n \in H^{-1}(\Omega)$ and it is divergence free: $\boldsymbol{\nabla}\cdot\boldsymbol{\epsilon}^n = 0$ and taking the difference of (33) and (18), reads:

$$\begin{aligned}\boldsymbol{e}^{n+1}-\boldsymbol{e}^n-\nu\delta t\Delta\boldsymbol{e}^{n+1} \quad = \quad & \delta t\boldsymbol{\epsilon}^{n+1}\delta t\mathbb{P}[(\boldsymbol{e}^n\cdot\nabla)\boldsymbol{w}(\boldsymbol{x},t_n) + \\ & (\boldsymbol{w}^n\cdot\nabla)\boldsymbol{e}^n - \frac{1}{2}\sum_{ij}\frac{\partial^2}{\partial x_i\partial x_j}(a_{ij}\boldsymbol{e}^n)],\end{aligned} \tag{34}$$

Thus, similar arguments to those used for the proof of Proposition 1 allow to show that:

$$\begin{aligned}\|\boldsymbol{e}^{n+1}\|^2_{L^2(\Omega)^d} - \|\boldsymbol{e}^n\|^2_{L^2(\Omega)^d} \quad + \quad & \frac{\delta t\nu}{2}\|\nabla\boldsymbol{e}^{n+1}\|^2_{L^2(\Omega)^{d\times d}} \\ \leq \quad & \delta t K(\boldsymbol{a},\nu)\|\nabla\boldsymbol{e}^n\|^2_{L^2(\Omega)^{d\times d}} \\ + \quad & \frac{4\delta t C^2(\Omega)}{\nu}2^{jd}\|\boldsymbol{w}^n\|^2_{L^2(\Omega)^d}\|\nabla\boldsymbol{e}^n\|^2_{L^2(\Omega)^{d\times d}} \\ + \quad & \frac{4\delta t C^2(\Omega)}{\nu}\|\boldsymbol{e}^n\|^2_{L^2(\Omega)^d}\|\nabla\boldsymbol{w}(t_n)\|^2_{L^{+\infty}(\Omega)^{d\times d}} \\ + \quad & 2\delta t\|\boldsymbol{\epsilon}^{n+1}\|_{H^{-1}(\Omega)}\|\boldsymbol{e}^{n+1}\|_{L^2(\Omega)^d}.\end{aligned} \tag{35}$$

Otherwise, using (18), we get:

$$\begin{aligned}-\nu\Delta\boldsymbol{w}(\boldsymbol{x},t_{n+1}) = -\partial_t\boldsymbol{w}(t_{\boldsymbol{x},n+1}) \quad - \quad & \mathbb{P}[(\boldsymbol{w}(\boldsymbol{x},t_{n+1})\cdot\nabla)\boldsymbol{w}(\boldsymbol{x},t_{n+1}) \\ - \quad & \frac{1}{2}\sum_{ij}\frac{\partial^2}{\partial x_i\partial x_j}(a_{ij}\boldsymbol{w}(\boldsymbol{x},t_{n+1}))],\end{aligned} \tag{36}$$

and

$$\begin{aligned}\boldsymbol{\epsilon}^{n+1} = -\partial_t\boldsymbol{w}(\boldsymbol{x},t_{n+1}) + & \frac{\boldsymbol{w}(\boldsymbol{x},t_{,n+1})-\boldsymbol{w}(\boldsymbol{x},t_n)}{\delta t} + \\ & \mathbb{P}[(\boldsymbol{w}(\boldsymbol{x},t_n)\cdot\nabla)\boldsymbol{w}(\boldsymbol{x},t_n) - (\boldsymbol{w}(\boldsymbol{x},t_{n+1})\cdot\nabla)\boldsymbol{w}(\boldsymbol{x},t_{n+1})] \\ & -\frac{1}{2}\mathbb{P}[\sum_{ij}\frac{\partial^2}{\partial x_i\partial x_j}(a_{ij}(\boldsymbol{x},t_n)\boldsymbol{w}(\boldsymbol{x},t_n) - a_{ij}(\boldsymbol{x},t_{n+1})\boldsymbol{w}(\boldsymbol{x},t_{n+1}))].\end{aligned}$$

Since we assume that $\boldsymbol{w}$ is regular enough, we have:

$$\begin{aligned}\boldsymbol{\epsilon}^{n+1} = \frac{1}{\delta t}\int_{t_n}^{t_{n+1}}(t-t_n)\partial_{tt}\boldsymbol{w}(\boldsymbol{x},t)dt - \mathbb{P}\left[\int_{t_n}^{t_{n+1}}\partial_t[(\boldsymbol{w}(\boldsymbol{x},t)\cdot\nabla)\boldsymbol{w}(t)]dt\right] \\ +\frac{1}{2}\sum_{ij}\mathbb{P}\left[\int_{t_n}^{t_{n+1}}\frac{\partial^2}{\partial x_i\partial x_j}(\partial_t[a_{ij}\boldsymbol{w}(\boldsymbol{x},t)])dt\right],\end{aligned} \tag{37}$$

and this gives:

$$\delta t \sum_n \|\epsilon^{n+1}\|^2_{H^{-1}(\Omega)} \leq \delta t^2 \|\partial_{tt} \boldsymbol{w}\|^2_{L^2(0,T;H^{-1}(\Omega))} + 3\delta t^2 \|\partial_t[(\boldsymbol{w}\cdot\nabla)\boldsymbol{w}]\|^2_{L^2(0,T;H^{-1}(\Omega))} + \frac{3\delta t^2}{2}\|\frac{\partial^2}{\partial x_i \partial x_j}(\partial_t[a_{ij}\boldsymbol{w}])\|^2_{L^2(0,T;H^{-1}(\Omega))},$$

and by definition we have:

$$\begin{aligned} \|\boldsymbol{e}^n\|^2_{L^2(\Omega)^d} &\leq 2\left(\|\boldsymbol{w}^n\|^2_{L^2(\Omega)^d} + \|\boldsymbol{w}(t_n)\|^2_{L^2(\Omega)^d}\right), \quad \|\nabla \boldsymbol{e}^n\|^2_{L^2(\Omega)^2} \\ &\leq 2\left(\|\nabla \boldsymbol{w}^n\|^2_{L^2(\Omega)^{d\times d}} + \|\nabla \boldsymbol{w}(t_n)\|^2_{L^2(\Omega)^{d\times d}}\right). \end{aligned} \tag{38}$$

From (35) and recalling that the right-hand terms of inequalities (38) are bounded by constants depending only on the initial data, there exists hence a suitable constant depending only on the initial data such that:

$$\|\boldsymbol{e}^{n+1}\|^2_{L^2} - \|\boldsymbol{e}^n\|^2_{L^2} + \frac{\delta t \nu}{4}\|\nabla \boldsymbol{e}^{n+1}\|^2_{L^2} \leq (\delta t + \delta x^{d/2}) C(\boldsymbol{w}^0, a_{ij}, \Omega). \tag{39}$$

Summation of (39) over n shows that:

$$\max_n \|\boldsymbol{e}^n\|^2_{L^2(\Omega)^d} \xrightarrow[\delta t\to 0, \delta x \to 0]{} 0. \tag{40}$$

and hence the semi-implicit temporal numerical scheme (18) converges.

3.4. Practical computational details of the divergence free numerical scheme. This section gives details on the implementation of the divergence-free wavelet numerical scheme in the especial case of the two space dimension: $\Omega = [0,1]^2$. We start by giving a brief reminder on the construction of a divergence-free wavelet basis. To construct divergence-free wavelet in $(L^2(\Omega))^2$, one needs first to have two multi-resolution analyses of $L^2(0,1)$ generated by the spaces V_j^1 and V_j^0 linked by differentiation:

$$\frac{d}{dx} V_j^1 = V_j^0.$$

In this case, it can be shown [**24**] that the associated wavelet generators ψ^1 and ψ^0 satisfy: $(\psi^1)' = 4\psi^0$. Then, the divergence-free wavelet generator is constructed by taking the curl of the scalar wavelet $\psi^1 \otimes \psi^1$:

$$\Psi^{div} = \mathbf{curl}[\psi^1 \otimes \psi^1] = 4\begin{pmatrix} \psi^1 \otimes \psi^0 \\ -\psi^0 \otimes \psi^1 \end{pmatrix}. \tag{41}$$

This divergence-free wavelet is contained in the standard multi-resolution analysis of $(L^2(\Omega))^2$ constituted by $\mathbf{V}_j = (V_j^1 \otimes V_j^0) \times (V_j^0 \otimes V_j^1)$, it provides hence fast divergence-free wavelet transform, see [**8, 15**]. This fast transform allows us to compute the divergence-free wavelet coefficients of $\boldsymbol{w}^n = (w_1^n, w_2^n)$ from those of its decomposition on the wavelet basis associated to $\mathbf{V}_j$ and conversely:

$$w_1^n = \sum_{|\boldsymbol{j}|\leq j, \boldsymbol{k}} d^{1,n}_{\boldsymbol{j},\boldsymbol{k}} \psi^1_{j_1,k_1} \otimes \psi^0_{j_2,k_2} \text{ and } w_2^n = \sum_{|\boldsymbol{j}|\leq j, \boldsymbol{k}} d^{2,n}_{\boldsymbol{j},\boldsymbol{k}} \psi^0_{j_1,k_1} \otimes \psi^1_{j_2,k_2}. \tag{42}$$

Moreover, in (18) to compute the coefficients of $\boldsymbol{w}^{n+1}$ from those of $\boldsymbol{w}^n$, one needs to invert the matrix corresponding to operator $(1 - \delta t \nu \Delta)$. To avoid computing this inverse at each time step, we use the method described in [**5**], which relies on the heat kernel operator factorization. This method is very close to the alternated direction implicit methods. Precisely, for a given $\alpha \in \mathbb{R}$, we have:

$$(1 - \alpha\Delta) \approx (1 - \alpha\frac{\partial^2}{\partial x^2})(1 - \alpha\frac{\partial^2}{\partial y^2}) + O(\alpha^2). \tag{43}$$

Thus, we only have to invert the matrix of the one dimension operator $(1 - \delta t\nu\frac{\partial^2}{\partial x^2})$, and this is done once before starting the time integration.

Now, taking $\psi^1_{j_1,k_1} \otimes \psi^0_{j_2,k_2}$ and $\psi^0_{j_1,k_1} \otimes \psi^1_{j_2,k_2}$ as test functions in (18), we get two linear systems on the wavelet coefficients matrices:

$$\mathcal{A}^1_{\delta t}[d^{1,n+1}_{\boldsymbol{j},\boldsymbol{k}}]\mathcal{A}^0_{\delta t} = \mathcal{M}^1[d^{1,n}_{\boldsymbol{j},\boldsymbol{k}}]\mathcal{M}^0 + \delta t\mathcal{M}^1[f^{1,n}_{\boldsymbol{j},\boldsymbol{k}}]\mathcal{M}^0 \tag{44}$$

and

$$\mathcal{A}^0_{\delta t}[d^{2,n+1}_{\boldsymbol{j},\boldsymbol{k}}]\mathcal{A}^1_{\delta t} = \mathcal{M}^0[d^{2,n}_{\boldsymbol{j},\boldsymbol{k}}]\mathcal{M}^1 + \delta t\mathcal{M}^0[f^{2,n}_{\boldsymbol{j},\boldsymbol{k}}]\mathcal{M}^1, \tag{45}$$

where $\mathcal{A}^i_{\delta t}$ are the stiffness matrix of the operator $(1 - \delta t\nu\frac{\partial^2}{\partial x^2})$ on the basis of $\{V^i_j\}_{i=0,1}$ and $\mathcal{M}^i$ are the mass matrices. These matrices are computed by solving en eigenvalue problem, see [**15**] and references therein. The source term s$[f^{1,n}_{\boldsymbol{j},\boldsymbol{k}}]$ and $[f^{2,n}_{\boldsymbol{j},\boldsymbol{k}}]$ are the matrices of the wavelet coefficients of the orthogonal projection onto the divergence-free function space of the non linear term and the anisotropic diffusion term:

$$\boldsymbol{f}^n = -\mathbb{P}[(\boldsymbol{w}^n \cdot \nabla)\boldsymbol{w}^n - \frac{1}{2}\sum_{ij}\frac{\partial^2}{\partial x_i \partial x_j}(a_{ij}\boldsymbol{w}^n)], \tag{46}$$

where the spatial derivatives are computed using finite difference method at the grid points.

The whole resolution method can be summarized as follows. Starting from $\boldsymbol{w}^0(x) = \boldsymbol{w}(\boldsymbol{x}, 0)$, compute its wavelet coefficients $[d^{1,0}_{\boldsymbol{j},\boldsymbol{k}}]$ and $[d^{2,0}_{\boldsymbol{j},\boldsymbol{k}}]$ on $\mathbf{V}_j = (V^1_j \otimes V^0_j) \times (V^0_j \otimes V^1_j)$ and for $0 \le n \le N$, repeat

STEP 1. Using finite difference method, compute the term

$$\tilde{\boldsymbol{f}}^n = -(\boldsymbol{w}^n \cdot \nabla)\boldsymbol{w}^n - \frac{1}{2}\sum_{ij}\frac{\partial^2}{\partial x_i \partial x_j}(a_{ij}\boldsymbol{w}^n)$$

STEP 2. Compute the divergence-free wavelet coefficients $[d^{div,n}_{\boldsymbol{j},\boldsymbol{k}}] = \mathbb{P}(\tilde{\boldsymbol{f}}^n)$ by:

$$\mathcal{M}^1[\tilde{f}^{1,n}_{\boldsymbol{j},\boldsymbol{k}}]\tilde{\mathcal{A}}^0 - (\tilde{\mathcal{A}}^0)^T[\tilde{f}^{2,n}_{\boldsymbol{j},\boldsymbol{k}}]\mathcal{M}^1 = \mathcal{M}^1[d^{div,n}_{\boldsymbol{j},\boldsymbol{k}}]\mathcal{R}^1 + \mathcal{R}^1[d^{div,n}_{\boldsymbol{j},\boldsymbol{k}}]\mathcal{M}^1$$

where $\mathcal{R}^1$ is the stiffness matrix of wavelet basis $\{\psi^1_{j,k}\}$, its terms correspond to $\langle(\psi^d_{j,k})', (\psi^1_{j',k'})'\rangle$ and the terms of $\tilde{\mathcal{A}}^0$ correspond to $\langle\psi^0_{j,k}, (\psi^d_{j',k'})'\rangle$.

STEP 3. Compute $[f^{1,n}_{\boldsymbol{j},\boldsymbol{k}}]$ and $[f^{2,n}_{\boldsymbol{j},\boldsymbol{k}}]$ from $[d^{div,n}_{\boldsymbol{j},\boldsymbol{k}}]$ using the change of basis between $\{(\psi^1_{\boldsymbol{j},\boldsymbol{k}})'\}$ and $\{\psi^0_{\boldsymbol{j},\boldsymbol{k}}\}$.

STEP 4. Find $[d_{\boldsymbol{j},\boldsymbol{k}}^{1,n+1}]$ and $[d_{\boldsymbol{j},\boldsymbol{k}}^{2,n+1}]$ solution of

$$\mathcal{A}_{\delta t}^{1}[d_{\boldsymbol{j},\boldsymbol{k}}^{1,n+1}]\mathcal{A}_{\delta t}^{0} = \mathcal{M}^{1}[d_{\boldsymbol{j},\boldsymbol{k}}^{1,n}]\mathcal{M}^{0} + \delta t\mathcal{M}^{1}\boldsymbol{f}_{1}^{n}\mathcal{M}^{0}$$

$$\mathcal{A}_{\delta t}^{0}[d_{\boldsymbol{j},\boldsymbol{k}}^{2,n+1}]\mathcal{A}_{\delta t}^{1} = \mathcal{M}^{0}[d_{\boldsymbol{j},\boldsymbol{k}}^{2,n}]\mathcal{M}^{1} + \delta t\mathcal{M}^{0}\boldsymbol{f}_{2}^{n}\mathcal{M}^{1}$$

As the matrices $\mathcal{A}_{\delta t}^{1}$ and $\mathcal{A}_{\delta t}^{0}$ are inverted once for all before starting the algorithm, step 4 corresponds thus only to a matrix multiplication. Let J be the maximal space resolution considered, then the theoretical complexity of Step 4 is $O(2^{3J})$. Step 2 is solved with a preconditioned conjugate gradient method, then its theoretical complexity is $O(2^{3J})$ and step 3 is a change of basis, whose complexity is linear. We deduce that the theoretical complexity of the method is about $O(2^{3J})$. In practice, this theoretical complexity can be improved using adaptive methods and coherent structure extraction methods, see [**6**, **38**].

4. Variational data assimilation

In this section we summarize the principle of variational data assimilation for the Navier-Stokes equations with uncertainties. We use a generic notation for the different inner products and associated norms; without loss of generality, we consider the discret framework where the evolution system is given by (18) supplemented with a source terme $\boldsymbol{f}$:

$$(I - \nu\delta t\Delta)\boldsymbol{w}^{n+1} = \boldsymbol{w}^{n} + \delta t\mathbb{P}[-\boldsymbol{w}^{n}\boldsymbol{\nabla}^{T}\boldsymbol{w}^{n} + \frac{1}{2}\sum_{i,j}\frac{\partial^{2}}{\partial x_{i}x_{j}}(a_{ij}^{n}\boldsymbol{w}^{n}) + \boldsymbol{f}]. \tag{47}$$

We suppose that (47) can be rewritten as:

$$\boldsymbol{w}^{n+1} = \mathbb{M}(\boldsymbol{w}^{n}, a_{ij}^{n}, \boldsymbol{f}), \tag{48}$$

where $\mathbb{M}$ is a non linear operator.

Basically, variational data assimilation methods aim at combining measured observations of the system $\boldsymbol{w}_{ob}^{k}$, $0 \leq k \leq N$ with the dynamical model (48) in order to improve the estimation of some parameters. Usually this is done by minimizing an energy functional with respect to control parameters. In this study, the considered control parameters are the initial condition $\boldsymbol{w}^{0}$, the source term $\boldsymbol{f}$ and/or the anisotropic diffusion coefficients a_{ij}. As a first simple example, if the initial condition is the only control parameter considered, a typical energy functional consists in:

$$\mathcal{J}(\boldsymbol{w}^{0}) = \frac{1}{2}\sum_{k=0}^{N}\|\mathbb{H}(\boldsymbol{w}^{k}) - \boldsymbol{w}_{ob}^{k}\|^{2}\delta t + \frac{\alpha}{2}\|\boldsymbol{w}^{0}\|^{2},$$

where $\mathbb{H}$ denotes the observation operator, maping the space of the solutions $\boldsymbol{w}^{k}$ to the measurements $\boldsymbol{w}_{ob}^{k}$ in the observations space. If the control is defined as the initial condition together with the source term, the energy functional reads now:

$$\mathcal{J}(\boldsymbol{w}^{0}, \boldsymbol{f}) = \frac{1}{2}\sum_{k=0}^{N}\|\mathbb{H}(\boldsymbol{w}^{k}) - \boldsymbol{w}_{ob}^{k}\|^{2}\delta t + \frac{\alpha}{2}\|\boldsymbol{w}^{0}\|^{2} + \frac{\beta}{2}\|\boldsymbol{f}\|^{2},$$

and if the control consists in the initial condition, the source term and the anisotropic diffusion, the energy functional is:

$$\mathcal{J}(\boldsymbol{w}^0, \boldsymbol{a}, \boldsymbol{f}) = \frac{1}{2}\sum_{k=0}^{N} \|\mathbb{H}(\boldsymbol{w}^k) - \boldsymbol{w}_{ob}^k\|^2 \delta t + \frac{\alpha}{2}\|\boldsymbol{w}^0\|^2 + \frac{\beta}{2}\|\nabla \cdot (\nabla \cdot \boldsymbol{a})\|^2 + \frac{\gamma}{2}\|\boldsymbol{f}\|^2.$$

The minimization of $\mathcal{J}$ is usually performed through a gradient descent method. For the sake of clarity, let us supposed that we have to control only the initial condition. Then, using directional derivative, it is easy to see that:

$$\langle \nabla \mathcal{J}(\boldsymbol{w}^0), \delta \boldsymbol{w}^0 \rangle = \sum_{k=0}^{N} \langle \mathbb{H}(\boldsymbol{w}^k) - \boldsymbol{w}_{ob}^k, \nabla_{\boldsymbol{w}} \mathbb{H}(\boldsymbol{w}^k) \cdot \nabla_{\boldsymbol{w}^0} \boldsymbol{w}^k \delta \boldsymbol{w}^0 \rangle \delta t + \alpha \langle \boldsymbol{w}^0, \delta \boldsymbol{w}^0 \rangle$$

Then, the gradient of $\mathcal{J}$ is defined on all the domain Ω. Thus, it is of great interest to have an efficient numerical method for the computation of this gradient. We used the adjoint method. Precisely, let us introduce the following langragiean function:

$$\mathcal{L}(\boldsymbol{w}^0, \lambda^{k+1}) = \mathcal{J}(\boldsymbol{w}^0) - \sum_{k=0}^{N-1} \langle \boldsymbol{w}^{k+1} - \mathbb{M}(\boldsymbol{w}^k, a_{ij}^k, \boldsymbol{f}), \lambda^{k+1} \rangle \tag{49}$$

Then, the problem of the minimization of $\mathcal{J}$ under the dynamical constraint (48) becomes a saddle point problem for the Lagrangian $\mathcal{L}$. The first-order optimality conditions lead respectively to the direct model (48) and to the following adjoint problem:

$$\begin{aligned} &\lambda^{N+1} = 0 \qquad \text{and for} \quad 0 \le k \le N \\ &\lambda^k - \nabla_{\boldsymbol{w}} \mathbb{M}^*(\boldsymbol{w}^k, a_{ij}^k, \boldsymbol{f}) \lambda^{k+1} = \delta t \nabla_{\boldsymbol{w}} \mathbb{H}^*(\boldsymbol{w}^k)(\mathbb{H}(\boldsymbol{w}^k) - \boldsymbol{w}_{ob}^k)\ , \end{aligned} \tag{50}$$

where the asterisk denotes the adjoint operator and a backward integration of this adjoint problem gives:

$$\frac{\partial \mathcal{L}}{\partial \boldsymbol{w}^0} = \alpha \boldsymbol{w}^0 + \lambda^0. \tag{51}$$

Remark that, if the λ^k are optimal, namely $\frac{\partial \mathcal{L}}{\partial \lambda^k} = 0$, it is easy to see that:

$$\frac{\partial \mathcal{L}}{\partial \boldsymbol{w}^0} = \nabla \mathcal{J}(\boldsymbol{w}^0). \tag{52}$$

This observation is used during the numerical computation of $\nabla \mathcal{J}$, to control the initial condition, the source term and the anisotropic diffusion of the proposed model in the next section, with image data observations.

5. Numerical results

To implement the model presented in section 2, the first step is to choose a viscosity tensor $\boldsymbol{a}$ that satisfies the incompressibility condition (5) and the ellipticity property (10). There are several choices. The simplest is to set $\boldsymbol{a}$ as a diagonal tensor: $a_{ij} = \mu \delta_{ij}$, with $\mu > 0$ a positive constant or $a_{ij} = \mu(\boldsymbol{x}, t)\delta_{ij}$, with $\mu > 0$ and $\Delta \mu = 0$. In this cases $\boldsymbol{a}$ is a Laplacian operator. The more general choice is

$$\boldsymbol{a} = \begin{pmatrix} \partial_{xy}^2 \chi & -\partial_{xx}^2 \chi \\ \partial_{yy}^2 \chi & -\partial_{xy}^2 \chi \end{pmatrix}, \tag{53}$$

where χ is a two variables regular function. By construction, this choice implies:

$$\nabla \cdot [\nabla \cdot \boldsymbol{a}] = 0.$$

However, to ensure (10), the ellipticity constraint must to be added in the optimization functional.

In this study, we choose the two simplest cases: $a_{ij} = c\delta_{ij}$ referred as case one and $a_{ij} = \mu(\boldsymbol{x}, t)\delta_{ij}$ referred as case two. The observation operator on which we rely is the optical flow operator:

$$\mathbb{H}(\boldsymbol{w}^k) - \boldsymbol{w}_{ob}^k := I_1^k(\boldsymbol{x} + \delta t\boldsymbol{w}^k) - I_0^k(\boldsymbol{x}),\ 0 \leq k \leq N,$$

with I_1^k and I_0^k two successive images and under the luminance invariance hypothesis we have:

$$I_1^k(\boldsymbol{x} + \delta t\boldsymbol{w}^k) - I_0^k(\boldsymbol{x}) = 0,\ 0 \leq k \leq N.$$

We made the control over the initial condition $\boldsymbol{w}_0(\boldsymbol{x})$, the anisotropic diffusion coefficients $a_{ij}(\boldsymbol{x}, t)$ and the source term $\boldsymbol{f}$. Then, the encoded functional to be minimized is:

$$\begin{aligned}\mathcal{J}(\boldsymbol{w}^0, \boldsymbol{a}, \boldsymbol{f}) &= \frac{1}{2}\sum_{k=1}^{N} \|I_1^k(\boldsymbol{x} + \delta t\boldsymbol{w}^k) - I_0^k(\boldsymbol{x})\|_{L^2}^2 \delta t + \\ &+ \frac{\alpha}{2}\|\boldsymbol{w}^0\|_{L^2}^2 + \frac{\beta}{2}\|\nabla \cdot (\nabla \cdot \boldsymbol{a})\|_{L^2}^2 + \frac{\gamma}{2}\|\boldsymbol{f}\|_{L^2}^2.\end{aligned}$$

Accordingly, the gradient of $\mathcal{J}$ is computed using the adjoint method as described in section 4 and we used a quasi-newton method for the optimization procedure, with MATLAB software.

To evaluate the accuracy of the approach, we work on a sequence of synthetic particle images associated to the numerical simulation of a 2D incompressible Navier-Stokes equations at $Re = 3000$, using the vorticity conservation equation and the Lagrangian equation for non-heavy particles transported by the flow (simulation details can be found in [**12**]). The image size is 256×256 pixels and the pixel grey levels have been normalized; a snapshot of the particle image sequence is displayed in Figure 1.

FIGURE 1. Example of used PIV image.

Estimated velocity and vorticity fields are evaluated based on the Root Mean Squared end-point Error (RMSE):

$$\text{RMSE} = \frac{1}{|\Omega|}\left(\sum_{\boldsymbol{x}\in\Omega}\left|\boldsymbol{w}(\boldsymbol{x}) - \boldsymbol{w}_{\text{ref}}(\boldsymbol{x})\right|^2\right)^{\frac{1}{2}}.$$

Figure 2 shows the plot of this RMS error both in terms of velocity field and its corresponding vorticity, on a sequence of five images. The spatial resolution for the model is $j = 6$ while image resolution is $j = 8$. To interpolate the velocity field at the finer image resolution, we used the wavelet interpolation operator. Here we used the Daubechies wavelet generator with 10 vanishing moments.

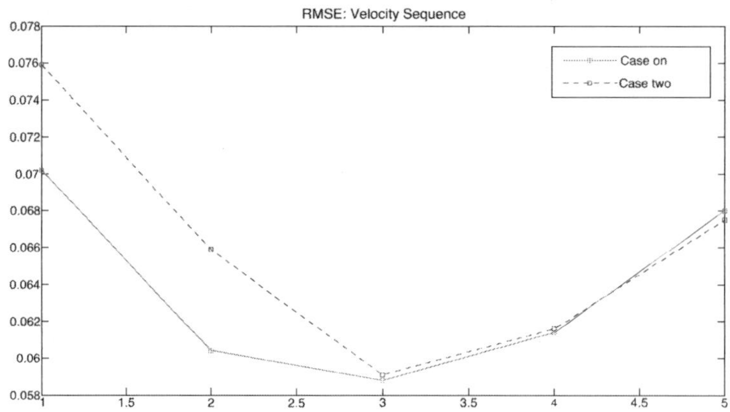

(a) RMSE on velocity

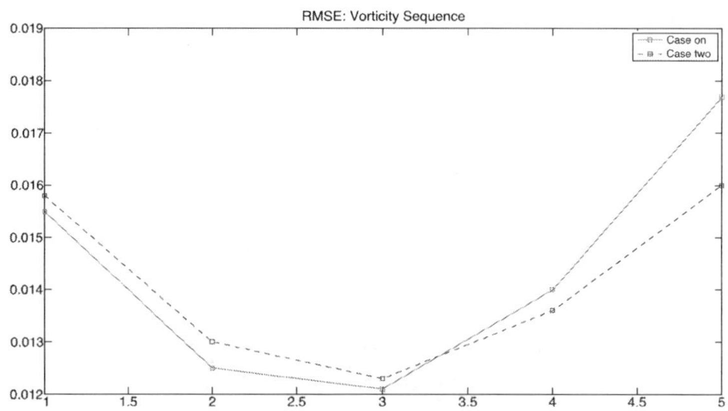

(b) RMSE on vorticity

FIGURE 2. Velocity RMSE (top) and vorticity RMSE (bottom).

Figure 3 and Figure 4 show the plot of the total energy spectrum of the solution computed with our model, superimposed with that of the ground-truth solution for case two (as case two includes indeed case one). As observed, the large scales have been well reproduced. This phenomena is also confirmed by the plots of the vorticity snapshots in Figures 5 and Figures 6, especially for $j = 6$. Moreover, we have a clear energy cutoff at fine scales for the proposed solution at $j = 7$, which is not the case for many classical large eddy simulation models [**13**]. It can be observed that case two, with its additional degree of freedom on the variance tensor leads to smoother vorticity solutions than case one.

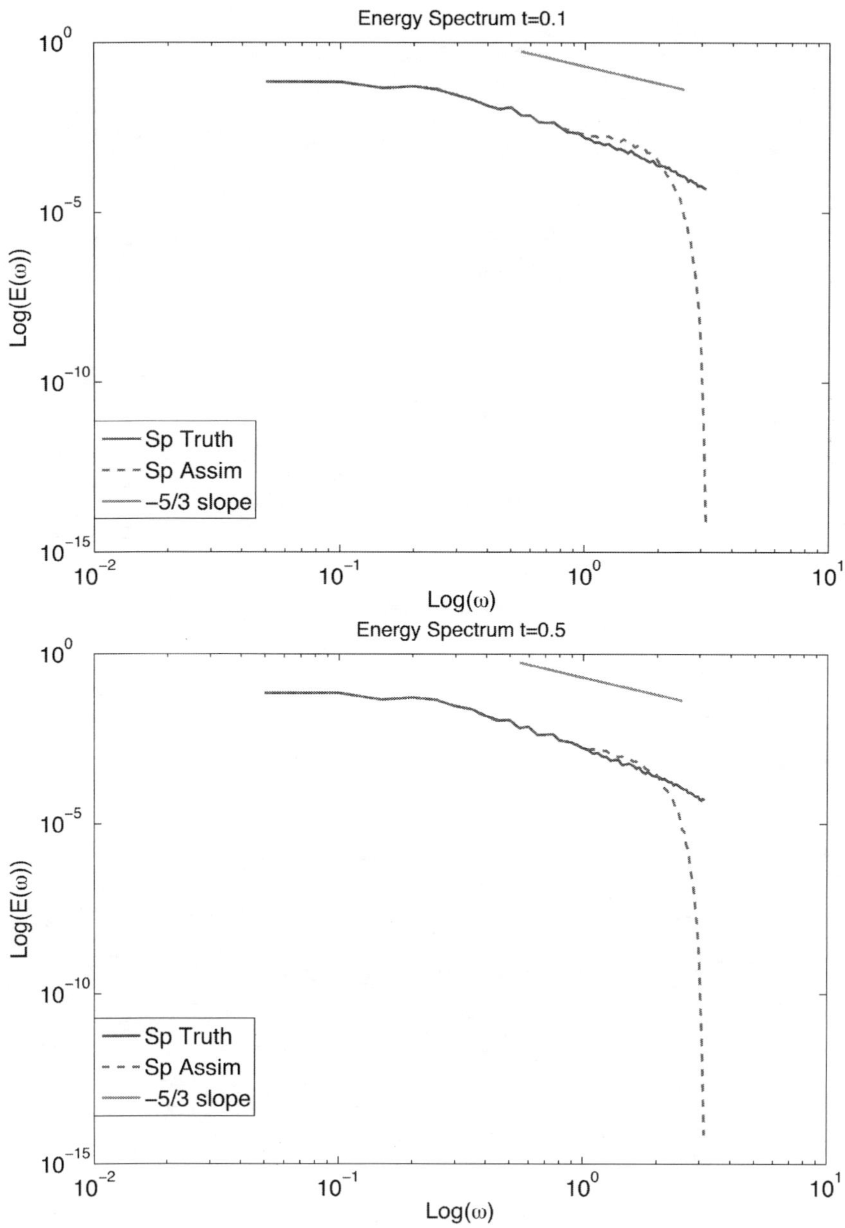

FIGURE 4. Total Energy Spectrum for the model at $j = 7$ (case two).

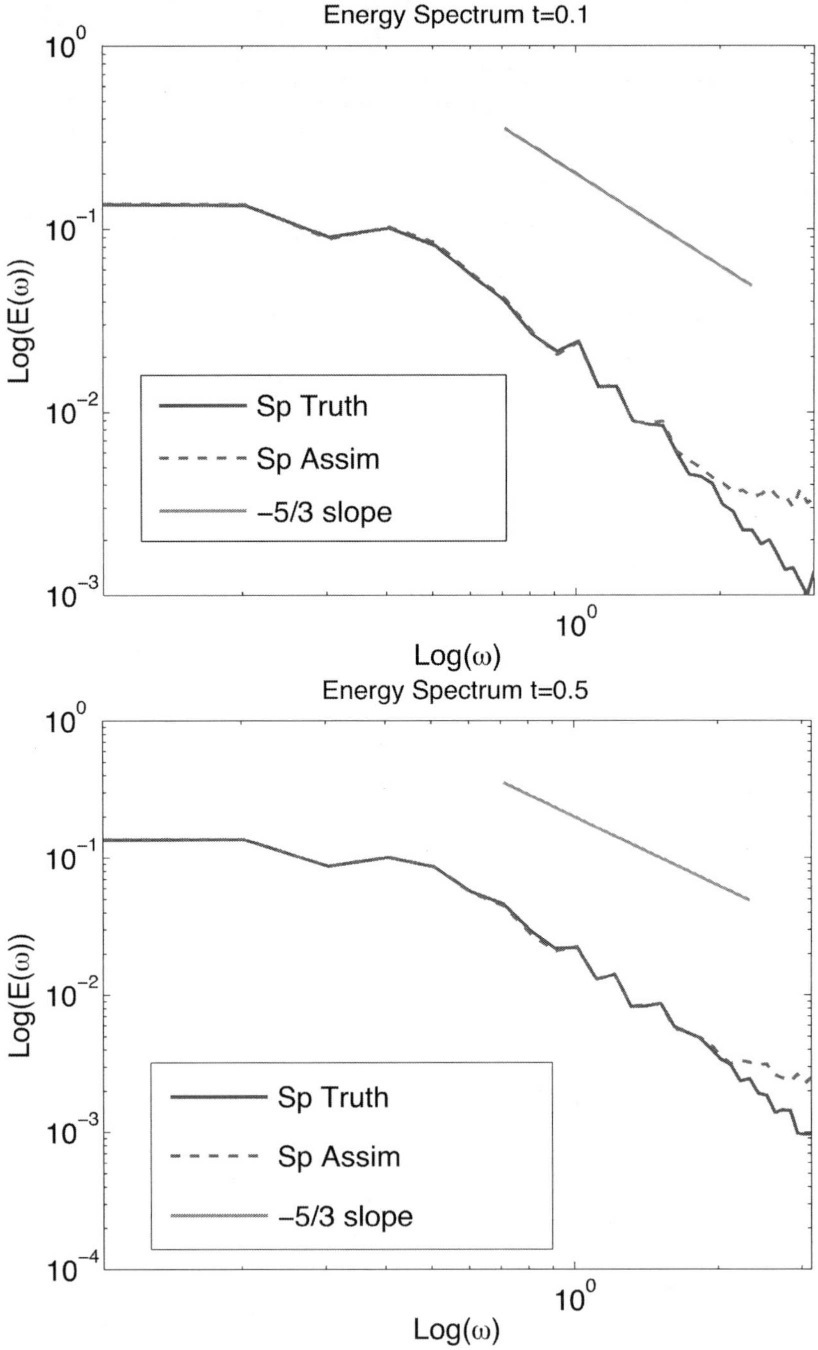

FIGURE 3. Total Energy Spectrum for the model at $j = 6$ (case two).

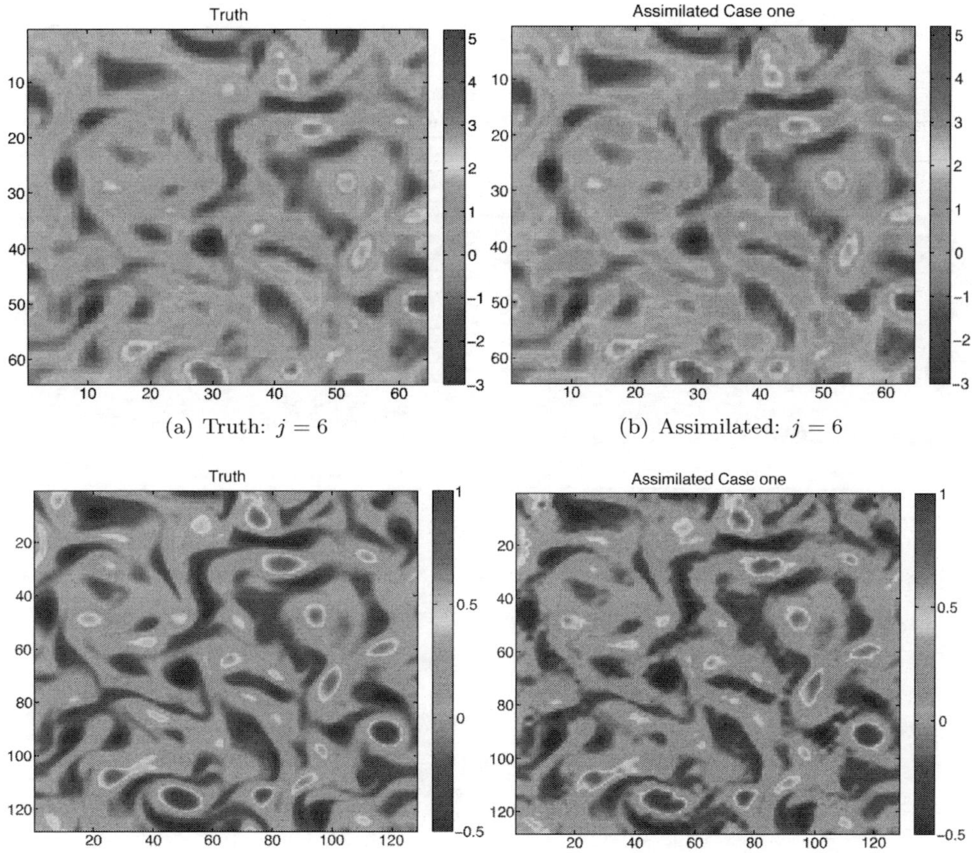

(a) Truth: $j = 6$ (b) Assimilated: $j = 6$

(c) Truth: $j = 7$ (d) Assimilated: $j = 7$

FIGURE 5. Vorticity snapshot plots for $j = 6$ top and $j = 7$ bottom (case one).

6. Conclusion

The present paper was devoted to image data assimilation in order to estimate a sub-grid viscosity model. The model was deduced under a stochastic uncertainty hypothesis on the flow. The main difficulty of the approach is the choice of the uncertainties covariance tensor. Two simple models were tested: a diagonal constant diffusion and an anisotropic diagonal diffusion. In each case, a simple divergence-free wavelet based numerical scheme was constructed and tested on synthetic particle image sequences in two space dimension. The results obtained are promising since they show a good large scale reproduction by the solution, especially for the total energy spectrum. Future works will deal with real data image sequences, with physical boundary conditions instead of periodic. In this case, it is well known that many turbulence models failed due to wall effect.

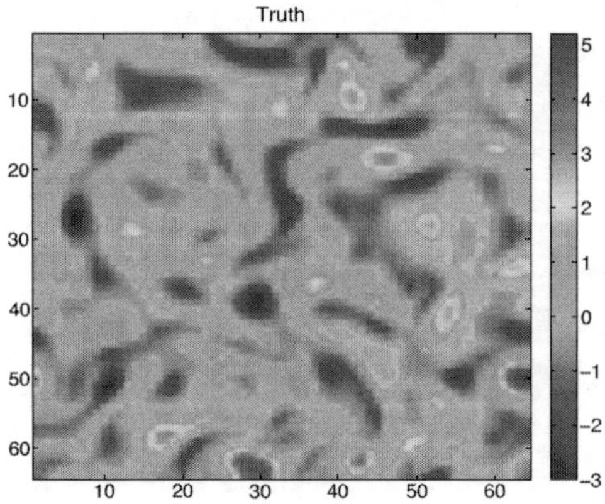

(a) Truth: $j = 6$

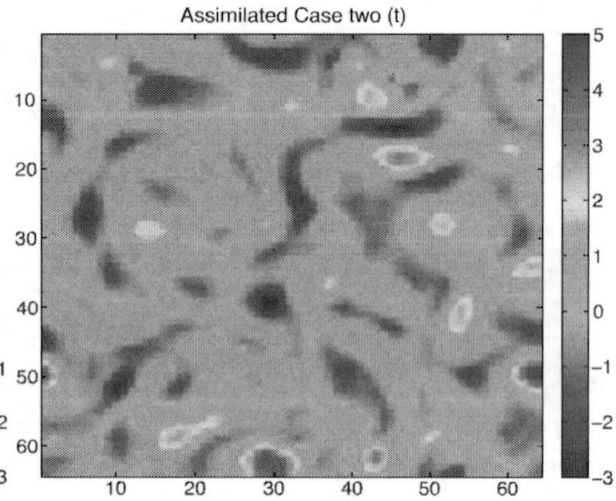

(b) Assimilated: $j = 6$

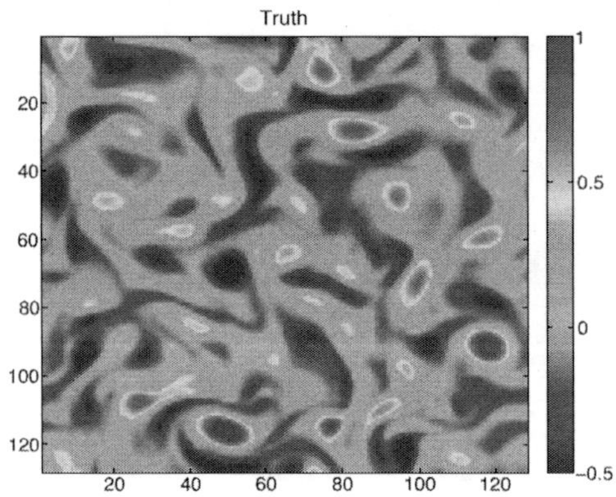

(c) Truth: $j = 7$

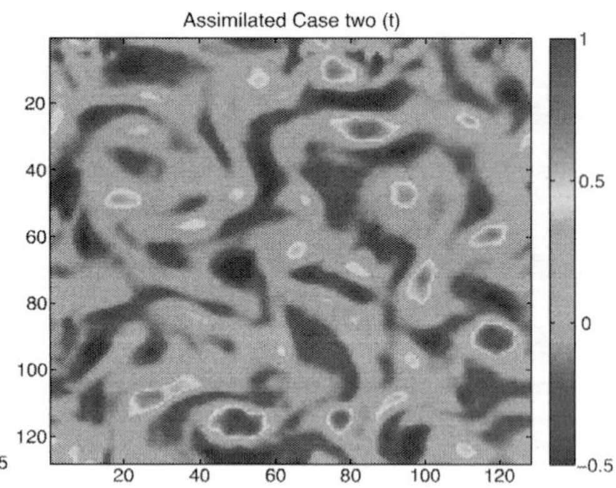

(d) Assimilated: $j = 7$

FIGURE 6. Vorticity snapshot plots for $j = 6$ top and $j = 7$ bottom (case two).

Appendix A. Proof of Lemma 1

To prove Lemma 1 it is sufficient to verify that if (28) holds, then by induction we must have:

$$\|\boldsymbol{w}^n\|_{L^2}^2 + \frac{\delta t}{2}(\nu - K(\boldsymbol{a},\nu)) \sum_{k=1}^{n} \|\nabla \boldsymbol{w}^k\|_{L^2}^2 \leq C(\boldsymbol{w}^0, \boldsymbol{a}), \tag{54}$$

where $K(\boldsymbol{a},\nu)$ is defined in (27) and $C(\boldsymbol{w}^0,\boldsymbol{a})$ any constant such that:

$$C(\boldsymbol{w}^0,\boldsymbol{a}) \geq \|\boldsymbol{w}^0\|_{L^2}^2 + \delta t^2 2^{jd} \|\boldsymbol{w}^0\|_{L^2}^2 \|\nabla \boldsymbol{w}^0\|_{L^2}^2 + \delta t K(\boldsymbol{a},\nu) \|\nabla \boldsymbol{w}^0\|_{L^2}^2. \tag{55}$$

Indeed, (26) gives this relation for $n = 1$. Suppose (54) true for order $n - 1$, then we have:

$$\delta t^2 2^{jd} \sum_{k=1}^{n-1} \|\boldsymbol{w}^k\|_{L^2}^2 \|\nabla \boldsymbol{w}^k\|_{L^2}^2 \leq \delta t^2 2^{jd} C(\boldsymbol{w}^0,\boldsymbol{a}) \sum_{k=1}^{n-1} \|\nabla \boldsymbol{w}^k\|_{L^2}^2, \tag{56}$$

and from (28) we get:

$$\delta t^2 2^{jd} \sum_{k=1}^{n-1} \|\boldsymbol{w}^k\|_{L^2}^2 \|\nabla \boldsymbol{w}^k\|_{L^2}^2 \leq \frac{\delta t}{2}(\nu - K(\boldsymbol{a},\nu)) \sum_{k=1}^{n} \|\nabla \boldsymbol{w}^k\|_{L^2}^2. \tag{57}$$

Otherwise, adding inequalities (26) for $k = 1, \cdots, n$; reads:

$$\begin{aligned} &\|\boldsymbol{w}^n\|_{L^2}^2 + \delta t(\nu - K(\boldsymbol{a},\nu)) \sum_{k=1}^{n} \|\nabla \boldsymbol{w}^k\|_{L^2}^2 \\ &\leq C(\boldsymbol{w}^0, \boldsymbol{a}) + \delta t^2 2^{jd} \sum_{k=1}^{n-1} \|\boldsymbol{w}^k\|_{L^2}^2 \|\nabla \boldsymbol{w}^k\|_{L^2}^2. \end{aligned} \tag{58}$$

Putting (57) in (58), we get the statements of (54) at order n and this proves Lemma 1.

References

[1] Bardina, J., Ferziger, J.H. & Reynolds W.C., Improved subgrid scale models for large eddy simulation, *AIAA*, **80**, 1357.

[2] A. Bensoussan and R. Temam, *Équations stochastiques du type Navier-Stokes* (French, with English summary), J. Functional Analysis **13** (1973), 195–222, DOI 10.1016/0022-1236(73)90045-1. MR0348841

[3] Berner, J. and Coauthors, Stochastic parameterization: Toward a new view of weather and climate models, *Bull. Amer. Meteor. Soc.*, 98, 565-588, 2017.

[4] Boussinesq, J., 1877, Essai sur la théorie des eaux courantes. Mémoires présentés par divers savants à l'Académie des Sciences, 23 (1): 1–680.

[5] P. Charton and V. Perrier, *A pseudo-wavelet scheme for the two-dimensional Navier-Stokes equation* (English, with English and Portuguese summaries), Mat. Apl. Comput. **15** (1996), no. 2, 139–160. MR1429545

[6] A. Cohen, *Numerical analysis of wavelet methods*, Studies in Mathematics and its Applications, vol. 32, North-Holland Publishing Co., Amsterdam, 2003. MR1990555

[7] G.-H. Cottet, D. Jiroveanu, and B. Michaux, *Vorticity dynamics and turbulence models for large-eddy simulations*, M2AN Math. Model. Numer. Anal. **37** (2003), no. 1, 187–207, DOI 10.1051/m2an:2003013. MR1972658

[8] E. Deriaz and V. Perrier, *Divergence-free and curl-free wavelets in two dimensions and three dimensions: application to turbulent flows*, J. Turbul. **7** (2006), Paper 3, 37, DOI 10.1080/14685240500260547. MR2207365

[9] F. Flandoli, *An introduction to 3D stochastic fluid dynamics*, SPDE in hydrodynamic: recent progress and prospects, Lecture Notes in Math., vol. 1942, Springer, Berlin, 2008, pp. 51–150, DOI 10.1007/978-3-540-78493-7_2. MR2459085

[10] Gawedzky, K. & Kupiainen, A., 1995, Anomalous scaling of the passive scalar. *Physical review letters* **75**, 3834–3837.

[11] Gent, P. & McWilliams, J., 1990, Isopycnal mixing in ocean circulation models. *J. Phys. Oceanogr.* **20**, 150–155.

[12] Heitz, D., Carlier, J. & Arroyo, G., 2007, Final report on the evaluation of the tasks of the workpackage 2, FLUID project deliverable 5.4. *Tech. rep.*, INRIA - Cemagref.

[13] S. K. Harouna and E. Mémin, *Stochastic representation of the Reynolds transport theorem: revisiting large-scale modeling*, Comput. & Fluids **156** (2017), 456–469, DOI 10.1016/j.compfluid.2017.08.017. MR3693232

[14] S. K. Harouna and V. Perrier, *Helmholtz-Hodge decomposition on $[0,1]^d$ by divergence-free and curl-free wavelets*, Curves and surfaces, Lecture Notes in Comput. Sci., vol. 6920, Springer, Heidelberg, 2012, pp. 311–329, DOI 10.1007/978-3-642-27413-8_20. MR2889785

[15] S. Kadri Harouna and V. Perrier, *Effective construction of divergence-free wavelets on the square*, J. Comput. Appl. Math. **240** (2013), 74–86, DOI 10.1016/j.cam.2012.07.029. MR2991109

[16] R. H. Kraichnan, *The structure of isotropic turbulence at very high Reynolds numbers*, J. Fluid Mech. **5** (1959), 497–543, DOI 10.1017/S0022112059000362. MR105963

[17] Kraichnan, R.,1968, Small-scale structure of a randomly advected passive scalar. *Phys. Rev. Lett.* **11**, 945–963.

[18] Kraichnan, R., 1970, Small-scale structure of a randomly advected passive scalar. *Phys. Rev. Lett.* **11**, 945–963.

[19] H. Kunita, *Stochastic flows and stochastic differential equations*, Cambridge Studies in Advanced Mathematics, vol. 24, Cambridge University Press, Cambridge, 1990. MR1070361

[20] J.-P. Laval, B. Dubrulle, and J. C. McWilliams, *Langevin models of turbulence: renormalization groups, distant interaction algorithms or rapid distortion theory?*, Phys. Fluids **15** (2003), no. 5, 1327–1339, DOI 10.1063/1.1564826. MR2003711

[21] Le Dimet, F.X. & Talagrand, O.,1986, Variational algorithms for analysis and assimilation of meteorological observations: theoretical aspects. *Tellus* **38A**, 97–110.

[22] Leith, C., 1971, Atmospheric predictability and two-dimensional turbulence. *J. Atmos. Sci.* 1971, **28**, 145–161.

[23] Leith, C., 1990, Stochastic backscatter in a subgrid-scale model: plane shear mixing layer. *Phys. of Fluids*, **2(3)**, 1521–1530.

[24] P. G. Lemarie-Rieusset, *Analyses multi-résolutions non orthogonales, commutation entre projecteurs et dérivation et ondelettes vecteurs à divergence nulle* (French, with English and French summaries), Rev. Mat. Iberoamericana **8** (1992), no. 2, 221–237, DOI 10.4171/RMI/123. MR1191345

[25] Lilly, D., 1966, On the application of the eddy viscosity concept in the inertial subrange of turbulence. *Tech. Rep.* 123. NCAR.

[26] Mason, P.J. and Thomson, D.J., 1989, Stochastic backscatter in large-eddy simulations of boundary layers, *J. of Fluid Mech.,* **242**, 51–78.

[27] N. J. Zabusky and M. V. Melander, *Three-dimensional vortex tube reconnection: morphology for orthogonally-offset tubes*, Phys. D **37** (1989), no. 1-3, 555–562, DOI 10.1016/0167-2789(89)90160-7. Advances in fluid turbulence (Los Alamos, NM, 1988). MR1024405

[28] E. Mémin, *Fluid flow dynamics under location uncertainty*, Geophys. Astrophys. Fluid Dyn. **108** (2014), no. 2, 119–146, DOI 10.1080/03091929.2013.836190. MR3223587

[29] R. Mikulevicius and B. L. Rozovskii, *Stochastic Navier-Stokes equations for turbulent flows*, SIAM J. Math. Anal. **35** (2004), no. 5, 1250–1310, DOI 10.1137/S0036141002409167. MR2050201

[30] Prandtl, L., 1925, Bericht über untersuchungen zur ausgebildeten turbulenz *Z. Angew. Math, Meth.* **5**, 136-139.

[31] V. Resseguier, E. Mémin, and B. Chapron, *Geophysical flows under location uncertainty, Part I Random transport and general models*, Geophys. Astrophys. Fluid Dyn. **111** (2017), no. 3, 149–176, DOI 10.1080/03091929.2017.1310210. MR3648562

[32] V. Resseguier, E. Mémin, and B. Chapron, *Geophysical flows under location uncertainty, Part II Quasi-geostrophy and efficient ensemble spreading*, Geophys. Astrophys. Fluid Dyn. **111** (2017), no. 3, 177–208, DOI 10.1080/03091929.2017.1312101. MR3648563

[33] V. Resseguier, E. Mémin, and B. Chapron, *Geophysical flows under location uncertainty, Part III SQG and frontal dynamics under strong turbulence conditions*, Geophys. Astrophys. Fluid Dyn. **111** (2017), no. 3, 209–227, DOI 10.1080/03091929.2017.1312102. MR3648564

[34] P. Sagaut, *Large eddy simulation for incompressible flows*, 3rd ed., Scientific Computation, Springer-Verlag, Berlin, 2006. An introduction; Translated from the 1998 French original; With forewords by Marcel Lesieur and Massimo Germano; With a foreword by Charles Meneveau. MR2174208

[35] Schumann, U., 1995, Stochastic backscatter of turbulence energy and scalar variance by random subgrid-scale fluxes. *Proc. R. Soc. Lond. A* **451**, 293–318 .

[36] Smagorinsky, J., 1963, General circulation experiments with the primitive equation: I. the basic experiment. *Monthly Weather Review* **91**, 99–165.

[37] R. Temam, *Navier-Stokes equations*, Revised edition, Studies in Mathematics and its Applications, vol. 2, North-Holland Publishing Co., Amsterdam-New York, 1979. Theory and numerical analysis; With an appendix by F. Thomasset. MR603444

[38] Vasilyev, O.V., De Stefano, G., Goldstein, D. & Kevlahan, N. K.-R., 2008, Lagrangian dynamic SGS model for Stochastic Coherent Adaptive Large Eddy Simulation *J. Turbulence* **11**, 1-14.

Laboratoire Mathématiques, Image et Applications (MIA), Université de La Rochelle, Avenue Michel Crépeau 17042 La Rochelle, France

Email address: souleymane.kadri_harouna@univ-lr.fr

INRIA Rennes - Bretagne Atlantique, Campus Universitaire de Beaulieu, 35042 Rennes cedex, France

Email address: etienne.memin@inria.fr

Selected Published Titles in This Series

757 **Kaïs Ammari and Stéphane Gerbi, Editors,** Identification and Control: Some New Challanges, 2020

755 **Marion Scheepers and Ondřej Zindulka, Editors,** Centenary of the Borel Conjecture, 2020

754 **Susanne C. Brenner, Igor Shparlinski, Chi-Wang Shu, and Daniel B. Szyld, Editors,** 75 Years of Mathematics of Computation, 2020

753 **Matthew Krauel, Michael Tuite, and Gaywalee Yamskulna, Editors,** Vertex Operator Algebras, Number Theory and Related Topics, 2020

752 **Samuel Coskey and Grigor Sargsyan, Editors,** Trends in Set Theory, 2020

751 **Ashish K. Srivastava, André Leroy, Ivo Herzog, and Pedro A. Guil Asensio, Editors,** Categorical, Homological and Combinatorial Methods in Algebra, 2020

750 **A. Bourhim, J. Mashreghi, L. Oubbi, and Z. Abdelali, Editors,** Linear and Multilinear Algebra and Function Spaces, 2020

749 **Guillermo Cortiñas and Charles A. Weibel, Editors,** K-theory in Algebra, Analysis and Topology, 2020

748 **Donatella Danielli and Irina Mitrea, Editors,** Advances in Harmonic Analysis and Partial Differential Equations, 2020

747 **Paul Bruillard, Carlos Ortiz Marrero, and Julia Plavnik, Editors,** Topological Phases of Matter and Quantum Computation, 2020

746 **Erica Flapan and Helen Wong, Editors,** Topology and Geometry of Biopolymers, 2020

745 **Federico Binda, Marc Levine, Manh Toan Nguyen, and Oliver Röndigs, Editors,** Motivic Homotopy Theory and Refined Enumerative Geometry, 2020

744 **Pieter Moree, Anke Pohl, L'ubomír Snoha, and Tom Ward, Editors,** Dynamics: Topology and Numbers, 2020

743 **H. Garth Dales, Dmitry Khavinson, and Javad Mashreghi, Editors,** Complex Analysis and Spectral Theory, 2020

742 **Francisco-Jesús Castro-Jiménez, David Bradley Massey, Bernard Teissier, and Meral Tosun, Editors,** A Panorama of Singularities, 2020

741 **Houssam Abdul-Rahman, Robert Sims, and Amanda Young, Editors,** Analytic Trends in Mathematical Physics, 2020

740 **Alina Bucur and David Zureick-Brown, Editors,** Analytic Methods in Arithmetic Geometry, 2019

739 **Yaiza Canzani, Linan Chen, and Dmitry Jakobson, Editors,** Probabilistic Methods in Geometry, Topology and Spectral Theory, 2019

738 **Shrikrishna G. Dani, Surender K. Jain, Jugal K. Verma, and Meenakshi P. Wasadikar, Editors,** Contributions in Algebra and Algebraic Geometry, 2019

737 **Fernanda Botelho, Editor,** Recent Trends in Operator Theory and Applications, 2019

736 **Jane Hawkins, Rachel L. Rossetti, and Jim Wiseman, Editors,** Dynamical Systems and Random Processes, 2019

735 **Yanir A. Rubinstein and Bernard Shiffman, Editors,** Advances in Complex Geometry, 2019

734 **Peter Kuchment and Evgeny Semenov, Editors,** Differential Equations, Mathematical Physics, and Applications, 2019

733 **Peter Kuchment and Evgeny Semenov, Editors,** Functional Analysis and Geometry, 2019

732 **Samuele Anni, Jay Jorgenson, Lejla Smajlović, and Lynne Walling, Editors,** Automorphic Forms and Related Topics, 2019

For a complete list of titles in this series, visit the AMS Bookstore at **www.ams.org/bookstore/conmseries/**.